KINETICS IN NANOSCALE MATERIALS

KINETICS IN NANOSCALE MATERIALS

KING-NING TU
ANDRIY M. GUSAK

Cover Design: Wiley
Cover Image: Courtesy of the authors

Published by John Wiley & Sons, Inc., Hoboken, New Jersey.
Published simultaneously in Canada.

Library of Congress Cataloging-in-Publication Data:

Tu, K. N. (King-Ning), 1937-
Kinetics in nanoscale materials / by King-Ning Tu, Andriy Gusak.
pages cm
Summary: "As the ability to produce nanomaterials advances, it becomes more important to understand how the energy of the atoms in these materials is affected by their reduced dimensions. Written by an acclaimed author team, Kinetics in Nanoscale Materials is the first book to discuss simple but effective models of the systems and processes that have recently been discovered. The text, for researchers and graduate students, combines the novelty of nanoscale processes and systems with the transparency of mathematical models and generality of basic ideas relating to nanoscience and nanotechnology"– Provided by publisher.
"Published simultaneously in Canada"–Title page verso.
Includes bibliographical references and index.
ISBN 978-0-470-88140-8 (hardback)
1. Nanostructured materials. 2. Chemical kinetics. 3. Nanostructured materials–Analysis. 4. Nanostructured materials–Computer simulation. I. Gusak, Andriy M. II. Title.
TA418.9.N35T8 2014
620.1′1599–dc23

2013042096

Printed in the United States of America.

10 9 8 7 6 5 4 3 2 1

CONTENTS

PREFACE

In the Department of Materials Science and Engineering at UCLA, three courses on kinetic processes in materials are being taught at the moment. The first course is MSE 131 on "Diffusion and Diffusion Related Phase Transformations," which is for upper undergraduate students. The textbook is "Phase Transformations in Metals and Alloys," 2nd edition, by D. A. Porter and K. E. Easterling, published by Chapman and Hall, London, 1992. The second course is MSE 223 on "Thin Film Materials Science," which is for first year graduate students. The textbooks are "Electronic Thin Film Science," by K. N. Tu, J. W. Mayer, and L. C. Feldman, published by Macmillan, New York, 1993, and "Electronic Thin Film Reliability," by K. N. Tu, published by Cambridge University Press, UK, 2011. The third course is MSE 201 on "Principle of Materials Science: Solid State Reactions," which is a mandatory course for Ph.D. students. It had been taught by Prof. Alan Ardell until his retirement in 2008. There is no textbook for this course, except the lecture notes by Prof. Ardell. One of the reasons that this book is written is to serve as the textbook for this course in the future. This book can also be used as a textbook for a kinetics course in the Department of Physics at Cherkasy National University, Cherkasy, Ukraine. Roughly speaking, MSE 131 covers mainly kinetics in bulk materials, MSE 223 emphasizes kinetics in thin films, and MSE 201 will focus on kinetics in nanoscale materials. It is worthwhile mentioning that kinetics in nanoscale materials is not completely new or very different from those in bulk and thin films. Actually, a strong link among them can be found, which is shown in this book. An example is the lower melting point of nanosize particles. In morphological instability of the solidification of bulk melt, the lower melting point of the tip of dendrite has been analyzed in detail.

Chapter 1 explains why the subject of kinetic processes in nanoscale materials is of interest. It begins with a discussion that the surface energy of a nanosphere is equal to its Gibbs–Thomson potential energy. This is implicit in the classical theory of homogeneous nucleation in bulk materials. Then, it is followed by several sections on some general kinetic behaviors of nanosphere, nanopore, nanowire, nanothin films, and nanomicrostructure in bulk materials. Specific topics on kinetics in nanoscale materials are covered by Chapter 2 on linear and nonlinear diffusion; Chapter 3 on Kirkendall effect and inverse Kirkendall effect; Chapter 4 on ripening among nano-precipitates; Chapter 5 on spinodal decomposition; Chapter 6 on nucleation events in bulk materials, thin films, and nanowires; Chapter 7 on contact reactions on Si: plane, line, and point contact reactions; Chapter 8 on grain growth in micro and nanoscales; Chapter 9 on self-sustained explosive reactions in nanoscale multilayered thin films; and Chapter 10 on formation and transformation of nanotwins in Cu. In the last two chapters, applications of nanoscale kinetics are emphasized by the explosive reactions for distance ignition or for local heating, and by nanotwinned Cu for interconnect and packaging technology for microelectronic devices.

In nanoscale materials, we encounter very high concentration gradient, very small curvature, very large nonequilibrium vacancies, very few dislocations, and yet very high density of grain boundaries and surfaces, and even nanotwins. They modify the driving force as well as the kinetic jump process. To model the nanoscale processes, our understanding of kinetic processes in bulk materials can serve as the stepping stone from where we enter into the nano region. On seeing the similarity between bulk and nanoscale materials, the readers can follow the link to obtain a better understanding of the kinetic processes in nanoscale materials. On seeing the difference, the readers will appreciate what modification is needed or what is new in the kinetic processes in nanoscale materials.

We would like to acknowledge that we have benefited greatly from the lecture notes by Prof. Alan Ardell on kinetics of homogeneous nucleation, spinodal decomposition, and ripening. We also would like to acknowledge that the second part of Chapter 2 on thermodynamic nonlinear effects on diffusion is taken from an unpublished 1986 IBM technical report written by Prof. Lydia Chiao in the Department of Physics at Georgetown University, Washington, DC. We apologize to the readers that because of our limited knowledge, we do not cover some of the very active and interesting topics of nanomaterials, such as the nucleation and growth of graphene on metal surfaces, VLS growth of nano Si wires, or interdiffusion in man-made superlattices. We hope that this book will help students and readers advance into these and other nanoscale kinetic topics in the future.

April 2014 *King-Ning Tu and Andriy M. Gusak*

CHAPTER 1

INTRODUCTION TO KINETICS IN NANOSCALE MATERIALS

1.1 INTRODUCTION

In recent years, a new development in science and engineering is nanoscience and nanotechnology. It seems technology based on nanoscale devices is hopeful. Indeed, at the moment the research and development on nanoscale materials science for nanotechnology is ubiquitous. Much progress has been accomplished in the processing of

Kinetics in Nanoscale Materials, First Edition. King-Ning Tu and Andriy M. Gusak.

nanoscale materials, such as the growth of silicon nanowires. Yet, we have not reached the stage where the nanotechnology is mature and mass production of nanodevices is carried out. One of the difficulties to be overcome, for example, is the large-scale integration of nanowires. We can handle a few pieces of nanowires easily, but it is not at all trivial when we have to handle a million of them. It is a goal to be accomplished. For comparison, the degree of success of nanoelectronics from a bottom-up approach is far from that of microelectronics from a top-down approach. In reality, the bottom-up approach of building nanoelectronic devices from the molecular level all the way up to circuit integration is very challenging. Perhaps, it is likely that a hybrid device will have a better chance of success by building nanoelectronic devices on the existing platform of microelectronic technology and by taking advantage of what has been developed and what is available in the industry.

The proved success of microelectronic technology in the past and now leads to expectations of both high yield in processing and reliability in the applications of the devices. These requirements extend to nanotechnology. No doubt, reliability becomes a concern only when the nanodevices are in mass production. We may have no concern about their reliability at the moment because they are not yet in mass production, but we cannot ignore it if we are serious about the success of nanotechnology.

On processing and reliability of microelectronic devices, kinetics of atomic diffusion and phase transformations is essential. For example, on processing, the diffusion and the activation of substitutional dopants in silicon to form shallow p–n junction devices require a very tight control of the temperature and time of fabrication. It is worth mentioning that Bardeen has made a significant contribution to the theory of atomic diffusion on our understanding of the "correlation factor" in atomic jumps. On reliability, the issue of electromigration-induced failures is a major concern in microelectronics, and the kinetic process of electromigration is a cross-effect of irreversible processes. Today, we can predict the lifetime of a microelectronic device or its mean-time-to-failure by conducting accelerated tests and by performing statistical analysis of failure. However, it is the early failure of a device that concerns the microelectronic industry the most. Thus, we expect that in the processing and reliability of nanoelectronic devices, we will have similar concerns of failure, especially the early failure, which tends to happen when the integration processes and the reliability issues are not under control. It is for this reason that the kinetics of nanoscale materials is of interest. If we assume that everything in nanoscale materials and devices is new, it implies that the yield and reliability of nanodevices is new too, which we hope is not completely true. In this book, we attempt to bridge the link between a kinetic process in bulk and the same process in nanoscale materials. The similarity and the difference between them is emphasized, so that we can have a better reference of the kinetic issues in nanodevices and nanotechnology.

To recall kinetic processes in bulk materials, we note that there are several kinds of phase changes in bulk materials in which the distance of diffusion or the size of phases are in nanoscale. Take the case of Guinier–Preston (GP) zones of precipitation, in which the thickness of GP zone is of atomic scale and the spacing between zones is of the order of 10 nm. In the case of spinodal decomposition, the wave length of decomposition is of nanometers. In homogeneous nucleation, the distribution of subcritical nuclei is a distribution of nanosize embryos. In ripening, a distribution of particles of nanoscale is assumed, and the analysis of ripening starts with the Gibbs–Thomson (GT) potential of these particles having a very small or nanoscale radius.

Furthermore, there are nanoscale microstructures in bulk-type materials. An example is the square network of screw dislocations in forming a small angle twist-type grain boundary. We can take two (001) Si wafer and bond them together face-to-face with a few degrees of misorientation of rotation, the dislocation network in the twist-type grain boundary forms one of the most regular two-dimensional nanoscale squares. Another example is a bulk piece of Cu that has a high density of nanotwins. One more example is a layer of nanosize grains formed by ball milling on the surface of a bulk piece of steel, which is called *surface mechanical attrition treatment* (*SMAT*) of nano-grains.

Our understanding of kinetic processes in bulk materials can serve as the stepping stone from where we enter into the kinetics in nanoregion. On seeing the similarity in kinetics between them, we can follow the similarity to reach a deeper level of understanding of the kinetic processes in nanoscale materials. On seeing the difference, we may appreciate what modification is needed in terms of driving force and/or kinetic process in nanoscale materials. In the early chapters of this book, several examples have been chosen for the purpose of illustrating the link between kinetic behaviors in bulk and in nanoscale materials, and in the later chapters a few cases of applications of nanoscale kinetics are given.

When we deal with nanoscale materials, we encounter very high gradient of concentration, very large curvature or very small radius, very large amount of nonequilibrium vacancies, very few dislocations, and yet very high density of surfaces and grain boundaries and, may be, nanotwins. They modify the driving force as well as the kinetic jump process. Indeed, the kinetic processes in nanoscale materials have some unique behavior that is not found in the kinetics of bulk materials. In this chapter of introduction, we present a few examples of nanoscale materials to illustrate their unique kinetic behavior. They are nanospheres, nanowires, nanothin films, and nanomicrostructures. More details will be covered in the subsequent chapters.

1.2 NANOSPHERE: SURFACE ENERGY IS EQUIVALENT TO GIBBS–THOMSON POTENTIAL

We consider a nanosize sphere of radius r. It has a surface area of $A = 4\pi r^2$ and surface energy of $E = 4\pi r^2 \gamma$, where γ is the surface energy per unit area and we assume that the magnitude of the surface energy per unit area γ is independent of r. We note that as surface energy is positive, the surface area (or the radius of the sphere) tends to shrink in order to reduce surface energy, which implies that the tendency to shrink exerts a compression or pressure to all the atoms inside the sphere. This pressure is called the *Laplace pressure*. The effect of the pressure is felt when we want to add atoms or remove atoms from the sphere because it will change the volume as well as the surface area. When we want to change the volume of the sphere under the Laplace pressure at constant temperature, we need to consider the work done and the work equals to the energy change, so that $pdV = \gamma dA$. The pressure can be calculated as

$$p = \gamma \frac{dA}{dV} = \gamma \frac{(d/dr)(4\pi r^2)}{(d/dr)((4/3)\pi r^3)} = \gamma \frac{8\pi r}{4\pi r^2} = \frac{2\gamma}{r} \tag{1.1}$$

However, we note that the work done by the Laplace pressure is different from the conventional elastic work done in a solid by a stress. The elastic work is given below,

$$E_{\text{elastic}} = V \cdot \int \sigma d\varepsilon = V \frac{1}{2} K \varepsilon^2 \tag{1.2}$$

To calculate the elastic work, we need to know at least the elastic bulk modulus K of the material (in case of homogeneous hydrostatic stress). On the other hand, the work done by Laplace pressure is due to the change in volume by adding or removing atoms under the Laplace pressure, and no modulus is needed.

We consider the case of adding an atom to a nanosphere, the Gibbs free energy ($G = U - TS + pV$) increases $p\Omega$, where U is internal energy, T is temperature, S is entropy, and Ω is atomic volume. By definition, $p\Omega$ is a part of the chemical potential of the nanosphere related to the change of its volume under the fixed external pressure. It is the change (increase) of Gibbs free energy due to the addition of one atom (or one mole of atoms, depending on the definition of chemical potential) to the nanosphere (see Section 2.2.3, on the definition of chemical potential). It is worth mentioning that adding an atom at constant temperature has effects on U, S, and p. This is because it adds a few more interatomic bonds to U, the configuration entropy increases because of more ways in arranging the atoms, and though it does not affect the external pressure, the Laplace pressure will decrease because of the increase in radius.

Here it is important to distinguish two alternative approaches to account for surface (capillary) effects:

1. Helmholtz free energy $F = U - TS$ of the limited system includes explicitly an additional free energy of the surface: $F = N \cdot f + \gamma \cdot A$, where f is a bulk free energy per atom, N is the number of atoms, A is an area of external boundary (in our case $A = 4\pi R^2$), γ is an additional surface free energy per unit area. In this case the "p" in the expression for Gibbs energy is just real external pressure of the thermal ambient, without any Laplace terms. In this case, $G = F + pV = N \cdot (f + p\Omega) + \gamma \cdot A = Ng + \gamma \cdot A$. Then the chemical potential $\mu = \partial G/\partial N = g + \gamma \cdot (\partial A/\partial N)$. Below we start with this case.
2. Alternatively, free energy $F = U - TS$ of the limited system may not include explicitly the surface energy but instead use some effective external pressure $p^{\text{ef}} = p + p_{\text{Laplace}}$. Then $\mu = \mu^{\text{bulk}} + p_{\text{Laplace}} \cdot \Omega$. If $p_{\text{Laplace}} = \gamma \cdot ((1/\Omega)(\partial A/\partial N))$, then the result will be the same.

To add the atom, if we imagine that the atomic volume Ω is "smeared" over the entire surface of the nanosphere as a very thin shell, it leads to the growth of the radius, dr, of the nanosphere as:

$$\Omega = dV = d\left(\frac{4}{3}\pi r^3\right) = 4\pi r^2 dr \Rightarrow dr = \frac{\Omega}{4\pi r^2}$$

so the work of Laplace pressure is $p_{\text{Laplace}} 4\pi r^2 \times dr = p_{\text{Laplace}}\Omega = 2\gamma\Omega/r$, where the product of $p_{\text{Laplace}} 4\pi r^2$ (force) and dr (distance) is the work done by the Laplace pressure. It is due to a surface change induced free energy change in the nanosphere, hence it should be added to the chemical potential of all the atoms belonging to the nanosphere. Thus, $p\Omega$ is the surface input into the chemical potential. We emphasize that this is an additional chemical potential energy of every atom in the nanosphere, not just the atoms on the surface, due to the surface effect. When r is small, this addition to chemical potential (GT potential), $2\gamma\Omega/r$, cannot be ignored.

Let us take the integral over the process of constructing the entire volume of the nanosphere by sequential adding of new spherical slices $4\pi r'^2 dr'$, and we obtain

$$\int_0^r p_{\text{Laplace}} dV = \int_0^r \frac{2\gamma}{r'} 4\pi r'^2 dr' = \gamma \cdot 4\pi r^2 = \Delta E_{\text{surface}} \qquad (1.3)$$

It means that the work done by Laplace pressure during the formation (growth) of the nanosphere is exactly equal to the surface energy. We have reached a very important conclusion that the surface energy ($4\pi r^2\gamma$) is equal to the sum of GT potential energy of all the atoms in the nanosphere, calculated as an integral over the evolution path of this sphere formation. (It is important to remember that in Eq. (1.3) the Laplace pressure under integral is not constant – it changes simultaneously with the growth of the sphere.) In other words, when we consider the GT potential, it means that all the atoms are the same, whether the atom is on the surface or within the nanosphere. We may say that from the point of view of GT potential, there is no surface atom, as all the atoms are the same, and hence there is no surface energy because the surface energy is being distributed to all the atoms.

To avoid possible misunderstanding, we emphasize that to form a nanosphere, we should add to the bulk energy an additional term of surface energy or the work of Laplace pressure, but not both of them. An example is in considering the formation energy of a nucleus in homogeneous nucleation, in which we include the surface energy of $4\pi r^2\gamma$ explicitly, see Eq. (1.11) or Eq. (6.1), so we do not need to add GT potential to all the atoms, even though the radius of a nucleus is very small. Another example is in ripening, in which the kinetic process is controlled by the mean-field concentration in equilibrium with particles having the mean radius, following the GT equation, but the surface energy of $4\pi r^2\gamma$ is implicit in the analysis, although the driving force comes from the reduction of surface energy. These two cases are covered in detail in later chapters.

As we can regard the hydrostatic pressure or Laplace pressure, p, as energy density or energy per unit volume, we might regard pV as the energy increase in a volume V under pressure. Strictly speaking, it is not completely correct. For example, the additional energy due to the existence of a surface is surface tension times the surface area: $\Delta E = \gamma 4\pi r^2$. However, the product $p_{\text{Laplace}} V$ is equal to $(2\gamma/r)(4/3)\pi r^3 = 2/3(\gamma \cdot 4\pi r^2)$. It is less by one-third from the surface energy of ΔE. As shown in Eq. (1.3), we need to take integration in order to obtain the correct energy.

We recall that when we consider the surface energy of a flat surface where the radius is infinite. In this case, the Laplace pressure is zero, so does the GT potential. Yet it does not mean the surface energy is zero. Instead, we use the number of broken bonds to calculate the surface energy of a flat surface by considering the cleavage of a piece of solid into two pieces having flat surfaces. In the case of a nanosphere, we simply use $4\pi r^2\gamma$ for its surface energy on the basis of GT potential. We recommend readers to analyze the above equations for the case when radius tends to infinity, spherical surface becomes more and more flat, Laplace pressure tends to zero, but total surface energy grows to infinity.

Next we might ask the question of a nonspherical particle, what is the chemical potential inside the nonspherical particle with curvature changing from one area of the surface to another area? In Appendix A, the concept of Laplace pressure is applied to nano-cubic particle and nano-disk particle, and the chemical potentials are given.

On the question of a hollow nanoparticle that has two surfaces, the inner and outer surfaces, the simple answer is that atoms at places with different curvatures possess different chemical potentials, and these potential differences or chemical potential gradient should enable surface and bulk diffusion to occur and lead to equalizing of curvatures. Nevertheless, the answer does not give a receipt of finding a spatial redistribution of chemical potential inside the particle if the temperature is low so that the smoothening proceeds only by surface diffusion. This gives us an example of the limit of applicability of thermodynamic concepts owing to slow kinetics: chemical potential is a self-consistent thermodynamic quantity assuming the condition of sufficiently fast diffusion kinetics. So, if diffusion is frozen at a low temperature, the driving force of chemical potential gradient has no response in such system or subsystem. We analyze the case of hollow nanospheres having two surfaces in a later section, assuming that atomic diffusion is fast enough for curvature change to happen.

1.3 NANOSPHERE: LOWER MELTING POINT

Nanosize will affect phase transition temperature besides pressure. Now we consider the melting of nanoparticles. Melting means transition from a crystalline phase to a liquid phase, where the crystalline phase is characterized by having a long range order (LRO). At the melting point, Gibbs free energy of the two phases is equal. The very notion of LRO for particles with the size of several interatomic distances or even several tens of nanometers becomes somewhat fuzzy, and the melting transition may become gradual within a temperature range, depending on the distribution of the nanoparticle size in the sample. Experimentally, we tend to measure the melting of a sample consisting of a large number of nanoparticles, rather than just one nanoparticle. Assuming that the melting temperature has an average value within a temperature range, we continue to define it as the temperature at which Gibbs free energy of the two phases is equal.

In Figure 1.1, a plot of Gibbs free energy versus the temperature of the liquid state and the solid state of a pure bulk phase having a flat interface is depicted by the two solid curves. We assume that the bulk sample has radius $r = \infty$. The two solid curves cross each other at the melting point of T_m $(r = \infty)$.

For solid nanoparticles of radius r, its Gibbs free energy curve is represented by one of the broken curves, and we note that the energy difference between the two curves of the solids is the GT potential energy of $p\Omega_s = 2\gamma_s\Omega_s/r_s$, where γ_s is the

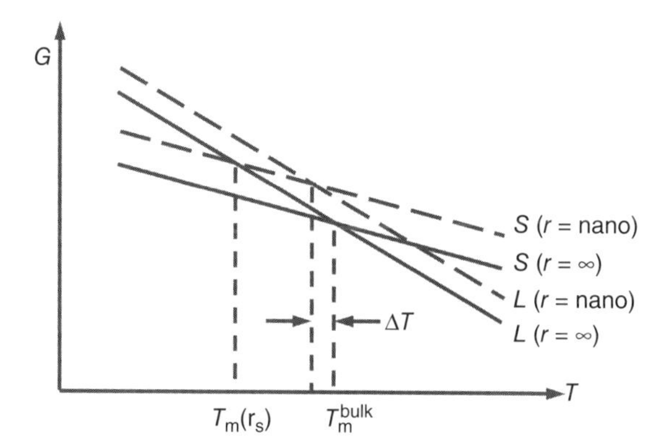

Figure 1.1 A plot of Gibbs free energy versus the temperature of the liquid state and solid state of a pure phase is depicted by the two solid curves. The two solid curves cross each other at the melting point of T_m. We assume that the solid state of a bulk sample has radius $r=\infty$. For solid and liquid nanoparticles of radius r_s and r_l, their Gibbs free energy curves are represented by the broken curves. The broken curves intersect at a lower temperature of T_m (nano), provided that we assume the surface energy of liquid is lower than that of the solid.

interfacial energy between the solid and the ambient and it is independent of size. Usually we are interested in the melting point of nanoparticles in air or vacuum. Strictly speaking, if this ambient is infinite and if it does not contain the vapor of atoms of the same nanoparticle, and if we have unlimited time for observation, eventually these particles will evaporate totally. But we are not interested in this process; instead, we want to know what happens with the nanoparticles at a much shorter time (typically less than seconds), for example, if it is heated to some constant temperature below T_m ($r=\infty$), will it melt? In this case, the actual concentration of atoms in the vapor phase is unimportant unless it influences significantly the surface tension.

For liquid nanoparticles of radius r, its Gibbs free energy curve is represented by the other broken curve, and we note that the energy difference between the two curves of the liquid is the GT potential energy of $p\Omega_\mathrm{l} = 2\gamma_\mathrm{l}\Omega_\mathrm{l}/r_\mathrm{l}$, where γ_l is the interfacial energy between the liquid and the ambient.

The solid-state curve of nanoparticle typically (if $2\gamma_\mathrm{s}\Omega_\mathrm{s}/r_\mathrm{s} > 2\gamma_\mathrm{l}\Omega_\mathrm{l}/r_\mathrm{l}$) intersects the liquid state curve of $r=\infty$ at a lower temperature, $T_\mathrm{m}(r_\mathrm{s})$, indicating that the melting point of the nanoparticles (if many nanoparticles melt simultaneously forming bulk liquid with formally infinite radius of surface) is lower than that of the bulk solid having a flat surface. How much lower in the melting point will depend on γ and r for the solid state and the liquid state. Here is an analysis.

First, we can write the equilibrium condition at the melting point of the nanosolid and liquid particles as

$$\mu_\mathrm{s}^\mathrm{bulk}\left(T_\mathrm{m}^\mathrm{bulk} + \Delta T\right) + \frac{2\gamma_\mathrm{s}\Omega_\mathrm{s}}{r_\mathrm{s}} = \mu_\mathrm{l}^\mathrm{bulk}\left(T_\mathrm{m}^\mathrm{bulk} + \Delta T\right) + \frac{2\gamma_\mathrm{l}\Omega_\mathrm{l}}{r_\mathrm{l}}. \tag{1.4}$$

Expanding the chemical potentials into Taylor series over ΔT including only the first order terms (for not very big size effect) and taking into account that the derivative of chemical potential over temperature is minus entropy, we obtain

$$\mu_{\rm s}^{\rm bulk}\left(T_{\rm m}^{\rm bulk}\right) - S_{\rm s}\Delta T + \frac{2\gamma_{\rm s}\Omega_{\rm s}}{r_{\rm s}} = \mu_{\rm l}^{\rm bulk}\left(T_{\rm m}^{\rm bulk}\right) - S_{\rm l}\Delta T + \frac{2\gamma_{\rm l}\Omega_{\rm l}}{r_{\rm l}}. \tag{1.5}$$

Then, taking into account the equality of the bulk chemical potentials for solid and liquid at the bulk melting temperature, the first term on both sides of the above equation cancels out. Using Clausius relation between the heat of transformation per atom $q_{\rm m}$ and entropy change per atom

$$\left[q_{\rm m} = \int_{\rm solid}^{\rm liquid} TdS = T_{\rm m}^{\rm bulk}\int_{\rm solid}^{\rm liquid} dS = T_{\rm m}^{\rm bulk}\cdot\left(S_{\rm l} - S_{\rm s}\right)\right],$$

we obtain:

$$S_{\rm l} - S_{\rm s} = \frac{q_{\rm m}}{T_{\rm m}^{\rm bulk}}$$

$$\frac{\Delta T}{T_{\rm m}^{\rm bulk}} = -\frac{\left((2\gamma_{\rm s}\Omega_{\rm s}/r_{\rm s}) - (2\gamma_{\rm l}\Omega_{\rm l}/r_{\rm l})\right)}{q_{\rm m}} = -\frac{2\gamma_{\rm s}\Omega_{\rm s}}{q_{\rm m}r_{\rm s}}\left(1 - \frac{\gamma_{\rm l}}{\gamma_{\rm s}}\frac{\Omega_{\rm l}}{\Omega_{\rm s}}\frac{r_{\rm s}}{r_{\rm l}}\right) \tag{1.6}$$

By taking into account the conservation of the number of atoms in the nanoparticle, $(4/3)\pi r_{\rm s}^3/\Omega_{\rm s} = (4/3)\pi r_{\rm l}^3/\Omega_{\rm l}$, we have finally:

$$\frac{\Delta T}{T_{\rm m}^{\rm bulk}} = -\frac{2\gamma_{\rm s}\Omega_{\rm s}}{q_{\rm m}r_{\rm s}}\left(1 - \frac{\gamma_{\rm l}}{\gamma_{\rm s}}\left(\frac{\Omega_{\rm l}}{\Omega_{\rm s}}\right)^{2/3}\right) \tag{1.7}$$

In Eq. (1.7), if we take $\gamma_{\rm l} = \gamma_{\rm s}$ and $\Omega_{\rm l} = \Omega_{\rm s}$, the bracket term becomes zero, it shows no temperature lowering. Typically, we can assume $\Omega_{\rm l} = \Omega_{\rm s}$, and thus we have to assume too $\gamma_{\rm s} > \gamma_{\rm l}$, as depicted in Figure 1.1. Taking the following reasonable values for a metal,

$$\gamma_{\rm s} = 1.5\,{\rm J/m^2},\ \ \gamma_{\rm l} = 1\,{\rm J/m^2},\ \ \Omega_{\rm s} = \Omega_{\rm l} = 10^{-29}\,{\rm m^3},\ \ q_{\rm m} = 2\times10^{-20}\,{\rm J},\ \ r = 10^{-8}\,{\rm m}.$$

We obtain $\Delta T/T_{\rm m}^{\rm bulk} \approx -0.05$, so that the absolute value of melting temperature lowering, ΔT, is about 50° for a metal having a melting point about 1000 K.

It is worth mentioning that the lowering of the melting point due to small radius of solids has been studied long ago in the analysis of morphological instability of solidification in the growth of dendritic microstructures in bulk materials. It is a rather well developed subject by Mullins and Sekerka, so we discuss here only the key issue in solidification very briefly [1]. In Figure 1.2, a schematic diagram of the solidification front having a protrusion is depicted. The heat is being conducted away from the liquid side. Thus we can assume the bulk part of the solid has a uniform temperature of $T_{\rm m}$, but the liquid has a temperature gradient so the liquid in front of the solid is undercooled. The tip of the protrusion has a radius r. If we assume the radius is large and we can ignore the effect of GT potential on temperature, the temperature along

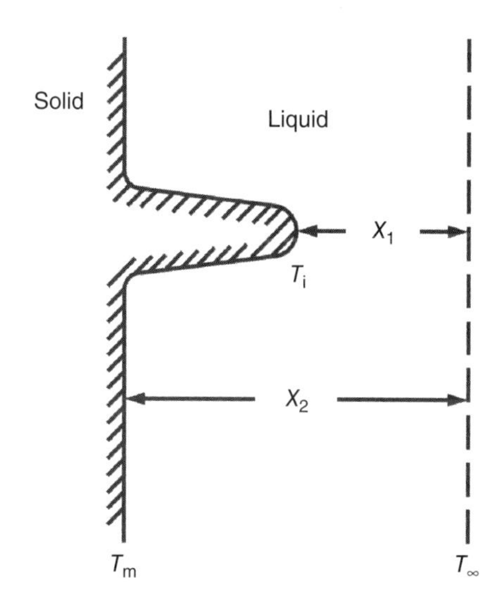

Figure 1.2 A schematic diagram of a solidification front having a protrusion.

the entire solid–liquid interface is T_m everywhere, including the tip. Now, in order to compare the temperature gradient in front of the tip and that in front of a point on the flat interface, we assume a uniform temperature T_∞ in the liquid, which is less than T_m, at a distance away from the front of solidification, as depicted in Figure 1.2. The temperature gradient in front of the tip of the protrusion is larger because $x_1 < x_2$.

$$\frac{T_m - T_\infty}{x_1} > \frac{T_m - T_\infty}{x_2} \tag{1.8}$$

The tip will advance into the undercooled liquid faster than the flat interface. Thus we have dendritic growth; in other words, the flat morphology of the growth front is unstable, and hence we have morphological instability.

However, if we assume now that the radius of the tip is of nanosize, we should consider the effect of GT potential on melting. In Figure 1.2, we assume that the melting point at the tip is T_i, and $T_i < T_m$. With respect to T_∞, the temperature gradient in front of the tip has changed. For comparison, we have now

$$\frac{T_i - T_\infty}{x_1} \Longleftrightarrow \frac{T_m - T_\infty}{x_2} \tag{1.9}$$

There is the uncertainty whether the gradient in front of the tip is larger or smaller than that in front of the flat surface. Because the radius of the tip tends to decrease with growth, the dentritic growth will persist. The optimal growth was found by solving the heat conduction equation and it occurs with the radius $r = 2r^*$, where r^* is the critical radius of nucleation of the solid in the liquid at T_∞. In the growth of thermal dendrites, it is well known that besides primary arms, there are secondary and tertiary arms.

Low melting point of nanospheres may have an important application in microelectronic packaging technology: to lower the melting of Pb-free solder joints. In flip chip technology, solder joints of about 100 μm in diameter are used to join Si chips to polymer-based substrate board. Owing to environmental concern, the microelectronic industry has replaced eutectic SnPb solder by the benign Pb-free solder. The latter, however, has a melting point about 220 °C, which is much higher than that of eutectic SnPb solder at 183 °C. The processing temperature or the so-called reflow temperature is about 30 °C above the melting point of the solder. The higher reflow temperature of Pb-free solder has demanded the use of dielectric polymer materials in the packaging substrate that should have a higher glass transition temperature. The use of polymer of higher glass transition temperature increases the cost of packaging. In addition, the higher reflow temperature also increases the thermal stress in the chip-packaging structure. Thus, solder paste of nanosize particles of Pb-free solder, the Sn-based solder, has been investigated for lowering the melting as well as the reflow temperature. Nevertheless, one of the complications that needs to be overcome is the fast oxidation of Sn nanoparticles in the solder paste.

1.4 NANOSPHERE: FEWER HOMOGENEOUS NUCLEATION AND ITS EFFECT ON PHASE DIAGRAM

Besides melting, other phase transformation properties of nanoscale particles can change with respect to bulk materials. We consider here the effect of nanoparticle size on homogeneous nucleation and then on phase diagrams. Generally speaking, in addition to pressure and temperature, GT potential will affect equilibrium solubility or composition, as shown by GT equation below

$$X_{\mathrm{B},r} = X_{\mathrm{B},\infty} \exp\left(\frac{2\gamma\Omega}{rkT}\right) \tag{1.10}$$

where $X_{\mathrm{B,r}}$ and $X_{\mathrm{B},\infty}$ are the solubility of a solute at the surface of a particle of radius r and ∞, respectively. As phase diagrams are diagrams of composition versus temperature, the equilibrium phase diagrams of bulk materials will be affected when it is applied to nanosize particles.

First, we consider the size effect on homogeneous nucleation in precipitation of an intermetallic compound phase, that is, nucleation within a nanoparticle of a supersaturated binary solid solution. We show that the homogeneous nucleation becomes very difficult and even suppressed.

In the precipitation of a supersaturated binary solid solution, we start from Figure 1.3, which is part of a bulk phase diagram of a two-phase mixture consisting of a practically stoichiometric compound "i," represented by the vertical line, and the boundary of the saturated solid solution, represented by the curved line 1, in Figure 1.3. When the solid solution is in the two-phase region, between the vertical line and the curved line, precipitation of the compound can occur by nucleation and growth.

The transformation starts from the formation of the critical nuclei of the compound phase in the supersaturated solution. For simplicity, we take the nuclei to be spherical. The change of the system's Gibbs free energy because of the nucleation of

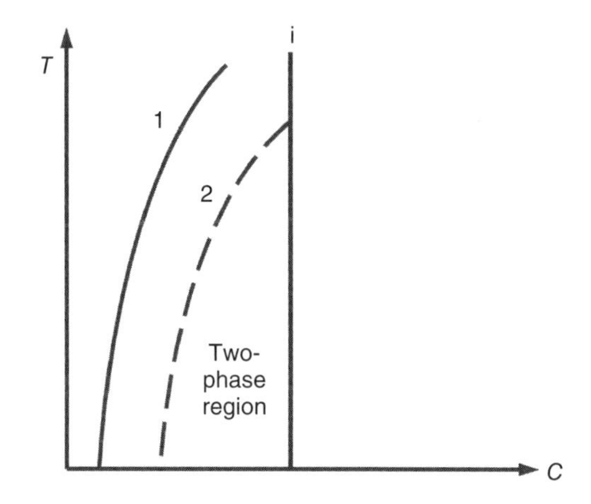

Figure 1.3 Part of a bulk phase diagram of a two-phase mixture consisting of a practically stoichiometric compound "i," the vertical line, and the saturated solid solution, the curved line 1. The broken curve represents the displacement of line 1 due to nanosize solid solution; it narrows down the two-phase region.

the compound sphere with radius r is

$$\Delta G(r) = -\Delta g \cdot \frac{(4/3) \cdot \pi r^3}{\Omega} + \gamma \cdot 4\pi r^2 \tag{1.11}$$

Here $((4/3) \cdot \pi r^3)/\Omega = n$ is the number of atoms in the spherical nucleus of radius r, Ω is atomic volume, Δg is a bulk driving force per one atom of the nucleus (the gain in energy per atom in the transformation), and γ is surface energy per unit area of the nucleus. The driving force, Δg, for macroscopic samples, can be calculated from the construction shown in Figure 1.4 and it is equal to

$$\Delta g = g_\alpha(\overline{C}) + (C_i - \overline{C}) \left. \frac{\partial g_\alpha}{\partial C} \right|_{\overline{C}} - g_i \tag{1.12}$$

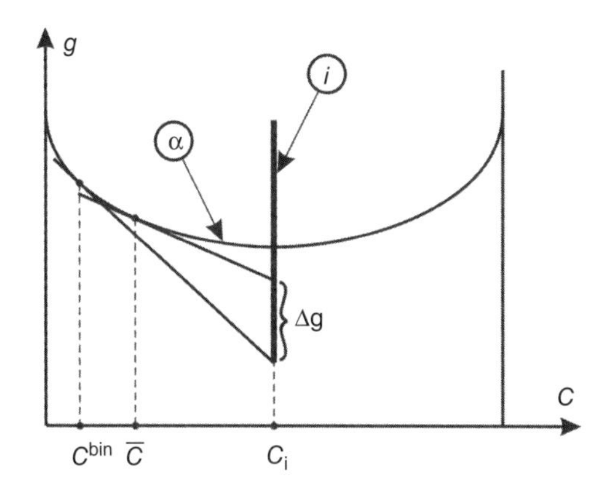

Figure 1.4 The driving force, Δg, for macroscopic samples, can be calculated from the tangent construction as shown. Qualitative concentration dependence of the Gibbs free energy per atom of parent phase α and of line compound phase i with $C_i = 1/2$. Composition C^{bin} of the bulk parent phase in the equilibrium with intermediate phase is determined by the common tangent. The driving force Δg of the bulk transformation is determined by the supersaturation magnitude.

Dependence of $\Delta G(r)$ on r has a maximum (nucleation barrier) at the critical size, at which the first derivative $d\Delta G(r)/dr$ is equal to zero. We obtain

$$r_{cr} = \frac{2\gamma\Omega}{\Delta g} \tag{1.13}$$

and the height of the nucleation barrier is

$$\Delta G^* = \Delta G(r_{cr}) = -\Delta g \cdot \frac{(4/3)\cdot \pi r_{cr}^3}{\Omega} + \gamma \cdot 4\pi r_{cr}^2 = \frac{1}{3}\gamma \cdot 4\pi r_{cr}^2 \tag{1.14}$$

The formation of the critical nucleus of the compound with a fixed composition, c_i, needs the fixed number of the solute atoms or B atoms as given below.

$$N_B^{cr} = C_i \cdot n_{cr} = C_i \cdot \frac{4\pi r_{cr}^3}{3\Omega} = C_i \cdot \frac{4\pi}{3\Omega} \cdot \left(\frac{2\gamma\Omega}{\Delta g}\right)^3 \tag{1.15}$$

where n_{cr} is the total number of atoms in the critical nucleus. If the precipitation proceeds within a limited volume (nanoparticle), we clearly need to consider the limitation due to the fact that a nanoparticle has a finite total number of B atoms;

$$N_B^{tot} = \overline{C} \cdot N = \overline{C}\frac{(4/3)\cdot \pi R^3}{\Omega}$$

where $\overline{C}$ is the fraction of B atoms in the nanoparticle, and R is the radius of the nanoparticle, and N is the total number of atoms in the nanoparticle of radius R. Thus, nucleation (and the transformation as a whole) becomes impossible if

$$C_i \cdot \frac{4\pi r_{cr}^3}{3\Omega} > \overline{C} \cdot \frac{4\pi R^3}{3\Omega} \Longleftrightarrow R < \left(\frac{C_i}{\overline{C}}\right)^{1/3} r_{cr} \tag{1.16}$$

For example, if $r_{cr} = 1\,\text{nm}$, $C_i = 1/2$, $\overline{C} = 0.02$, then nanoparticle of sufficiently small size, $R <\approx 3\,\text{nm}$, cannot have homogeneous nucleation as considered above. Moreover, we expect that even for larger sizes, when nucleation is theoretically possible, the barrier will be high, making the probability of transformation practically impossible. In order to make nucleation possible, we need to increase the concentration of B atoms in the nanoparticle, in turn, we have to move the phase boundary of nanoparticles to a much higher concentration. Combining with a lower melting point of the nano compound phase, we show in Figure 1.3 the phase boundary of nanoparticles by the broken curve, line 2. The two-phase region for nanoparticles is actually narrower than that of the bulk phase.

The above consideration is based on homogeneous nucleation; however, we have to consider heterogeneous nucleation in nanoparticles. Nevertheless, it is known that the crystallization temperature in small droplets of high purity water can be lowered because of the suppression of heterogeneous nucleation as well as the difficulty of homogeneous nucleation. If we assume a high-purity nanoparticle and also assume that the surface energy of the compound phase is higher than that of the nano solid solution phase, the heterogeneous nucleation can be ignored.

1.5 NANOSPHERE: KIRKENDALL EFFECT AND INSTABILITY OF HOLLOW NANOSPHERES

Hollow nanoparticles of CoO or Co_3S_4 were formed when Co nanoparticles were annealed in oxygen or sulfur atmosphere, respectively [2]. The formation of the hollow nanoparticles was explained on the basis of Kirkendall effect by assuming that the out-diffusion of Co is faster than the in-diffusion of oxygen or sulfur during the annealing. We recall that the Kirkendall effect was originally observed in bulk diffusion couple of Cu and CuZn [3]. Markers of Mo wire were placed at the original interface between the Cu and the CuZn. After interdiffusion, the markers were found to have moved into CuZn, indicating that the Zn atomic flux (J_A) is greater than that of Cu atomic flux (J_B). The unbalance of the two atomic fluxes in the interdiffusion has to be balanced by a flux of vacancies,

$$J_V = J_A - J_B$$

which is directed toward the faster diffusing component. (Here we took the absolute values of the fluxes, to show explicitly that the vacancy flux is the difference of two atomic fluxes by absolute value.) Thus, to have void formation within a nanoparticle, we should place the faster diffusing component inside.

The flux of vacancy may or may not lead to void formation. When the vacancy concentration is assumed to be equilibrium everywhere in the bulk diffusion couple, no void forms. Indeed, in Darken's analysis of interdiffusion, there is no void formation because he has assumed that vacancy is in equilibrium everywhere in the diffusion couple. This is because the nucleation of a void requires the supersaturation of vacancy "vapor" or, in other words, the nonequilibrium vacancies.

In a nanosphere, the confinement of vacancies within the spherical shell structure will enable vacancies to accumulate and reach the supersaturation needed to nucleate a void. However, when we consider interdiffusion in a nanosphere, besides Kirkendall effect, we need to consider inverse Kirkendall effect, which is discussed in Section 1.5. Now we consider the role of curvature or the GT effect on the stability of a hollow nanosphere.

Figure 1.5a is a schematic diagram of the cross-section of a hollow nanosphere of a pure phase, in which r_1 and r_2 are the inner and outer radius, respectively. If we consider GT potential of both surfaces, an atom as well as a vacancy in the hollow sphere will be driven to diffuse by the potential gradient between the two surfaces. Here we take three approaches to consider the instability issue.

First, the inner surface has a negative curvature, but the outer surface has a positive curvature. The chemical potential of atoms near r_1 and r_2 are

$$\mu_1 \left(= \mu_0 + \frac{2\gamma}{-r_1}\Omega \right) < \mu_2 \left(= \mu_0 + \frac{2\gamma}{r_2}\Omega \right) \tag{1.17}$$

where μ_0 refers to the chemical potential of atoms in bulk materials. Under the potential gradient, atoms will diffuse from the outer surface to the inner surface, and the vacancies will diffuse in the opposite direction, leading to the elimination of the void

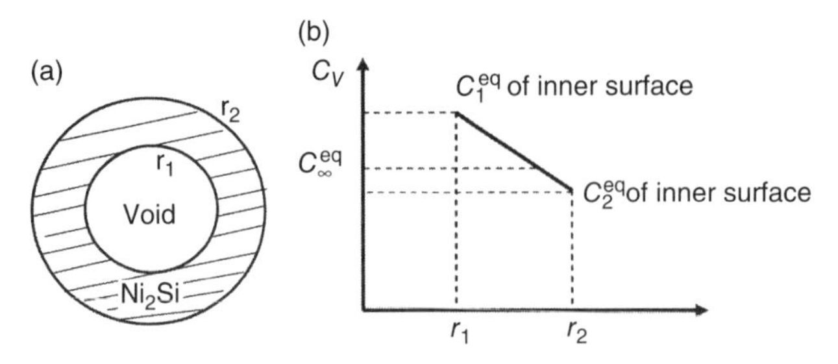

Figure 1.5 (a) A schematic diagram of the cross-section of a hollow nanosphere, in which r_1 and r_2 are the inner and outer radius, respectively. (b) A schematic diagram depicts the vacancy gradient across the shell of the hollow nanosphere.

and to transform the hollow nanosphere to a compact nanosphere finally. Figure 1.5b depicts the vacancy gradient across the shell of the hollow nanosphere. While vacancy concentration gradient inside the shell must be hyperbolic, to be shown later, we approximate it by a straight line because the distance is very short.

The second way is to look at the vacancy gradient by considering the radial stress potential difference between the inner and the outer surface. Because surface energy is positive, both surfaces tend to shrink. The tendency of shrinking of the inner surface exerts a radial tensile stress in the region near the inner surface. Following the Nabarro–Herring model of stress potential in analyzing creep, under tension, it is easier to form vacancies in the tensile region because the formation of a vacancy requires breaking bonds and it is easier to break those bonds that are already stretched under tension. Thus there are more vacancies near the inner surface with respect to the equilibrium vacancy concentration in a region without hydrostatic pressure or stress. In comparison, the outer surface exerts a radial compressive stress to atoms nearby, so there are fewer vacancies. Thus, the vacancy concentration gradient as shown in Figure 1.5b will lead to the diffusion of vacancies from the inner surface to the outer surface.

To avoid misunderstanding, one should remember that even in case of tensile stresses at inner surface and compressive stresses at external surface, the gradient of hydrostatic stress (one-third of the trace of stress tensor) inside spherical or cylindrical layer is zero. (Radial component changes, two tangential components change, and their sum is the same in each point inside shell.) There is no hydrostatic stress gradient inside shell; there is just difference of vacancy concentrations at the boundaries: vacancies diffuse from the inner boundary to the external one not because of some mechanical force, but due to entropic reasons – they diffuse from place with higher concentration (in the vicinity of inner boundary) to the place with lower concentration (in the vicinity of external boundary).

The third way is to examine the work done to form a vacancy in the inner as well as the outer surfaces. In this case, we first consider the formation of a vacancy near the inner surface. We remove an atom near the inner surface in order to leave a vacancy there, and we place the atom on the inner surface, meaning a shrinking of the inner surface (decrease of surface energy). Indeed, if we imagine that the atom of volume Ω is taken from the "bulk" and is "smeared" over the inner spherical surface of radius

r_1, then this radius should be reduced by $dr_1 = -\Omega/4\pi r_1^2$. The corresponding change of surface energy or work done for the formation of such a vacancy is $d(\gamma \cdot 4\pi r_1^2) = \gamma 8\pi r_1 dr_1 = \gamma 8\pi r_1 \cdot (-\Omega/4\pi r_1^2) = -2\gamma\Omega/r_1$. Similarly, we can obtain the change of surface energy or work done due to the formation of a vacancy near the outer surface, and it is positive and equal to $2\gamma\Omega/r_2$. It means that it is easier to form a vacancy near the inner surface than that in the outer surface. The equilibrium vacancy concentration near the inner and the outer surfaces are, respectively,

$$C_V^{eq}(r_1) = \exp\left(-\frac{(E_V^{\infty} - (2\gamma\Omega/r_1))}{kT}\right) = C_V^{eq}\exp\left(\frac{2\gamma\Omega}{kTr_1}\right) \quad \text{and}$$

$$C_V^{eq}(r_2) = \exp\left(-\frac{(E_V^{\infty} + (2\gamma\Omega/r_2))}{kT}\right) = C_V^{eq}\exp\left(-\frac{2\gamma\Omega}{kTr_2}\right) \tag{1.18}$$

where C_V^{eq} is the equilibrium vacancy concentration in the bulk. Thus, the vacancy concentration gradient as shown in Figure 1.5b, will lead to the diffusion of vacancies from the inner surface to the outer surface.

The overall result is in agreement with an energy consideration based on the total surface area. By conservation of volume, we have for the transformation of a hollow sphere to a solid sphere,

$$\frac{4}{3}\pi r_2^3 - \frac{4}{3}\pi r_1^3 = \frac{4}{3}\pi r_0^3$$

where r_0 is the radius of the solid sphere. The reduction in surface area will be

$$4\pi r_0^2 < 4\pi(r_2^2 + r_1^2)$$

In the hollow shell structure as shown in Figure 1.5a, we cannot define what is the equilibrium concentration of vacancies. As there are two surfaces with different potential to serve as references for source and sink of vacancies, no equilibrium vacancy concentration can be given. System will reach equilibrium vacancy concentration only after reaching equilibrium shape, which is after the collapse of a hollow shell into a compact particle. This is unique in a hollow nanosphere. If we assign a vacancy concentration corresponding to the equilibrium vacancy concentration with respect to a planar surface in a bulk sample, C_V^{eq}, it is between the vacancy concentration at r_1 and r_2.

If we anneal the hollow nanosphere at a high temperature to enhance diffusion, the void at the core of the nanosphere will disappear. To estimate the time scale for the filling of a hollow nanosphere, a single elemental phase is assumed for simplicity. If the inner and outer radii of the shell are not too small ($r_i \gg (2\gamma\Omega/kT) \equiv \beta$), then the exponents in the above equations can be expanded, so that we can take the vacancy concentration near the inner surface, r_1, and the outer surface, r_2, respectively, to be

$$C_v(r_1) = C_v^{eq}\left(1 + \frac{\beta}{r_1}\right)$$
$$C_v(r_2) = C_v^{eq}\left(1 - \frac{\beta}{r_2}\right) \tag{1.19}$$

where $\beta = 2\gamma\Omega/kT$, and γ is the surface energy per unit area, Ω is atomic volume, and kT has the usual meaning. We show here that we can use the vacancy concentrations in Eq. (1.19) as the boundary conditions for the diffusion equation in the nanosphere.

In spherical coordinates and if we assume a steady state process, the diffusion equation can be expressed as

$$\nabla^2 C = \frac{\partial^2 C}{\partial r^2} + \frac{2}{r}\frac{\partial C}{\partial r} = \frac{1}{r^2}\frac{\partial}{\partial r}\left(r^2 \frac{\partial C}{\partial r}\right) = 0 \tag{1.20}$$

which means

$$r^2 \frac{\partial C}{\partial r} = \text{const} = -B, \quad dC = -B\frac{dr}{r^2}$$

By integration, we obtain the solution of the diffusion equation to be $C(r) = B/r + A$. By using Eq. (1.19) as boundary conditions, we have

$$C_v(r_1) = \frac{B}{r_1} + A = C_v^{eq}\left(1 + \frac{\beta}{r_1}\right)$$
$$C_v(r_2) = \frac{B}{r_2} + A = C_v^{eq}\left(1 - \frac{\beta}{r_2}\right)$$

By solving the last two equations for A and B, we have

$$B = C_v^{eq}\beta\frac{r_2 + r_1}{r_2 - r_1}$$
$$A = C_v^{eq}\left(1 - \frac{2\beta}{r_2 - r_1}\right)$$

So we obtain the concentration profile of vacancies as [4],

$$C_v(r) = C_v^{eq}\beta\left(\frac{r_2 + r_1}{r_2 - r_1}\right)\frac{1}{r} + C_v^{eq}\beta\left(-\frac{2}{r_2 - r_1}\right) + C_v^{eq} \tag{1.21}$$

Knowing $C_v(r)$, we can calculate its first derivative at r_1. Then, using Fick's first law of diffusion, we obtain the total flux of vacancies, J, leaving (or atoms arriving at) at the spherical surface of r_1. The volume of the void is $V = (4/3)\pi(r_1)^3$, the number of atoms needed to fill the void is $N = V/\Omega$, where Ω is the atomic volume. We can take $N = JA't$, where A' the surface area of the void and t is time, but A' is shrinking with time; instead, we can take $N = \overline{J}t$, where $\overline{J}$ is an average total flux during shrinking and t is the shrinkage time, and the time needed to fill the void can be estimated roughly to be

$$t \cong \frac{kT}{A\gamma D\Omega} r_1^3 \tag{1.22}$$

where A is a constant of the order of 10, and $D \approx D_V C_V^{eq}$ is the self-diffusion coefficient of the atoms in the nanosphere.

If we take Au as an example because we know its surface energy and self-diffusivity very well, $\gamma = 1400\,\text{erg/cm}^2$ and $D = 0.1 \times \exp -(1.8\,\text{eV})/\text{kT}\,\text{cm}^2/\text{s}$. Assume a hollow nanosphere having $r_1 = 30\,\text{nm}$ and $r_2 = 60\,\text{nm}$, it will take about 5×10^3 s at 400 °C to transform the hollow nanosphere to a solid nanosphere. In the case where the hollow particles has $r_1 = 3\,\text{nm}$ and $r_2 = 6\,\text{nm}$, the required transformation time is only a few seconds.

We can simplify the above analysis, without solving the diffusion equation in spherical coordinates, by assuming that the vacancy flux is steady because the thickness of the nanoshell is extremely small, and thus the vacancy flux is given as

$$\begin{aligned} J_V &= -D_V \frac{\Delta C_V}{\Delta r} = -D_V C_V^{eq} \frac{(1 + (\beta/r_1)) - (1 - (\beta/r_2))}{r_1 - r_2} \\ &= -D_V C_V^{eq} \beta \frac{((1/r_1) + (1/r_2))}{r_1 - r_2} \\ &= D_V C_V^{eq} \frac{2\gamma\Omega}{kT} \left(\frac{1}{r_1} + \frac{1}{r_2} \right) \frac{1}{\Delta r} \end{aligned} \tag{1.23}$$

where $\beta = 2\gamma\Omega/kT$ and $\Delta r = r_2 - r_1$. Then we assume this average vacancy flux will remove the void in time "t" and we take $(4/3)(\pi r_1^3)(1/\Omega) = J_V(4\pi r_1^2)t$. We reach the same conclusion as that given by Eq. (1.22).

We can have several kinds of nano hollow spheres; they are a pure element, an intermetallic compound phase, an alloy phase, and a coaxial bilayer structure. We have discussed here the kinetic behavior of nano hollow spheres of a pure element. In the following section, we consider nano hollow spheres of an alloy or solid solution phase for the consideration of inverse Kirkendall effect. Then we study the nano hollow spheres having a coaxial bilayer structure for the consideration of interdiffusion. More detailed studies are presented in Chapter 3.

1.6 NANOSPHERE: INVERSE KIRKENDALL EFFECT IN HOLLOW NANO ALLOY SPHERES

If the hollow nanosphere is an alloy phase, the vacancy diffusion as discussed in the previous section will induce the inverse Kirkendall effect. The classic Kirkendall effect of interdiffusion in a diffusion couple of A and B showed that when the flux of A is not equal (by absolute value) to the counter flux of B, a vacancy flux will be generated to balance the interdiffusion. The inverse Kirkendall effect refers to the effect when a preexisting vacancy flux (generated by some external force) affects the

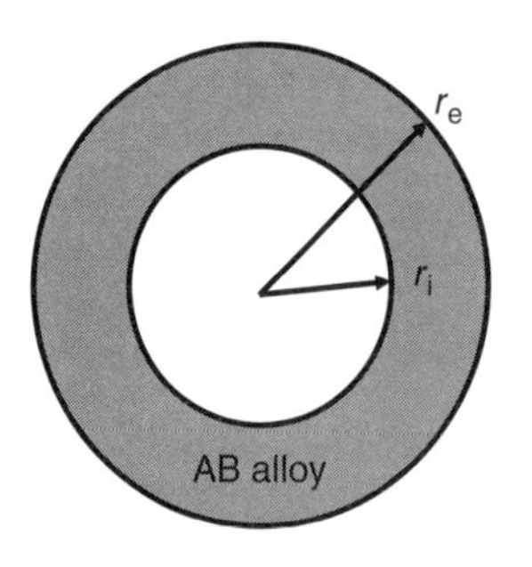

Figure 1.6 A schematic diagram depicts a homogeneous AB alloy with a hollow shell structure. A vacancy gradient exists and the vacancy diffusion from the inner side to the outer side, and consequently, will lead to dealloying when the intrinsic diffusion of A and B is different.

interdiffusion of A and B in a homogeneous alloy of AB. A classic example is the irradiation of a homogeneous alloy in a nuclear reactor. Under irradiation, segregation in a homogeneous AB alloy occurs and the alloy becomes inhomogeneous because the irradiation has produced excess vacancies and in turn a flux of vacancy in the alloy. The diffusion of the vacancies has led to the interdiffusion and segregation of A and B in the homogeneous alloy.

We consider a homogeneous AB alloy with a hollow shell structure, as shown in Figure 1.6. When such a hollow nanoshell is annealed at a constant temperature, dealloying or segregation of A and B occurs. This is different from thermomigration or the Soret effect, which occurs when a homogeneous alloy is annealed in a temperature gradient. The segregation in the nano shell alloy takes place isothermally, so there is no temperature gradient. Following GT effect, there will be a higher vacancy concentration near the inner shell surface than that near the outer shell surface. Because of the vacancy concentration gradient, a vacancy flux exists, and the diffusion of vacancies will affect the diffusion of A and B atoms. If we assume that the intrinsic diffusivity of A and B are different, it leads to dealloying or segregation in the hollow alloy shell. The faster diffusing species will segregate to the inner shell and create a gradient of chemical potential to retard the vacancy flux. The diffusion of A and B is uphill. Provided that the vacancy potential is larger than the counterpotential of dealloying, the hollow alloy shell will eventually transform to a solid nanosphere in order to reduce the total surface area, but the rate is typically slower than that of a pure phase. The kinetic analysis is presented in Chapter 3.

1.7 NANOSPHERE: COMBINING KIRKENDALL EFFECT AND INVERSE KIRKENDALL EFFECT ON CONCENTRIC BILAYER HOLLOW NANOSPHERE

When we consider the interdiffusion of A and B in a planar two-layer structure as shown in Figure 1.7a, we have only the Kirkendall effect when the atomic fluxes of A and B are unequal. When we bend the planar bilayer into a nano shell having a coaxial bilayer structure, as shown in Figure 1.7b, the Kirkendall effect and inverse Kirkendall effect coexist [5]. How they interact with each other is not straightforward and it requires a careful analysis.

If A is the outer layer and B is the inner layer and if the flux of A, J_A, is bigger than the flux of B, J_B, the balancing vacancy flux, J_V, due to Kirkendall effect, will

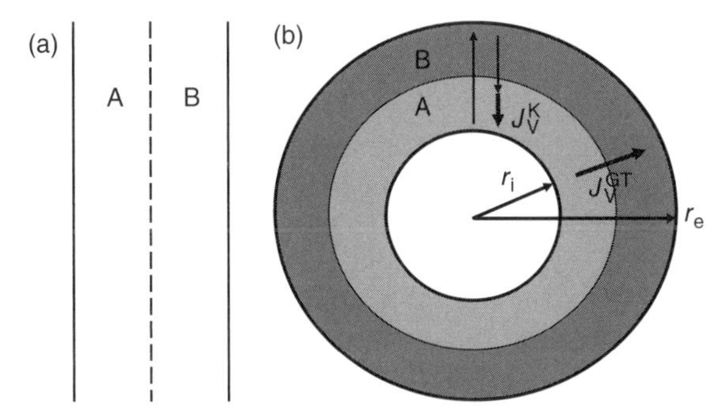

Figure 1.7 (a) A schematic diagram depicts the interdiffusion of A and B in a planar bilayer structure, in which Kirkendall effect occurs. If we assume the flux of A is larger than the flux of B, a vacancy flux will accompany B as indicated by the arrow. (b) If we bend the planar bilayer into a coaxial nano shell structure, the Gibbs–Thomson potential will create a vacancy flux outward. Now we have both the Kirkendall effect and the inverse Kirkendall effect. Depending on whether A is the outer layer or B is the outer layer, the two vacancy fluxes can go in the same direction or in the opposite direction.

diffuse outward, which will accompany the vacancy flux due to GT effect. Conversely, if $J_A < J_B$, the two vacancy fluxes will move in the opposite directions. Then if we have A as the inner layer and B as the outer layer, the case will be different. A detailed analysis of the interaction between the Kirkendall effect and the inverse Kirkendall effect in nanospheres is presented in Chapter 3.

1.8 NANO HOLE: INSTABILITY OF A DONUT-TYPE NANO HOLE IN A MEMBRANE

In biological materials, we often encounter a nano hole in a membrane. The stability of the nano hole is of interest. To analyze the kinetic problem, we consider the geometry and energy of formation of a nano hole in a sufficiently large area (formally infinite) membrane. Again, it is a kinetic problem concerning both a negative curvature and a positive curvature.

By conservation of matter, we assume the edge of the hole is like the one in a donut. The cross-section of such a hole is depicted in Figure 1.8a. To form the hole, we imagine that it is transformed from a homogeneous membrane, and its cross-section is depicted in Figure 1.8b. We consider that the volume of $\pi(r + 2R)^2 H$ of the membrane is transformed to the donut-type of hole, where r is the radius of the hole (negative curvature) and R is the radius of the donut ring (positive curvature) and H is the thickness of the membrane. By conservation of volume,

$$\pi(r + 2R)^2 H \cong \pi R^2 2\pi(r + R) \tag{1.24}$$

In the above equation, we have approximated the volume of the donut ring (toroid) as the volume of a cylinder of cross-section of πR^2 and height of $2\pi(r + R)$.

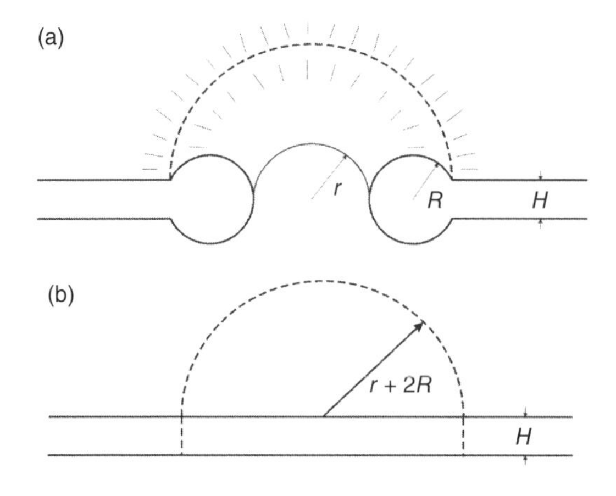

Figure 1.8 (a) The cross-section of a donut-type hole. (b) Regarding the formation of the hole, we imagine that it is transformed from a circular disc in a homogeneous membrane and its cross-section is depicted as a semicircular disc.

Equation (1.24) can be reduced to a quadratic equation for r as function of R.

$$r^2 - 2\left(\frac{\pi R^2}{H} - 2R\right) r + 4R^2 - \frac{2\pi R^3}{H} = 0$$

The solution of r is

$$r = R\left(\frac{\pi R}{H} - 2\right) + \sqrt{R^2\left(\frac{\pi R}{H} - 2\right)^2 + 2R\left(\frac{\pi R}{H} - 2\right)} \tag{1.25}$$

Next, we analyze the change of surface energy as a result of the formation of the nano hole.

$$\frac{1}{\gamma}\Delta G^{\text{surface}} \cong -2\pi(r + 2R)^2 + (2\pi R - H)2\pi(r + R) \tag{1.26}$$

In the above equation, again we have approximated the surface of the donut ring as that of a cylinder of $2\pi R[2\pi(r + R)]$.

To proceed, we take nondimensional parameters, $x \equiv R/H$ and $y \equiv r/H$. We convert Eqs (1.25) and (1.26) respectively into

$$y = x\left[(\pi x - 2) + \sqrt{(\pi x - 2)^2 + 2(\pi x - 2)}\right] \tag{1.27}$$

$$\frac{\Delta G^{\text{surface}}}{2\gamma\pi H^2} \cong (2\pi x - 1)(y + x) - (y + 2x)^2 \tag{1.28}$$

Next, we substitute y, expressed by Eq. (1.27), into Eq. (1.28), we obtain

$$\frac{\Delta G^{\text{surface}}}{2\gamma\pi H^2} \cong f(x) = -x^2\left(\pi x + \sqrt{(\pi x)^2 - 2\pi x}\right)^2 + (2\pi x - 1)x\left(\pi x - 1 + \sqrt{(\pi x)^2 - 2\pi x}\right) \tag{1.29}$$

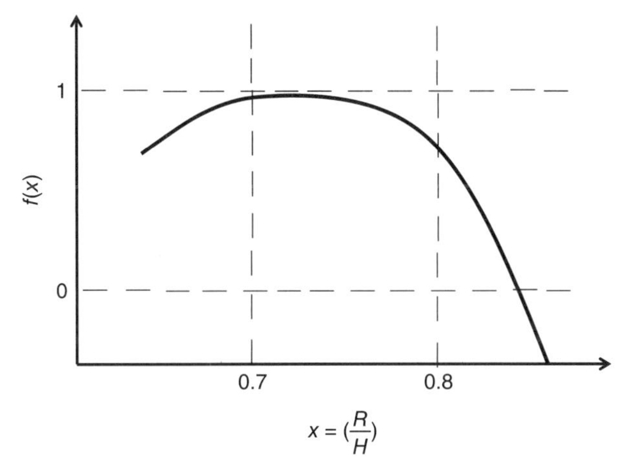

Figure 1.9 A plot of $f(x) = \Delta E^{\text{surface}}/2\pi\gamma H^2$ versus $x = R/H$. The function $f(x)$ has a maximum in the interval of $(2/\pi, \infty)$. The maximum is at $x^{\text{crit}} = R^{\text{crit}}/H \cong 0.69$. Or, at $y^{\text{crit}} = r^{\text{crit}}/H \cong 0.53$.

When we plot the $f(x)$ versus x, we find that this function has a maximum in the interval of $(2/\pi, \infty)$, as shown in Figure 1.9.

The maximum is at $x^{\text{crit}} = R^{\text{crit}}/H \cong 0.69$. Or, at $y^{\text{crit}} = r^{\text{crit}}/H \cong 0.53$.

It means that a small hole with radius less than the critical one will shrink, but large holes will grow. This behavior is similar to the homogeneous nucleation of subcritical and overcritical nuclei.

Physically, we may interpret the above finding by using GT potential of negative and positive curvatures. In the inner surface of donut ring, we have $\mu = \mu^{\text{planar}} + \gamma\Omega((1/R) - (1/r))$, and it is a competition between the positive curvature and the negative curvature when the ring grows or shrinks. If the radius of the hole is very small, the negative curvature wins, which means the Gibs–Thomson part of chemical potential at the surface becomes negative. Then there is the driving force to drive atoms into the hole, and the hole starts to shrink. If the hole is large, in comparison to the membrane thickness, the positive curvature wins, and the chemical potential will be positive. It will provide a driving force to take atoms away from the surface of the hole, and the hole will start to grow.

Furthermore, if the membrane is under biaxial stress, the picture will remain the same, except that the critical size of the hole will be shifted. Under tensile stress, the critical size of the hole will be less.

The dependence of hole stability on membrane thickness is of interest, especially when the thickness is monoatomic as in graphene. A hole in graphene, except a vacancy, will grow when desorption occurs.

1.9 NANOWIRE: POINT CONTACT REACTIONS BETWEEN METAL AND SILICON NANOWIRES

Silicon nanowires have received much attention recently; its processing, characterization, and properties have been studied widely. The most popular method of growing Si nanowires is the vapor–liquid–solid (VLS) method. Because the growth direction

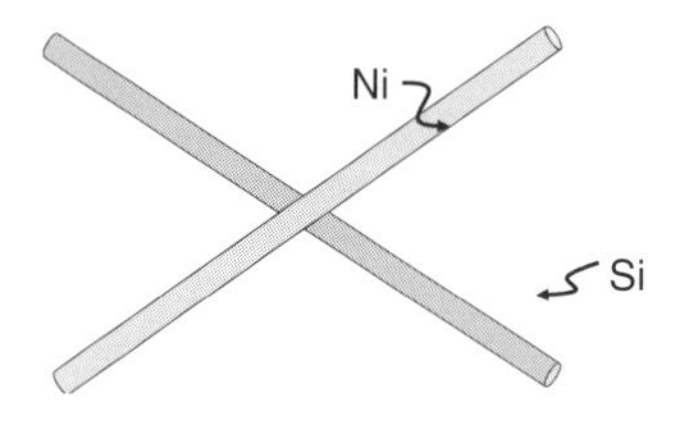

Figure 1.10 Schematic diagram of two nanowires crossing each other to form a point contact.

of Si nanowire is [111], it has created the most challenging issue in the integration of Si nanowires on a [100] oriented Si wafers. This is because in microelectronic complementary metal–oxide–semiconductor (CMOS) technology, Si wafers and Si chips are [100] oriented. Recently, an array of [100] oriented Si nanowires have been obtained by etching a [100] oriented Si wafer.

It is no doubt that manipulating a large number of nanowires into an ordered structure for device application is a very challenging issue. However, it is possible to arrange a small number of nanowires to produce sensor devices, especially medical devices. Hence was arranged a study of nano silicide formation in Si nanowires to form a heterostructure of silicide/Si/silicide, in which the silicides serve as metallic contacts or electrodes to the middle Si, which acts as the sensor [6, 7]. Silicide is a metal–Si intermetallic compound, and C-54 $TiSi_2$, $CoSi_2$, and NiSi have been used widely as source and drain as well as gate contacts on microelectronic Si devices.

When two nanowires cross and touch each other, the contact they make is a point contact, as depicted in Figure 1.10. How chemical reaction occurs across a point contact is of fundamental interest. High resolution transmission electron microscopy has been used to study point contact reactions between a Si nanowire and a metal nanowire of Ni, Co, or Pt [8, 9]. The metal atoms will diffuse through the point contact and transform the Si nanowire into a nanowire of silicide. The kinetic rate of point contact reactions is supply-limited, which is different from the well-known diffusion-limited and interfacial-reaction-limited chemical reactions in bulk and in thin film materials.

We can regard the point contact as zero-dimension contact. When a nanowire is placed on a flat solid surface or on a thin film pad, the contact is a line contact, which is one-dimensional, provided that the interface of the line contact is smooth. When we deposit a thin film on another thin film or when we make a bulk diffusion couple, the contact is a plane contact, which is two-dimensional. Then in a random mixing of two phases, the contact becomes three-dimensional. In Chapter 7, we present a systematic discussion of interfacial chemical reaction in these contacts.

1.10 NANOWIRE: NANOGAP IN SILICON NANOWIRES

In a heterostructure of silicide/Si/silicide, when the length of the middle Si is of nanoscale, we define it as a nanogap structure. Most of silicides are metallic and can serve as electrodes for the Si nanogap.

Figure 1.11 is a schematic diagram of two Ni nanowires crossing a Si nanowire. On annealing to 700 °C, Ni diffuses from the point contact into the Si and forms silicide from both ends of the Si nanowire. Because the Si nanowire has a finite length, the two growth fronts approach each other from the two ends. We can stop the growth fronts before they touch each other by terminating the annealing and we obtain a nanogap structure as shown in Figure 1.12a–c. In Figure 1.12c, we allowed the growth front to join and we have transformed a silicon nanowire completely to a silicide nanowire. Figure 1.13 is a high resolution transmission electron microscopic lattice image of the epitaxial interface between the Si and NiSi. The epitaxial relations have been determined to be

Si(1 1 1)//NiSi(31-1) and Si[1-10]//NiSi[1-12]

The misfit between the Si and NiSi is 5.6%. However, we were unable to find misfit dislocation in the epitaxial interface.

Figure 1.14 shows high resolution transmission electron microscopic images of nanogap of Si down to 2.2 nm length. As the lattice plane images are (111) planes of silicon, we can measure the (111) interplanar spacing in the nanogap and compare it to that of (111) in a strain-free Si wafer. It is found that the Si in the nanogap is highly strained, up to 11% of compression in the nanogap Si of length of 2.2 nm. The high strain is reasonable. This is because across the epitaxial interface, the strain along the

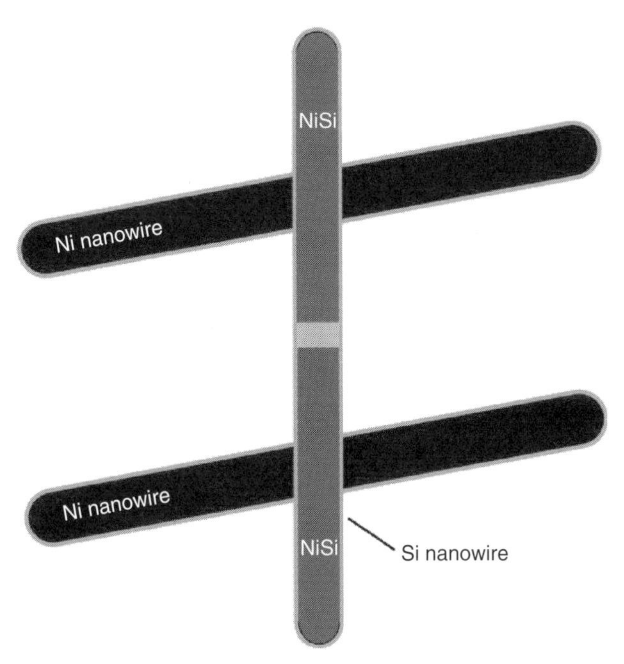

Figure 1.11 A schematic diagram of two Ni nanowire crossing a Si nanowire. On annealing to 700 °C, Ni will diffuse into the Si and form silicide from both ends of the Si nanowire. *Source:* Reproduced with permission from *Nano Letters* 2007.

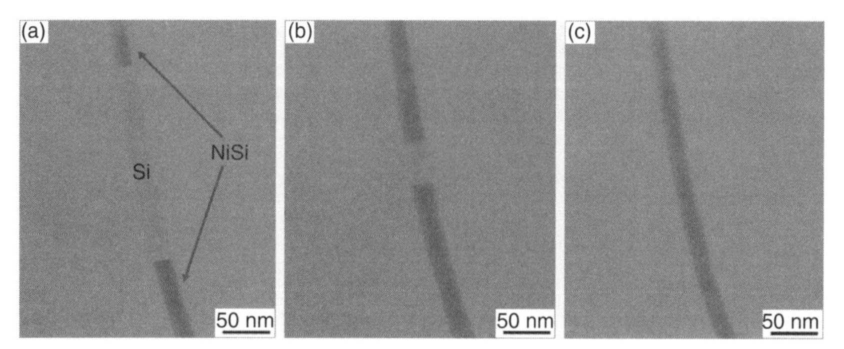

Figure 1.12 (a–c) Hetero-structure of NiSi/Si/NiSi. (a) SEM image of a heterostructure of NiSi/Si/NiSi. (b) The two growth fronts of silicide were stopped before they could touch each other by stopping the annealing and we obtain a nanogap structure. (c) We allowed the growth front to join and we have transformed a silicon nanowire to a silicide nanowire. *Source:* Reproduced with kind permission from *Nano Letters* 2007.

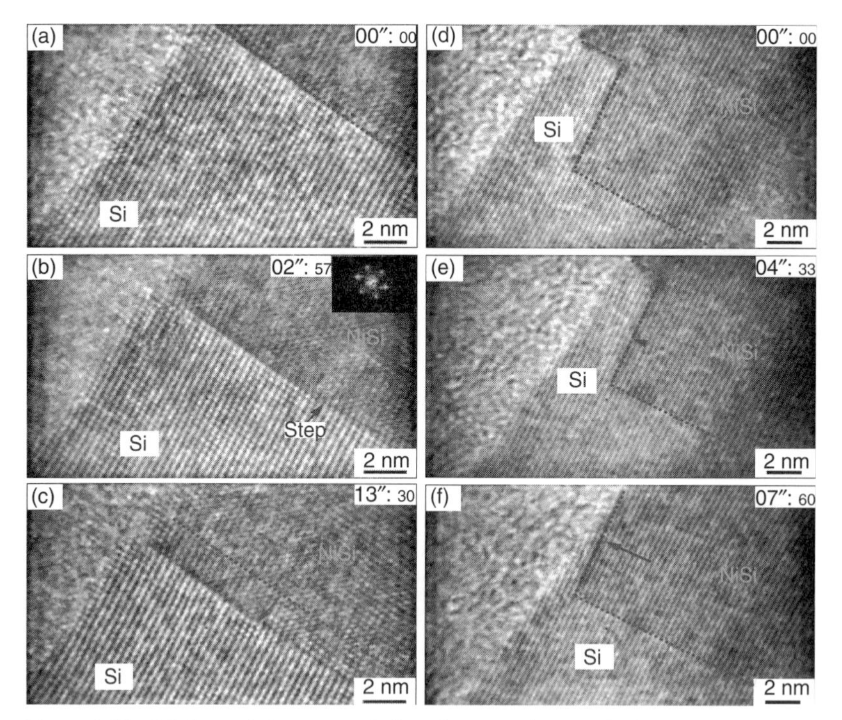

Figure 1.13 High resolution transmission electron microscopic lattice images of the epitaxial interface between the Si and NiSi. No misfit dislocation was found across the interface. *Source:* Reproduced with kind permission from *Nano Letters* 2007.

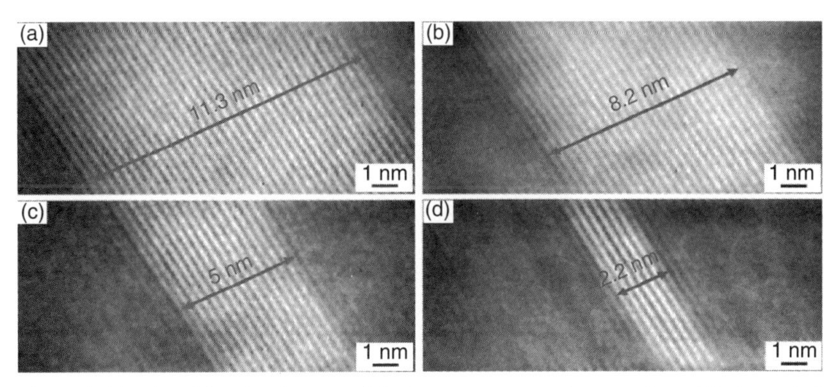

Figure 1.14 (a–d) Lattice images of the hetero-structure down to 2 nm. High resolution transmission electron microscopic images of nanogap of Si down to 2.2 nm length. *Source:* Reproduced with kind permission from *Nano Letters* 2007.

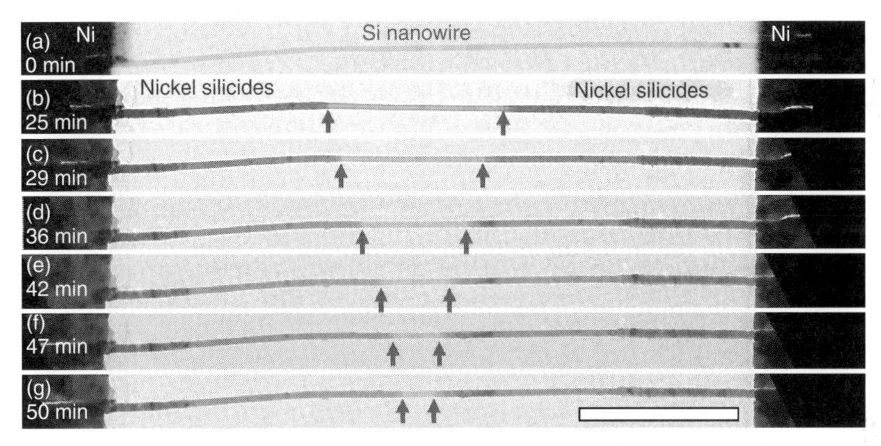

Figure 1.15 A series of *in situ* TEM images of the growth of $NiSi_2$ from two Ni pads and the narrowing of the middle silicon segment. The arrows indicate the silicide/Si reaction fronts. The scale bar is 1 μm. *Source:* Reproduced with kind permission from *Nano Letters* 2012.

two major axes of Si on the Si interface is 5.6% in tension, and under the assumption of constant unit cell volume, the strain in the direction normal to the epitaxial interface is about 11% in compression.

Figure 1.15 shows a series of *in situ* TEM images of the growth of $NiSi_2$ from two Ni pads and the narrowing of the middle silicon segment. The arrows indicate the silicide/Si reaction fronts. The scale bar is 1 μm [10].

Besides NiSi/Si/NiSi, heterostructures of PtSi/Si/PtSi, $CoSi_2$/Si/$CoSi_2$, and Ni_2Si/Si/Ni_2Si have been obtained. It is easier to place a Si nanowire across two metal pads and to anneal the pad/wire/pad structure to form nanogap of Si. Figure 1.16 shows a scanning electron microscopic image of a Si nanowire on two Ni pads followed by an annealing to form the heterostructure of Ni_2Si/Si/Ni_2Si. The nanogap in Figure 1.16 has a length of 8 nm.

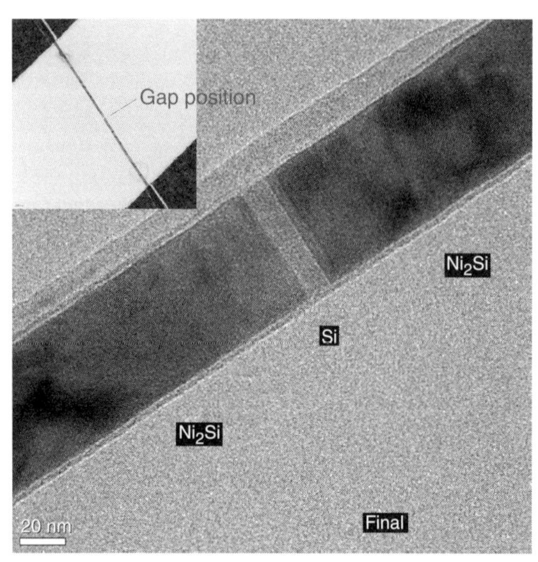

Figure 1.16 Scanning electron microscopic image of a Si nanowire on two Ni pads and followed by an annealing to form the heterostructure of $Ni_2Si/Si/Ni_2Si$. The middle Si gap is 8 nm. *Source:* Reproduced with kind permission from *Nano Letters* 2012.

When a gate oxide and gate electrode is prepared on the nanogap of Si, we have a field-effect transistor and we can study the transistor behavior of the nanogap. On the other hand, we can etch the Si away so that we have an empty gap between the two metallic silicide electrodes. When a biological cell is placed in the empty gap, it is possible to study the electrical properties of the cell.

1.11 NANOWIRE: LITHIATION IN SILICON NANOWIRES

Lithium ion batteries are being developed as energy storage devices in portable electronics as well as in transportation vehicles. Electrochemical reaction between lithium and the electrodes is the most challenging problem for the success of the technology. The reaction controls the capacity, cyclability, and reliability of the battery. Silicon has been found to be a promising anode material with an extremely high theoretical capacity, especially in nanowires-based electrodes [11, 12]. However, lithiation of Si at room temperature is found to involve solid-state amorphization [13] accompanied by extremely large volume expansion and stress generation, resulting in the fracture of the Si electrode. The diffusion and reaction of Li and Si occurs at room temperature and leads to the formation of amorphous Li_xSi alloy. Nanowires of Si have been investigated for better reliability, and no doubt more study is needed.

1.12 NANOWIRE: POINT CONTACT REACTIONS BETWEEN METALLIC NANOWIRES

In thin film solar cell applications, electrodes are required. Transparent and conductive metal oxide thin film such as indium tin oxide (ITO) electrodes have been developed and used widely. For large area flexible organic thin film solar cells and light-emitting diodes, the oxide electrodes tend to crack on bending. Therefore, transparent conductive electrodes consisting random meshes of silver nanowires have been investigated and reported [14, 15]. In the mesh, a high density of point contacts occur between Ag nanowires and these contacts are essential for electrical connection. On thermal annealing at 200 °C for an hour or on passing a high current density at room temperature, many point contacts have been found broken by the breaking of the silver nanowires and the formation of droplets.

It is well known in very-large-scale integration of Si technology that electromigration in the Cu interconnect occurs by surface diffusion at 100 °C under a current density of 10^5 amp/cm^2. Thus, surface diffusion on Ag nanowires can occur very fast at 200 °C.

Figure 1.17 depicts a quarter of the cross-sectional view of point contact reaction between two Ag nanowires. Besides the radius of the two nanowires, we need to consider the very small negative radius following formation of a neck between the two. The negative radius will enlarge by surface diffusion as the ripening reaction continues. As ripening proceeds, thinning will occur in the nanowires. The distance between the thinning region and the point contact will be influenced by the effective surface diffusion distance. It is known that the cylinder has larger surface energy than the a sphere of the same volume, so that cylindrical form is not stable – it is metastable. If, for some reasons – stochastic or deterministic, the cylindric wire becomes thinner in some place, the GT potential of the atoms at the neck, similar to the nano donut hole problem, has two terms

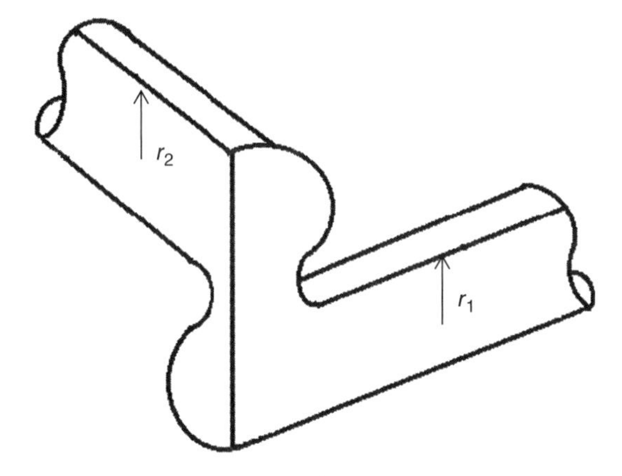

Figure 1.17 A quarter of the cross-sectional view of point contact reaction between two Ag nanowires.

of opposite signs (convex with radius r_{neck} in the plane perpendicular to cylinder axis and concave in the plane along cylinder axis with radius R_{well}): $\mu^{GT}(\text{neck}) = \gamma \cdot ((1/r_{neck}) - (1/R_{well}))$. At the same time, the GT-potential at the cylinder far from neck is just $\mu^{GT}(\text{neck}) = \gamma \cdot (1/R_{cylinder})$. If the neck becomes so narrow that $\gamma \cdot ((1/r_{neck}) - (1/R_{well})) > \gamma \cdot (1/R_{cylinder})$, or $(1/r_{neck}) > (1/R_{well}) + (1/R_{cylinder})$, it will become absolutely unstable. Most probably, the diffusion of atoms from cylinders to the "contact neck" from both sides may make the necks of both cylinders, or at least one of them, less than critical. Eventually, the thinning will lead to breakage by forming a droplet that has a bigger radius than the nanowires. Because of the formation of the neck having a negative radius, ripening can occur between two Ag nanowires of the same radius, and the driving force is the reduction of surface energy.

When a current is passing through the point contact, we need to consider the effect of current crowding [16] as well as joule heating on the point contact reaction. In order to reduce the contact resistance, because of a large number of them, contact reaction occurs to change the contact configuration in order to facilitate the passing of the current. In essence, the radius of the turn from the upper nanowire to the lower nanowire, or vice versa, will change by mass transport because of surface diffusion.

1.13 NANO THIN FILM: EXPLOSIVE REACTION IN PERIODIC MULTILAYERED NANO THIN FILMS

A periodic multilayered thin film of nano thickness is energetically unstable if the two kinds of thin films in the periodic structure can react to form an intermetallic compound [17–20]. When the period is of nano thickness, the very high rate of interfacial reaction can generate a large amount of heat that enables a self-sustained explosive reaction. An example is the multilayered Al/Ni/Al/Ni thin films having a period in nanometer. When the multilayer is disturbed by an electric pulse or by a mechanical shock of the impact of a sharp needle, for example, explosive reaction occurs. The reaction releases a sufficient amount of heat that can cause self-propagation of the reaction front from the point of explosion to the rest of the multilayered thin film and transforms it completely to intermetallic compound. The released heat can also serve as a heat source to trigger other chemical reactions, so the multilayered nano thickness thin films can have applications as ignition materials.

Figure 1.18 is a schematic diagram of a multilayered Al/Ni nano thin film. In the diagram, an Al layer of $2l$ thickness is sandwiched between two Ni layers, each of them has a thickness of l, which we assume to be 10 nm. By repeating the sandwiched structure, we can obtain a multilayered Al/Ni/Al/Ni structure, having a period of Al/Ni of $4l = 40$ nm in thickness. The reason of having the sandwiched Ni/Al/Ni structure is that we assume interdiffusion and reaction occurs by the diffusion of Ni into Al. This is because the melting temperature of Al at 660 °C is much lower than that of Ni over 1400 °C. We note that both Al and Ni are face-centered cubic metals. Thus, at a given reaction temperature, there are much more equilibrium vacancies in Al than that in Ni, so Ni diffuses into Al. No doubt, we can also consider a sandwiched structure of Al/Ni/Al for analysis.

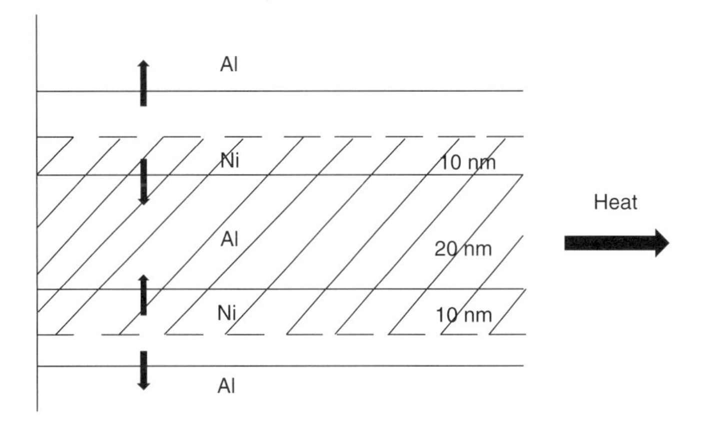

Figure 1.18 A schematic diagram of a multilayered Al/Ni nano thin film.

If we assume, in the explosive reaction, the reaction temperature is close to the melting point of Al, so atomic diffusivity in Al will be near 10^{-8} cm^2/s. Assuming a diffusion-controlled explosive reaction, the time needed for interdiffusion and reaction in the multilayered structure will be

$$\tau = \frac{l^2}{D} = \frac{(10^{-6}\text{cm})^2}{10^{-8}\text{cm}^2/\text{s}} \approx 10^{-4}\,\text{s} \tag{1.30}$$

The reaction is assumed to be exothermic or no heat will be transferred out in the normal direction to the two interfaces, in the y-direction, shown in Figure 1.18. Thus the heat generated by the reaction between Al and Ni will be transferred only in the x-direction. During this time, the heat transfer will travel a distance λ in the x-direction, where

$$\lambda = \sqrt{a^2\tau} = \sqrt{\frac{a^2 l^2}{D}} \tag{1.31}$$

where a^2 is thermal diffusivity in the heat transfer equation below,

$$\frac{\partial T}{\partial t} = a^2 \frac{\partial^2 T}{\partial x^2} \tag{1.32}$$

When two boundary conditions are given, the solution of the above equation expresses the temperature distribution in a sample. The unit of a^2 is cm^2/s, same as that of atomic diffusivity, D. In heat transfer, the heat flux density equation is given as

$$J_{\text{heat}} = -\kappa \frac{\partial T}{\partial x} \tag{1.33}$$

where J_{heat} is heat flux and its unit is joule/cm^2-s and κ is thermal conductivity and its unit is joule/K cm s. Often the unit of thermal conductivity is given as watt/(K cm). If we change thermal energy as joule to electrical energy as eV, and we divide the electrical energy by time, we have watt. It is known that $a^2 = \kappa/C_p\rho$, or $\kappa = a^2C_p\rho$, where C_p is heat capacity per unit mass and its unit is joule/K-m_0 and $\rho = m_0/\Omega$ is mass density, where m_0 is atomic mass and Ω is atomic volume. For metals, if Dulong–Petit law holds, we have $C_p = 3k/m_0$. Thus, $a^2 = \Omega\kappa/3k$. For metals, typically, we have $a^2 = 0.1-1\ \text{cm}^2/\text{s}$. In the above Eq. (1.32), we have assumed

$$\tau = \frac{l^2}{D} = \frac{\lambda^2}{a^2} \tag{1.34}$$

By taking $\tau = 10^{-4}$ s in Eq. (1.34), we find that $\lambda \approx \sqrt{a^2\tau} \approx 10^{-2}$ cm, which is the distance of heat transfer. So, the velocity of explosive propagation or flame velocity can be estimated to be

$$v = \frac{\lambda}{\tau} = \frac{\sqrt{a^2D}}{l} \approx 100\,\text{cm/s} \tag{1.35}$$

Furthermore, if melting occurs, the diffusivity will be $10^{-5}\ \text{cm}^2/\text{s}$, and the velocity will be about 10–100 m/s. The calculated velocity is in good agreement with the measured velocity of reaction propagation, which is reviewed in Chapter 9. The velocity is found to be inversely proportional to the layer thickness and proportional to the square root of the product of thermal diffusivity and atomic diffusivity (interdiffusion coefficient).

Explosive reactions or self-propagating high-temperature synthesis (SHS) reactions were discovered in 1960s, first for powder mixtures and later for multilayered thin films or multifoils. Vast applications in materials joining and in medical applications and military service have been found. Intensive heating during SHS makes multifoils a good candidate for far-away ignitors as well as a limited local heating to avoid large thermal stress, for example, in flip chip solder joining to reduce thermal stress between chip and its packaging substrate. More details on the selection of bilayers and kinetics of explosive reaction are presented in Chapter 9.

1.14 NANO MICROSTRUCTURE IN BULK SAMPLES: NANOTWINS

Recently, a high density of nanotwins in bulk Cu has been prepared by pulse current electroplating. The nanotwinned Cu has been reported to have ultra-high strength, about 10 times stronger than the large (coarse) grained Cu, and yet it maintains a normal ductility and electrical conductivity [21, 22]. The combined properties are unique so the nanotwinned Cu has received a wide attention. This is because how to strengthen a metal has been one of the key issues in metallurgical research. Typically, a hardened material tends to be brittle. There were four known mechanisms of strengthening a metal or an alloy; they are work hardening, solution hardening, precipitation or dispersion hardening, and grain size reduction or the Hall–Petch effect. Now, nanotwin is the fifth one. It is attractive because it does not make the materials brittle.

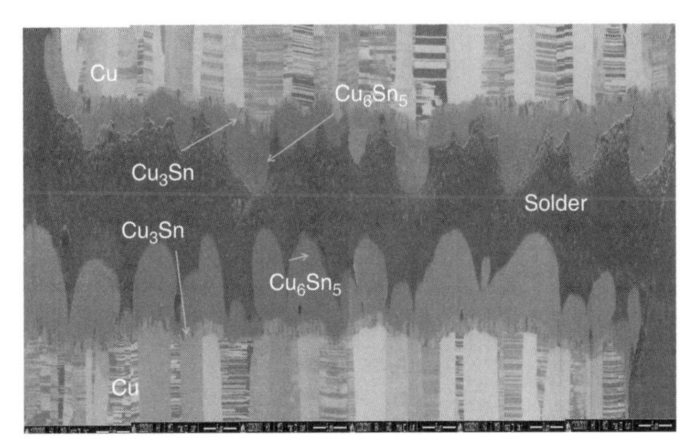

Figure 1.19 Scanning electron microscopic image of the cross-section of a microbump solder joint in 3D IC packaging. From top to bottom, the layers are Cu/Cu_3Sn/Cu_6Sn_5/SnAg/Cu_6Sn_5/Cu_3Sn/Cu. The diameter of the microbump is about 20 μm and the thickness of each Cu layer is about 10 μm. We note that in the Cu layer, there is a high density of nanotwins.

Figure 1.19 is a scanning electron microscopic image of the cross-section of a microbump solder joint in three-dimensional integrated circuit (3D IC) packaging. From the top to the bottom, the layers are Cu/Cu_3Sn/Cu_6Sn_5/SnAg/Cu_6Sn_5/Cu_3Sn/Cu. The diameter of the microbump is about 20 μm and the thickness of each Cu layer is about 10 μm. We note that in the Cu layer, there is a high density of <111> oriented nanotwins. The (111) twin plane of nanotwins is parallel to the free surface. The intermetallic compounds of Cu–Sn are formed by reflow reaction between the Cu and the molten SnAg solder.

In the application of nanotwinned Cu as under-bump-metallization in microbump packaging, the oriented twin plane in the Cu can affect the unidirectional growth of the Cu–Sn intermetallic compound during the reflow reaction. Consequently, in arrays of a large number of microbumps on a Si chip surface in 3D IC packaging technology, we can obtain the same microstructure in every microbump. In other words, the microstructure of a large number of microbumps is not random and can be controlled to have the same microstructure by using the oriented nanotwinned Cu. More details are presented in Chapter 10.

When the high density of <111> oriented nanotwin Cu is annealed at 450 °C for a few minutes, it transforms into unusually large grains of <100> oriented Cu by abnormal grain growth. Figure 1.20a is a SEM image of a large <100> oriented grain in the center surrounded by a large number of much smaller <111> oriented nanotwinned grains in the matrix, after an annealing at 450 °C for 5 min. Figure 1.20b is a SEM image of the same sample after another 5 min at 450 °C. The <100> grain has grown so big that only a part of it is shown on the left-hand side of the image. X-ray diffraction verifies that it is <100> oriented.

The <100> oriented grains are larger than 100 μm. Thus, we can prepare an array of a large number of <100> oriented single crystal of Cu as underbump-metallization (UBM) for 3D IC packaging. More details are presented in Chapter 10.

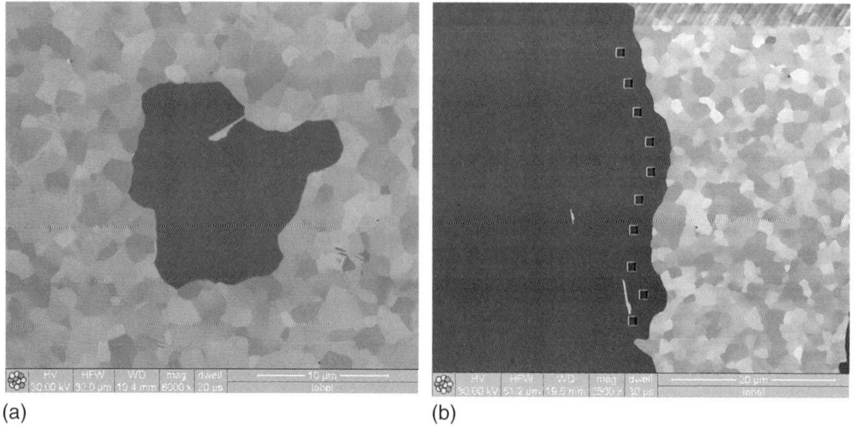

Figure 1.20 (a) SEM image of a large <100> oriented grains in the center surrounded by a large number of much smaller <111> oriented nanotwinned grains. The transformation occurred after an annealing at 450 °C for 5 min. (b) After the annealing at 450 °C for 10 min, the abnormal <100> grain has grown much larger. Only a part of it is shown in the left-hand side of the image. The square markers in the <100> were markers for velocity measurement. Yet, not much growth was found in subsequent annealing. *Source:* Reproduced with kind permission from Chih Chen, National Chiao Tung University, Taiwan.

1.15 NANO MICROSTRUCTURE ON THE SURFACE OF A BULK SAMPLE: SURFACE MECHANICAL ATTRITION TREATMENT (SMAT) OF STEEL

Surface modification of a bulk sample has been studied widely for a long time. We can transform the surface layer of a bulk sample to have a different microstructure and/or different composition from the bulk. The obvious case is the formation of a native surface oxide on a metal, such as Al and Ti. When the oxide is protective, it resists the corrosion of the metal. In microelectronic industry, surface processing of adding and removing a monoatomic layer from a semiconductor single crystal surface are the most essential processing steps, and much effort has been dedicated to the study of homo and hetero epitaxial growth of semiconductor thin films on elemental and compound semiconductors. Another example is atomic layer deposition of gate oxide thin films. Furthermore, we can use ion implantation to convert the surface layer of a Si wafer from crystalline to amorphous, or we can codeposit a metal and Si to form an amorphous silicide layer on a Si wafer followed by annealing to convert the silicide from amorphous to crystalline. More commonly, we can deposit a metal thin film on a Si wafer and then, by chemical interfacial reaction, form a silicide layer on the Si wafer to serve as source/drain and gate contacts.

We note that the native oxide tends to be amorphous, and homo and hetero epitaxial growth require the growth of a single crystal epitaxial layer. For certain applications, we prefer the grain size in the surface layer as large as possible, which is nontrivial because it is known that the grain size in thin films is limited typically by the film thickness. When the film thickness is nanometer, the average grain size is also nanometer. This is especially the case in the depositing of a polycrystalline Si film on a large area glassy substrate for solar cell applications; much effort has been taken to enlarge the grain size in the poly-Si, even by using laser to melt the Si to enhance crystal growth. However, in this chapter, we limit our discussion to the formation of a nanocrystalline layer on the surface of a bulk sample. Recently, it has been found that a nanocrystalline surface layer on a bulk sample of metals and alloys can have better mechanical properties against fracture and wear as well as corrosion.

The use of mechanical deformation by ball milling can cause a gradual and multidirectional plastic damage of the surface layer of a bulk piece of metal, alloy, or steel and lead to the formation of a surface layer up to 50 μm thick of nanoscale grains. This is called *SMAT* [23–25]. The word "attrition" means the deformation is occurring gradually. The surface layer of nanosize grains, due to severe plastic deformation, has stored a large amount of excess energy (entropy) and a high density of crystalline defects of dislocations and subgrains and grain boundaries. The process of strain-induced grain refinement may include grain rotation. During the process of SMAT, the temperature of the surface layer may have increased substantially, which can lead to recrystallization. Yet, the recrystallized grains will be damaged again and again because the process of SMAT is continuous. It has been reported that the excess energy and defects of the surface layer enable rapid diffusion of nitrogen, so nitridation can occur at a much lower temperature than that in a bulk sample without the nano layer. The nitride surface layer can provide protection against corrosion and wear of the steel. The surface layer is hardened, but the bulk remains tough; these are the combined unique requirements of a sword, for instance.

Diffusion of Cr in nanocrystalline α-Fe produced by SMAT was measured at temperatures of 573, 613, and 653 K for 1440, 600, and 120 min, respectively. In the top surface layer of 5 μm thick of the α-Fe, the average grain size is about 10–25 nm. The diffusivity was found to be about 10^{-14} to 10^{-11} cm^2/s, which is about seven to nine orders of magnitude faster than lattice diffusion in α-Fe and about four to five orders of magnitude faster than grain boundary diffusion in α-Fe. For a rough estimate, if we take $x = 5$ μm and $t = 600$ min, we have D ($\approx x^2/t$) to be about 10^{-11} cm^2/s. The measured activation energy for Cr diffusion in the nanocrystalline α-Fe is about 197 ± 21 kJ/mol, which is comparable to that of grain boundary diffusion in α-Fe, about 218 kJ/mol. But the preexponential factor, $D_0 = 14 \times 10^4$ cm^2/s is much higher than those of lattice diffusion at 2.5 cm^2/s and of grain boundary diffusion at 6×10^2 cm^2/s for α-Fe. The large preexponential factor indicates a very large effect of entropy.

Besides the large preexponential factor, the much enhanced diffusivity of Cr in the nanocrystalline Fe was suggested owing to a large volume fraction of nonequilibrium grain boundaries having a high density of dislocations and a considerable amount of triple points in the nanocrystalline layer.

REFERENCES

1. (a) Mullins WW, Sekerka RF. Stability of a planar interface during solidification of a dilute binary alloy. Journal of Applied Physics 1964;35:444–451. (b) Fleming MC. *Solidification Processing*. New York: McGraw-Hill; 1974.

2. Yin Y, Rioux RM, Endonmez CK, Hughes S, Somorjai GA, Alivisatos AP. Formation of hollow nanocrystals through the nanoscale Kirkendall effect. Science 2004;304:711–714.
3. Smigelkas AD, Kirkendall EO. Zinc diffusion in alpha brass. Trans AIME 1947;171:130–142.
4. Tu KN, Goesele U. Hollow nanostructures based on the Kirkendall effect: design and stability considerations. Appl Phys Lett 2005;86:093111-1–093111-3.
5. (a) Gusak AM, Zaporozhets TV, Tu KN, Goesele U. Kinetic analysis of the instability of hollow nanoparticles. Philos Mag 2005;85(36):4445–4464. (b) Gusak AM, Tu KN. Interaction between Kirkendall effect and inverse Kirkendall effect in nanoscale particles. Acta Mater 2009;57:3367–3373. (c) Gusak AM, Zaporozhets TV. Hollow nanoshell formation and collapse in binary solid solutions with large range of solubility. J Phys Condens Mat 2009;21:415303–415313.
6. Patolsky F, Timko BP, Zhang G, Lieber CM. Nanowire-based nanoelectronic devices in the life sciences. MRS Bull 2007;32:142–149.
7. Huang Y, Duan XF, Cui Y, Lauhon LJ, Kim KH, Lieber CM. Logic gates and computation from assembled nanowire building blocks. Science 2001;294:1313–1317.
8. Lu K-C, Tu KN, Wu WW, Chen LJ, Yoo B-Y, Myung NV. Point contact reactions between Ni and Si nanowires and reactive epitaxial growth of axial nano-NiSi/Si. Appl Phys Lett 2007;90:253111–253111-3.
9. Lu K-C, Wu W-W, Wu H-W, Tanner CM, Chang JP, Chen LJ, Tu KN. In-situ control of atomic-scale Si layer with huge strain in the nano-heterostructure NiSi/Si/NiSi through point contact reaction. Nano Lett 2007;7(8):2389–2394.
10. Tang W, Dayeh SA, Picraux ST, Huang JY, Tu KN. Ultrashort channel silicon nanowire transistors with nickel silicide source/drain contacts. Nano Lett 2012;12:3979–3985.
11. Chan CK, Peng H, Liu G, McIlwrath K, Zhang XF, Huggins RA, Cui Y. High performance lithium battery anodes using silicon nanowires. Nat Nanotechnol 2008;3:31–35.
12. Liu XH, Wang JW, Huang S, Fan F, Huang X, Liu Y, Krylyuk S, Yoo J, Dayeh SA, Davydov AV, Mao SX, Picraux ST, Zhang S, Li J, Zhu T, Huang JY. In situ atomic scale imaging of electrochemical lithiation in silicon. Nat Technol 2012;7:749–756.
13. Herd S, Tu KN, Ahn KY. Formation of an amorphous Rh–Si alloy by interfacial reaction between amorphous Si and crystalline Rh thin films. Appl Phys Lett 1983;42:597–599.
14. Lee J-Y, Connor ST, Cui Y, Peumans P. Solution-processed metal nanowire mesh transparent electrodes. Nano Lett 2008;8:689–692.
15. Garnett EC, Cai W, Cha JJ, Mahmood F, Connor ST, Greyson Christoforo M, Cui Y, McGehee MD, Brongersma ML. Self-limited plasmonic welding of silver nanowire junctions. Nat Mater 2012;11:241–249.
16. Tu KN, Yeh CC, Liu CY, Chen C. Effect of current crowding on vacancy diffusion and void formation in electromigration. Appl Phys Lett 2000;76:988–990.
17. Ma E, Thompson CV, Clevenger LA, Tu KN. Self- propagating explosive reaction in Al/Ni multilayer thin films. Appl Phys Lett 1990;57:1262–1264.
18. Clevenger LA, Thompson CV, Tu KN. Explosive silicidation in nickel/amorphous-silicon multilayer thin films. J Appl Phys 1990;67:2894–2898.
19. De Avillez RR, Clevenger LA, Thompson CV, Tu KN. Quantitative investigation of Ti/amorphous-Si multilayer thin film reactions. J Mater Res 1990;5:593–600.
20. Tong M, Sturgess D, Tu KN, Yang J-M. Explosively reacting nanolayers as localized heat sources in solder joints – a microstructural analysis. Appl Phys Lett 2008;92:144101–144101-3.
21. Lu L, Shen YF, Chen XH, Qian LK, Lu K. Ultrahigh strength and high electrical conductivity in copper. Science 2004;304:422–426.
22. Xu D, Kwan WL, Chen K, Zhang X, Ozolins V, Tu KN. Nanotwin formation in copper thin films by stress/strain relaxation in pulse electrodeposition. Appl Phys Lett 2007;91:254105–254105-3.
23. Lu J, Lu K. Chapter 14: Surface nanocrystallization (SNC) of materials and its effect on mechanical behavior. In: *Comprehensive Structure Integrity*. Vol. 8. 2003. p 495–528.
24. Tong WP, Tao NR, Wang ZB, Lu J, Lu K. Nitriding iron at lower temperature. Science 2003;299:686–688.
25. Wang ZB, Tao NR, Tong WP, Lu J, Lu K. Diffusion of chromium in nanocrystalline iron produced by means of surface mechanical attrition treatment. Acta Mater 2003;51:4319–4329.

PROBLEMS

1.1. On Laplace pressure of a sphere, show three ways to derive $p = 2\gamma/r$, where γ is the surface energy per unit area and r is the radius of the sphere.

1.2. Consider the homogeneous nucleation of a circular disc of monolayer thickness on a surface. We smear an atom and add it to the circumference to the disc. Calculate the chemical potential of atoms in the disc by adding one atom to the disc.

1.3. Surface input into chemical potential exists not only for curved interfaces but also for faceted crystals with flat facets. Let crystal has the form of rectangular parallelepiped with the sides a, b, c and corresponding faces 100, 010, and 001 all of which have the same tension γ. What will be the surface chemical potentials at the faces ab, bc, and ca respectively? Atomic volume is Ω. Show that the chemical potentials will be equal if $a = b = c$.

1.4. Please find and derive the Wulff rule for the ratio of surface areas at equilibrium crystal shape.

1.5. On a surface step of height "h," the upper corner has a positive radius of r and the lower corner has a negative radius of r. On annealing at temperature T, surface diffusion occurs and it transports matter from the upper corner to the lower corner, resulting in a round-off of the corners. Assume surface diffusivity is D, calculate the time it takes to change the radius from r to $h/2$.

1.6. On a flat surface, there are two local deviations: a hemispherical dimple and a hemispherical bump of the same absolute value radius R. They are separated by a distance L between their centers. Estimate the rate of shrinkage of both deviations by vacancy diffusion on annealing at T. Assume that the vacancy diffusivity is D and the surface energy of the dimple surface is γ. To simplify the problem, we assume that the process with dimple and the bump will maintain their hemispherical shape and the flat surface reminds unchanged.

1.7. In bulk crystal growth, the shape and facet of a crystal obey Wulff's plot. On nanowire growth, because they tend to have a very high aspect ratio, it seems that we may not use Wulff's plot to explain the shape of nanowires. Why not?

1.8. Consider a hollow nanosphere of Au having an outer and inner radius of 6 and 3 nm, respectively. If we anneal it at 600 °C, it will transform to a solid sphere. Calculate the time needed to do so. Please note that the smallest radius of a void is not zero; it is a vacancy and you can take the diameter of a vacancy to be 0.3 nm.

1.9. We consider the amorphization of a nanoalloy particle. Let nanoparticle of binary nanoalloy has average atomic fraction $\overline{X}$ of component B, which is just in the middle of two stoichiometric compositions of intermetallic compounds 1 and 2: $\overline{X} = (X_1 + X_2)/2$.

Assume that at a given temperature and pressure, the amorphous phase is the metastable phase so that in the macroscopic sample, it is favorable to form a mixture of stable compounds instead of the amorphous phase: $g_{am}(\overline{X}) > (g_1 + g_2)/2$.

It is intuitively clear that in sufficiently small particle, decomposition to form a mixture of the stable compounds may become unfavorable because the additional energy of forming the interface may become larger than the gain in bulk free energy as compared to that in forming the amorphous alloy. Find the critical size of nanoparticle, in which amorphization instead of decomposition becomes favorable.

For simplicity, take all surface tensions of all interfaces as the same: $\gamma_{1/\mathrm{Vac}} = \gamma_{2/\mathrm{Vac}} = \gamma_{1/2} = \gamma_{\mathrm{am}/V} \equiv \gamma$.

1.10. It is well known that the nuclei of atoms behave themselves, in many respects, as the droplets of nuclear, electrically charged liquid consisting of protons and neutrons. Fission of uranium or transuranian elements is the result of competition between repulsive electric forces and capillary forces trying to make the surface energy smaller. In **Fission** (characteristic for massive large nuclei), electrostatic repulsion eventually wins. To the contrary, nuclear **fusion** is actually a coagulation, merging of two nuclear droplets into one. The driving force of fusion is the decrease of total surface energy. In reaction $H_1^2 + H_1^3 \rightarrow He_2^4 + n$, the energy about 17 MeV is released. Taking the radius of one nucleon (proton or neutron) about $R_0 \approx 10^{-15}$ m, and the radius of arbitrary nucleus $R \approx R_0 A^{1/3}$, estimate the surface tension of the nuclear matter.

CHAPTER 2

LINEAR AND NONLINEAR DIFFUSION

2.1 INTRODUCTION

Atomic diffusion is the basic kinetic process in bulk, thin films, and nanomaterials [1–5]. In classical metallurgy, a blacksmith inserts a bar of iron into a charcoal furnace to allow the gas phase of carbon to diffuse into the iron. The time of diffusion in the charcoal furnace is short, so the blacksmith has to take out the red-hot bar and hammer it in order to homogenize carbon in the bar. This process of "heat and beat" is carried out repeatedly to enhance diffusion of carbon and to homogenize carbon in iron so that the Fe–C alloy or steel can be formed.

In microelectronic technology, pure Si is not useful until electrically active n-type and p-type dopants can be incorporated substitutionally into the Si lattice by diffusion. The fundamental structure of a transistor, that is, the p–n junction is obtained by a nonuniform distribution of both n- and p-type dopants in Si in order to

Kinetics in Nanoscale Materials, First Edition. King-Ning Tu and Andriy M. Gusak.

achieve the built-in potential that guides the motion of electrons and holes in the junction. Thus the diffusion of dopants in Si has been a very important processing step in microelectronics, especially in making shallow junction devices. In those nanoelectronic devices to be built using nanowires of Si, the electrodes will be nanosilicide as we have discussed in Chapter 1 on point contact reactions, and the diffusion of metal atoms in the nanowire of Si is the key processing step.

Although diffusion is an old method, modern technology nevertheless makes use of it. The diffusion of Cu in Si is of concern due to the poisoning effect of Cu in forming deep trap states in Si. The diffusion of Cu into Sn to form Cu_6Sn_5 intermetallic compound at room temperature can induce spontaneous Sn whisker growth, and Sn whisker growth is a major reliability issue in satellites. In energy technology, the diffusion of Li in Si is of interest because of the potential application of nanowires of Si as electrodes in rechargeable Li-based battery. Yet the very large volume change in dissolving a large amount of Li in Si is a serious reliability issue.

In bulk and thin film materials, when the diffusion distance is over 100 nm, we can assume that the diffusion is governed by Fick's first and second laws, in which the linear diffusion is defined by having the atomic flux, J, to be proportional to the driving force, dC/dx or $d\mu/dx$, where C and μ are concentration and chemical potential, respectively. In nanoscale materials, when the diffusion distance is below 100 nm, the diffusion may be different from that in bulk and thin film materials because of the very small distance of diffusion. In a very small distance, the chemical potential gradient or the driving force of diffusion can be very large, when atomic flux is proportional to a high order of the driving force so that high order effect or nonlinear effect may exist. In this chapter, we discuss the connection between linear and nonlinear microscopic diffusions in a crystalline lattice. We consider the high order effects from both kinetic and thermodynamic viewpoint.

2.2 LINEAR DIFFUSION

Fick's first law on atomic flux is given below

$$J = -D\frac{\partial C}{\partial x}$$

where the atomic flux, J, has the unit of number of atoms per unit area per unit time, or #atom/(cm^2 s), D is atomic diffusivity and has the unit of cm^2/s, and C is atomic concentration and has the unit of #atom/cm^3. Fick's first law has the same form as Fourier's law of thermal conduction and Ohm's law of electrical conduction. Although the form is the same, the conduction or transport mechanism is different. For example, on atomic diffusion, we ask what is the activation energy of diffusion? This is because of the defect mechanism in atomic diffusion.

It is worth noting that there is a negative sign in Fick's first law. It is conventional because we typically draw the concentration curve from upper left corner to lower right corner in the coordinates of x versus C, in which C decreases from left to right and x also increases from left to right, so the slope is negative. The outcome is a downhill diffusion in which both J and D are positive.

To review the concept of atomic diffusion in bulk crystalline solids, we begin with the vacancy mechanism of diffusion in a face-centered cubic metal. We make the following assumptions in order to develop the analytical model.

(1) It is a thermally activated unimolecular process. Unimolecular process means that we consider the jumping of a single atom in the diffusion process, unlike the diffusion in gas phases, in which it occurs by the collision of two molecules, and unlike the chemical reactions in liquid solutions, which are typically bimolecular processes, such as rock salt formation, where the collision of ions of Na and Cl takes place.

(2) It is a defect mediated process following Kirkendall effect. Here the defect is a vacancy or vacant lattice site. It is known that very small elements such as hydrogen and carbon can diffuse interstitially in metals. We do not discuss interstitial defects here. We note that vacancy is a thermodynamic equilibrium defect, thus there is always an equilibrium concentration of vacancy in the solid to facilitate atomic diffusion. Atomic diffusion occurs by the exchange of lattice site positions between a vacancy and one of its nearest neighbor atoms. In other words, diffusion occurs by the jumping of an atom into the neighboring vacant lattice site.

(3) The thermally activated process requires the diffusing atom to possess enough activation energy to pass through an activated state before it can take the vacant site. The distribution of the activated state is assumed to obey Boltzmann's equilibrium distribution function on the basis of transition state theory. Hence, the Boltzmann distribution function is used to describe the probability of diffusion jumps.

(4) It is assumed that the probability of reverse jumps is large due to small driving force in solid-state diffusion, so we have to consider reverse jump processes. In other words, the diffusion process is near-equilibrium or not far from the equilibrium state.

(5) Statistically, atomic diffusion obeys the principle of random walk.

(6) A long range directional diffusion of atoms requires a driving force or a chemical potential gradient.

2.2.1 Atomic Flux

Atomic flux is represented by the parameter J, which has the unit of #atom/(cm^2 s). We can have three expressions of the unit as shown below:

$$\frac{\#\text{atom}}{\text{cm}^2} \times \frac{1}{\text{s}} = \frac{\#\text{atom}}{\text{cm}^3} \times \frac{\text{cm}}{\text{s}} = \frac{\#\text{atom}}{\text{cm}^4} \times \frac{\text{cm}^2}{\text{s}}$$

In the first expression, we can regard #atom/cm^2 as areal concentration of atoms on an atomic plane and 1/s as the jump frequency of atoms jumping away from the atomic plane. If we divide the areal concentration by an interplanar spacing and multiply the frequency by the interplanar spacing (jump distance), we obtain the second expression or the relation of $J = Cv$, where C is the concentration of atoms per unit volume and v is drift velocity. In the third expression, we note that #atom/cm^4 can be regarded as (#atom/cm^3)/cm, which is the concentration gradient, and cm^2/s is the unit of diffusivity, and we obtain the relation of $J = D(dC/dx)$. How do we link the second expression to the third expression is shown below.

2.2.2 Fick's First Law of Diffusion

We consider a simple one-dimensional case of diffusion and represent the minimum potential energy of a row of atoms by the periodic pair potential curves between every two atoms, as shown in Figure 2.1a. We assume that at positions A and B, there is an atom and a vacancy, respectively, for diffusion consideration.

At equilibrium, which is represented by a horizontal base line of the potential curve, the atom at A is attempting to jump over the potential energy barrier with the attempt frequency, ν_0, to exchange position with the neighboring vacancy at B. The successful exchange or the exchange jump frequency is given below on the basis of Boltzmann's distribution function.

$$\nu = \nu_0 \exp\left(\frac{-\Delta G_{\mathrm{m}}}{kT}\right) \tag{2.1}$$

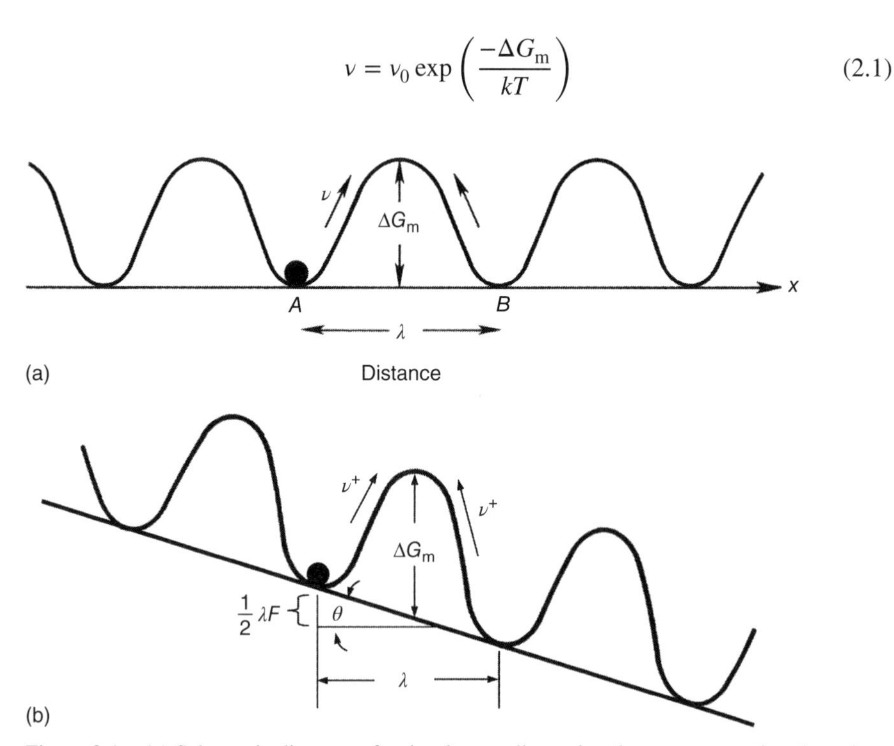

Figure 2.1 (a) Schematic diagram of a simple one-dimensional case representing the pair potential energy of a row of atoms by combining the pair potential curves between every two atoms. (b) To have a directional diffusion, we must introduce a driving force to drive the diffusion in a given direction. This can be represented by tilting the base line of the potential energy. The tilting introduces a gradient of the potential energy, which is the driving force of diffusion.

where ν_0 is the attempt frequency or Debye frequency of atomic vibration, ν is the exchange frequency, and ΔG_m is the activation energy of motion (saddle point energy).

We note that there is a reverse jump at the same attempt frequency. Such diffusion in an equilibrium state will lead to random walk of the vacancy in the lattice. The jumping of the atom to the right is equivalent to the jumping of the vacancy to the left.

To have a directional diffusion or a diffusional flux, we must introduce a driving force to drive the diffusion in a given direction. This can be represented by tilting the base line of the potential energy with a tilt angle of θ as shown in Figure 2.1b. The tilting introduces a gradient of the potential energy; the tilt is from left to right. It has raised the potential energy by the amount of $\Delta\mu$ to help the atom jump into the vacancy. If we assume the interatomic distance or the jump distance is λ, by taking $\Delta x = \lambda/2$, we have

$$\Delta\mu = \Delta x(\tan\theta) = \frac{\lambda}{2}\tan\theta \tag{2.2}$$

In this equation, we can take "$\tan\theta$" as a force because energy equals distance times force. Actually, the force is small or θ is small, and we can take $\tan\theta = \theta$. We recall that force, F, in a potential field of μ is given as potential gradient or the slope of the potential energy curve,

$$F = -\frac{\Delta\mu}{\Delta x} = \tan\theta$$

Note the negative sign and the negative slope. Thus, $\Delta\mu = \lambda F/2$. Under the driving force F (a mathematical expression of F is presented later), the forward jump is increased by $\Delta\mu$ as shown below,

$$\nu^+ = \nu_0 \exp\left(\frac{-\Delta G_\mathrm{m} + \Delta\mu}{kT}\right) = \nu\exp\left(+\frac{\lambda F}{2kT}\right) \tag{2.3}$$

The reverse jump is decreased by

$$\nu^- = \nu_0 \exp\left(\frac{-\Delta G_\mathrm{m} - \Delta\mu}{kT}\right) = \nu\exp\left(-\frac{\lambda F}{2kT}\right) \tag{2.4}$$

And the net frequency is

$$\nu_n = \nu^+ - \nu^- = 2\nu\sinh\left(\frac{\lambda F}{2kT}\right) \tag{2.5}$$

Now, we take the "condition of linearization," that

$$\frac{\lambda F}{kT} \ll 1$$

So we can assume the approximation of $\sinh x = x$, when x is very small. Later, when we consider diffusion in nanoscale distance, where the condition of linearization is no longer true, we cannot take $\sinh x = x$; instead, a higher order term must be added, which is the kinetic origin of nonlinear diffusion, which is discussed later.

In the linear condition, the net jump frequency, ν_n, is "linearly" proportional to the driving force F as shown below, which is the base of linear diffusion.

$$\nu_n = \nu \frac{\lambda F}{kT} \tag{2.6}$$

We can define a drift velocity, υ which equals to net jump frequency times jump distance. The physical meaning of the drift velocity is that under the driving force, all the equilibrium vacancies in the solid will exchange position at the net frequency in the opposite direction of the force, and it will result in a flux of atoms moving with velocity υ in the direction of the force.

$$\upsilon = \lambda \nu_n = \frac{\nu \lambda^2}{kT} F \tag{2.7}$$

The atomic flux, J, which has the unit of number of atoms/cm^2 s, as discussed in the last section, is given as

$$J = C\upsilon = C \frac{\nu \lambda^2}{kT} F = CMF \tag{2.8}$$

where $M = \nu\lambda^2/kT$. In Eq. (2.8), C is the concentration and its unit is number of atoms/cm^3, and υ has the unit of cm/s. Here we note that C represents only those atoms that exchange position with the jumping vacancies, or rather that C represents the vacancy concentration, not the concentration of all the atoms in the volume of cm^3. If we prefer to consider all the atoms in the volume of cm^3, we need to multiple it by the probability of the formation of a vacancy, which is covered later when we discuss atomic diffusivity in Section 2.2.7.

It is worth mentioning that in the literature on diffusion, especially on interdiffusion in an alloy, often the fraction of concentration of a component in the alloy is used, for example, C_A and C_B as the fraction of component A atoms and B atoms in an AB alloy, respectively. Indeed, in Darken's analysis of Kirkendall effect, the equation of marker velocity as well as the equation of interdiffusion coefficient are expressed in terms of the gradient of fraction of the component atoms in the alloy, for example, $X_B = C_B/C$ is the fraction of B atoms in the AB alloy, so X_B is a number and has no unit. Readers are referred to Section 3.1.2 in Chapter 3 on Darken's analysis. Fraction also occurs in the literature on spinodal decomposition.

However, when C_B is defined as atomic fraction, we need to use NC_B or C_B/Ω to represent the number of B component atoms per unit volume, and $N = 1/\Omega$ is defined as the number of atoms per unit volume, where Ω is atomic volume. Or we can define $N = N_A$ as the number of atoms per mole if N_A is the Avogadro number. Throughout this book, we define $C = N/V$, where N is the total number of a certain component atoms in the volume V, and the unit of C is number of atoms/cm^3, not a fraction, except when it is defined specifically. We use X to represent fraction.

On the other hand, when it is convenient to use C to represent fraction in this book or in other textbooks, as in the chemical potential of ideal dilute solution,

$\mu = kT\ln C$, it will be made clear. For a pure element, we have $C = 1/\Omega$ or $C\Omega = 1$. Then in an alloy of AB, which has atomic fraction of A to be X_A and fraction of B to be X_B, we can take $X_A/\Omega = C_A$ and $X_B/\Omega = C_B$.

Now, let us go back to Eq. (2.8), in which the atomic flux J is "linearly" proportional to the driving force, F. Because the drift velocity $\upsilon = MF$, where $M = \nu\lambda^2/kT$, M is defined as atomic mobility, the unit of mobility is cm²/joule s, or cm²/eV s, where eV is the product of electric charge and volt, eV is electric energy. As kT is energy, we can give kT the unit of joule or eV.

Now if we assume AB is an ideal dilute solid solution, we have chemical potential $\mu_B = kT\ln X_B$, which is discussed in the next section. Then we calculate

$$F = -\frac{\partial \mu_B}{\partial x} = -\frac{\partial \mu_B}{\partial X_B}\frac{\partial X_B}{\partial x} = -\frac{kT}{X_B}\frac{\partial X_B}{\partial x} \tag{2.9}$$

Then we have from Eq. (2.8),

$$J_B = C_B\frac{\nu\lambda^2}{kT}F = \frac{X_B}{\Omega}\frac{\nu\lambda^2}{kT}\left(-\frac{kT}{X_B}\frac{\partial X_B}{\partial x}\right) = -\nu\lambda^2\left(\frac{\partial(X_B/\Omega)}{\partial x}\right) = -D_B\left(\frac{\partial C_B}{\partial x}\right)$$

Hence, we have obtained Fick's first law of diffusion

$$\frac{J}{-(\partial C/\partial x)} = D = \nu\lambda^2 \tag{2.10}$$

where D is the diffusion coefficient (or tracer diffusivity) in units of cm²/s. Then, $M = D/kT$.

In Appendix B, we discuss the interdiffusion coefficient, $\widetilde{D}$, and we show that

$$-\frac{J}{(\partial C/\partial x)} = \widetilde{D} = CMf'' \tag{2.11}$$

where f'' is the second derivative of Helmholtz free energy per unit volume of the solid solution. The sign of interdiffusion coefficient takes the sign of f'', so it can be positive or negative. In the spinodal region, because f'' is negative, $\widetilde{D}$ is negative and the diffusion is uphill.

At the inflection point, f'' is zero, $\widetilde{D} = 0$, but $D = \nu\lambda^2 \neq 0$. It means that at the inflection point, atoms can still diffuse by random walk, but there is no chemical force to drive the atoms in a particular direction. Thus, an amorphous alloy having a composition near the inflection point is kinetically more stable. Whether or not phase transformation can occur at the inflection point is an interesting question; it deserves a numerical simulation analysis.

2.2.3 Chemical Potential

We recall that in a potential field, the physical meaning of driving force "F" is generally defined as a potential gradient,

$$F = -\frac{\partial \mu}{\partial x}$$

We note that the negative sign is a convention. For atomic diffusion, μ is the chemical potential of an atom (or a mole of atoms) in the diffusion field, and it is defined at constant temperature and constant pressure to be

$$\mu = \left(\frac{\partial G}{\partial C}\right)_{T,p}$$

where G is Gibbs free energy per unit volume and C is concentration. Physically, the differential means that when we change a very small amount of C by ΔC, the change in G will be ΔG, which at constant temperature and pressure is given by

$$\Delta G = \left(\frac{\partial G}{\partial C}\right) \Delta C$$

In diffusion, we diffuse an atom of A into an AB alloy, so the change in ΔC is one atom. In the last equation, when we take $\Delta C = 1$, we have $\Delta G = \mu$. Thus we may regard chemical potential as the change in free energy of an alloy when its composition is changed by adding or removing one atom.

We show below that in an ideal dilute solution, $\mu = kT \ln X_A$, where X_A is a concentration fraction and has no unit. Ideal solution is defined by having $\Delta H_{mix} = 0$ and ΔS_{mix} = configuration change of entropy (ideal mixing). Thus the Gibbs free energy of the solution is

$$\Delta G_{mix} = \Delta H_{mix} - T\Delta S_{mix} = -T(X_A \ln X_A + X_B \ln X_B) \tag{2.12}$$

where X_A and X_B are the concentration fraction of A and B in the solution. For dilute solution, we can assume $X_B \ll 1$, then we obtain

$$\mu = \frac{d\Delta G_{mix}}{dC} = kT \ln X_B \tag{2.13}$$

Again, we note that in Eq. (2.13), very often it is expressed as the chemical potential of an ideal and dilute solution by $kT \ln C$, where C should be a number and has no unit. On the other hand, in Fick's first law, Eq. (2.10), we use C to represent concentration and it has unit.

From the point of view of diffusion, besides chemical potential of ideal dilute solution, we have already mentioned, in Section 1.2, the Gibbs–Thomson chemical potential of curvature, $\mu = 2\gamma\Omega/r$. We have considered the energy change in adding one atom to a nanosphere by smearing the atomic volume of Ω over the surface of the nanosphere, and we obtain the energy change to be $\mu = 2\gamma\Omega/r$, which by definition is the chemical potential. Furthermore, in constant temperature creep, we have from Nabarro–Herring model that the hydrostatic stress or the normal stress can be taken as energy density, in which case the energy change per atom or per vacancy is $\mu = \sigma\Omega$, where σ is normal stress. Later, in Chapter 5 on spinodal decomposition, we discuss the chemical potential of an inhomogeneous solid solution. These are the four types of chemical potential that we found often in dealing with diffusion related kinetics.

2.2.4 Fick's Second Law of Diffusion

In the above, we have derived Fick's first law under a constant driving force, so the atomic flux is steady and is assumed to be constant with position and time because the diffusivity is assumed to be unchanged at a constant temperature. However, in most problems of diffusion, the atomic flux changes with position and time even though the temperature is constant. To handle such a nonsteady state problem, we need to derive Fick's second law of diffusion, which is a continuity equation from the "principle of conservation of mass" and we need to use Gauss theorem of flux divergence. We review briefly the idea of flux divergence below.

We consider a simple cube of six square surfaces and we consider six fluxes of atoms, passing normally through the six surfaces of the cube, shown in Figure 2.2. In a period of Δt, the number of atoms entering one of the square surfaces, A_1, will be

$$\Delta N_1 = -J_1 A_1 \Delta t \tag{2.14}$$

We note that there is a rule of sign of the flux entering or leaving an enclosed volume. Flux is a vector and the surface normal is another vector. For an outgoing flux, the angle between the flux vector and the surface normal vector is 0°. Because $\cos 0° = +1$, the sign is positive. For an entering flux, the angle between the two vectors is 180°. As $\cos 180° = -1$, the sign is negative.

Atomic flux J is defined as number of atoms/cm^2 s, or number of atoms passing through a unit area in a unit time, thus JAt gives the number of atoms. If we sum all the atomic fluxes, J_i, passing through each surface of area, A_i, for a given period of time, Δt, or we sum all the atoms coming in and out of the cubic volume, we have

$$\sum_{i=1}^{6} J_i A_i \Delta t = \sum_{i=1}^{6} \Delta N_i = \Delta N$$

where ΔN is the net change of the total number of atoms inside the cubic box, which we note can be positive (net gain) or negative (net loss). If the volume of the cubic box is V and C is the concentration in the box, the change in concentration is

$$\Delta C = \pm \frac{\Delta N}{V}$$

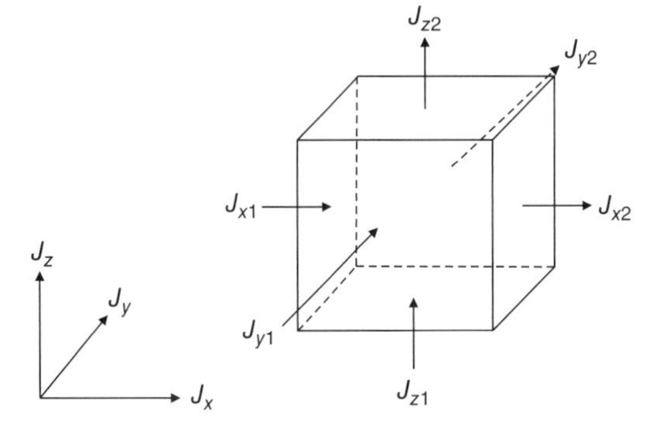

Figure 2.2 A simple cubic of six square surfaces and six fluxes of atomic passing through the six surfaces of the cubic.

The sign in the above equation is determined physically, by the net change whether it gains or loses. From the last two equations, we obtain

$$\frac{dN}{dt} = V\frac{dC}{dt} = \sum_{i=1}^{6} J_i A_i \tag{2.15}$$

According to Gauss theorem, the last term should equal to the divergence inside the cubic volume, so we have

$$\sum_{i=1}^{6} J_i A_i = (\nabla \cdot J)V \tag{2.16}$$

where V is the volume of the cubic. In general, Gauss theorem applies to an arbitrary shape of a volume of V enclosed by an area of A,

$$\sum_{i} J_i A_i = \oint_A \vec{J} \cdot \vec{n}\, dA = (\nabla \cdot J)V \tag{2.17}$$

Combining the last three equations, we have

$$V\frac{dC}{dt} = \oint_a \vec{J} \cdot \vec{n}\, dA = (\nabla \cdot J)V \tag{2.18}$$

so the well-known continuity equation is obtained. Typically there is a negative sign in the equation.

$$\frac{dC}{dt} = -\nabla \cdot J \tag{2.19}$$

As we have mentioned, whether we have a positive or negative sign depends on whether the net change of atoms inside the cubic is positive or negative. Conventionally, we assume the out-flux is larger than the in-flux, which is shown in Section 2.2.5, so the concentration inside the cubic decreases with time and we have a negative size. As Gauss theorem assumes that the fluxes and derivatives are continuous functions, the equation is called *continuity equation*.

There are actually three equations in Eq. (2.18). The first is the Gauss equation relating the second term and the third term. The second is the continuity equation relating the first term and the third term. The third is the growth equation relating the first term and the second term and it assumes conservation of mass, or the conservation of the number of atoms ($\Delta N = -JA\Delta t$), or the conservation of the number of vacancies when we consider the growth of a void. We illustrate in Figure 2.3 the interrelationship of the three equations. Often we encounter the growth equation in solid-state reactions, such as the growth of a precipitate. This is because we have used the small cube in Figure 2.2 to consider flux divergence, and it is best for gas states because the concentration can change within the cubic. On the other hand, if the cubic is a solid, for example, a pure metal, the concentration of the cubic is $C = 1/\Omega$, which is constant and cannot change, where Ω is atomic volume. If the concentration cannot change (except by alloying), the cubic has to grow when a flux of atoms comes to it.

$$V\frac{dC}{dt} = \oint \vec{J}\cdot\vec{n}dA$$

$$(\nabla\cdot J)V$$

Figure 2.3 Three equations that relate to each other in obtaining Fick's second law.

So we need the growth equation. We encounter growth equation frequently in phase transformations.

As we have already derived the flux equation of Fick's first law,

$$J = -D\frac{\partial C}{\partial x}$$

in the one-dimensional case, then

$$\frac{\partial C}{\partial t} = \frac{\partial}{\partial x}D\left(\frac{\partial C}{\partial x}\right)$$

and if D is independent of position

$$\frac{\partial C}{\partial t} = D\left(\frac{\partial^2 C}{\partial x^2}\right) \tag{2.20}$$

This is Fick's second law of diffusion. Equation (2.20) is a one-dimensional diffusion equation. It is worth noting that in this equation, the C can have any unit or no unit because C is in both sides of the equation; its unit can be canceled. On the other hand, in Fick's first law, C must have a unit. Furthermore, D has the unit of cm^2/s and it is clear from the variation of C on t and on x in the equation.

2.2.5 Flux Divergence

We use Cartesian coordinates to consider the flux divergence in a small cubic volume, as shown in Figure 2.4. If we use a vector to represent the flux, J, we can decompose the vector into three components, J_x, J_y, and J_z. The amount of material flowing into the cube through the surface x_1 of area of $dydz$ per unit time is equal to

$$\overrightarrow{J_{x_1}}\cdot\overrightarrow{x_1} = J_{x_1}x_1\cos 180^\circ = -J_{x_1}dydz$$

Similarly, the amount of material flowing out of the cubic box through the surface x_2 (opposite to x_1) is equal to

$$\overrightarrow{J_{x_2}}\cdot\overrightarrow{x_2} = \left(J_{x_1} + \frac{\partial J_x}{\partial x}dx\right)x_2\cos 0^\circ = J_{x_1}dydz + \frac{\partial J_x}{\partial x}dxdydz$$

If we add these together, the net flux out of the box in the x direction becomes

$$(J_{x_2} - J_{x_1})dydz = \left(\frac{\partial J_x}{\partial x}\right)dxdydz$$

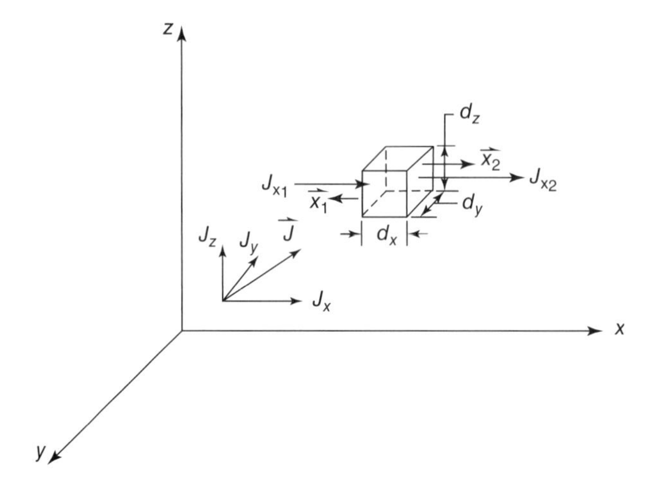

Figure 2.4 The flux divergence in a small cube volume in the Cartesian coordinates.

We note that we have used a continuous and differentiable function of J between x_1 and x_2, so that

$$J_{x_2} = J_{x_1} + \frac{\partial J_x}{\partial x} dx \tag{2.21}$$

Furthermore, J_{x_2} is larger than J_{x_1}. As a consequence, more materials are flowing out the cubic box. If we follow this approach in the y and z directions, we have

$$(J_{y_2} - J_{y_1}) dxdz = \left(\frac{\partial J_y}{\partial x}\right) dxdydz$$

$$(J_{z_2} - J_{z_1}) dxdy = \left(\frac{\partial J_z}{\partial x}\right) dxdydz$$

If we sum all of them together, we have

$$\sum_{i=1}^{6} J_i A_i = \left(\frac{\partial J_x}{\partial x} + \frac{\partial J_y}{\partial x} + \frac{\partial J_z}{\partial x}\right) dV \tag{2.22}$$

where $dV = dxdydz$. Now, instead of a cube, consider an arbitrary volume. An arbitrary volume can always be cut up into small cubes, and the flux going out from a square surface of any one cube is equal to the flux going into the neighboring cube. This occurs across all the internal surfaces, so that they all cancel. The only fluxes we have to consider are those on the outer surface. For an arbitrary volume V bounded by area A, the summation can be expressed as an integral,

$$\int_A J \cdot n dA = (\nabla \cdot J) V \tag{2.23}$$

This is the well-known Gauss theorem, and the right-hand side is defined as the divergence of the flux, where

$$\nabla \cdot J = \frac{\partial J_x}{\partial x} + \frac{\partial J_y}{\partial x} + \frac{\partial J_z}{\partial x} \tag{2.24}$$

Now this quantity (divergence of J) times, V, must equal to the change of the total number of atoms inside the volume due to mass conservation. It needs a negative sign because we derived this quantity under the conditions that the out-flux is greater than the in-flux and that there is no source within the volume; we lose material from this volume. Therefore it must equal to the time rate of the decrease of concentration as $C = N/V$. This is the continuity equation in differential form,

$$-(\nabla \cdot J) = \frac{\partial C}{\partial t} \tag{2.25}$$

It describes a non-steady-state flux flow. It is a very well-known equation in fluid mechanics, as well as here in diffusion.

2.2.6 Tracer Diffusion

An example of three-dimensional diffusion is the case of a drop of ink spreading out in a glass of water. A two-dimensional example would be a drop of gasoline that spreads out on the surface of a pond. Here we consider a one-dimensional problem

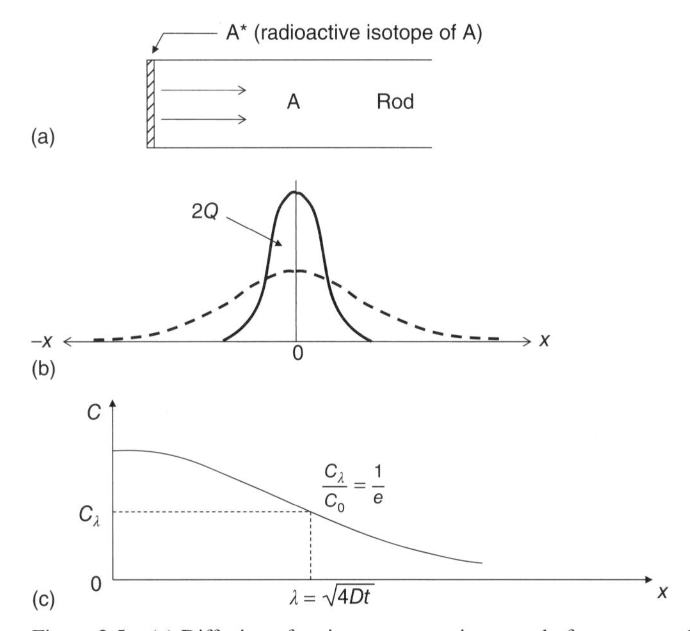

Figure 2.5 (a) Diffusion of an isotope tracer into a rod of a pure metal. (b) The composition profile for the one-dimensional symmetrical diffusion case with a total amount of tracer of $2Q$. (c) Concentration as a function of distance, where $C_\lambda/C_0 = 1/e$ at $\lambda = \sqrt{4Dt}$.

using Eq. (2.20). We take a long rod of a pure metal and plate a small amount of its isotope, a tracer, on the end surface, as shown in Figure 2.5a. We shall determine the concentration profile of the isotope diffusing into the rod as a function of time and temperature. It is a long distance diffusion in bulk dimension. The standard method of setting up the problem is to put a reflecting mirror at the end surface, and thus change the problem into a symmetrical one as shown in Figure 2.5b. The solution is then a standard one,

$$C(x,t) = \frac{Q}{(\pi Dt)^{1/2}} \exp\left(\frac{-x^2}{4Dt}\right) \tag{2.26}$$

The constant Q is the fixed amount of isotope material plated on the end surface of the sample and it satisfies the following condition. The initial conditions of the problem are that

$$\text{at } x = 0, C \to Q/Ad \text{ as } t \to 0$$

$$\text{for } |x| > 0, C \to 0 \text{ as } t \to 0$$

where A is the cross-section of the sample and d is the width of the isotope on the sample surface. We should have

$$\int_0^{\infty} C dx = Q$$

Two important concentration values are

$$C(0,t) = C_0 \frac{Q}{(\pi Dt)^{1/2}} \quad \text{at } x = 0$$

$$C(\lambda_D, t) = C_\lambda = \frac{C_0}{e} \text{ at } x = \lambda_D = (4Dt)^{1/2} \tag{2.27}$$

The last equation shows that, the position where the ratio, C_λ/C_0, of a local concentration to concentration at the source point is $1/e$, which occurs always at $x^2 = 4Dt$, as shown in Figure 2.5c. Here we present one solution that shows such a unique relation of diffusion, and many other solutions of the diffusion equation using different boundary conditions always involve x^2 proportional to Dt. This proportionality is one of the most important relationships of diffusion. If a kinetic process is controlled by diffusion, it must obey this relationship, $x^2 \sim Dt$, except in ripening, which is discussed in Chapter 4.

On the basis of the solution in Eq. (2.26), we can measure diffusivity, for example, from knowing the concentration profile, shown in Figure 2.5c. Because we know the time of diffusion, we plot $\ln C$ versus x^2, and the slope equals $1/(4Dt)$ according to Eq. (2.26), and we obtain D at the temperature of the experiment at a given time. When we measure D as a function of temperature, at four to five temperatures, we can plot $\ln D$ versus $1/kT$ and the slope will give us the activation energy of the diffusion, Q. We can express the diffusion coefficient as

$$D = D_0 \exp\left(\frac{-Q}{kT}\right) \tag{2.28}$$

where D_0 is the preexponential factor and Q is the activation energy of diffusion.

2.2.7 Diffusivity

In Eq. (2.10), we have derived Fick's first law, in which $D = \lambda^2 \nu$. In the derivation, we have assumed that the diffusing atom has a neighboring vacancy. For the majority of atoms in the lattice, this is not true because they do not have a vacancy as a nearest neighbor. In general, we can define the probability of an atom having a neighboring vacancy in the solid as

$$\frac{n_v}{n} = \exp\left(-\frac{\Delta G_f}{kT}\right) \tag{2.29}$$

where n_v is the total number of vacancy in the solid, and n is the total number of lattice sites in the solid, and ΔG_f is the Gibbs free energy of formation of a vacancy. In a face-centered cubic metal, as a lattice atom has 12 nearest neighbors, the probability for a particular atom to have a vacancy as a neighbor is

$$n_c \frac{n_v}{n} = n_c \exp\left(-\frac{\Delta G_f}{kT}\right)$$

where $n_c = 12$ is the number of nearest neighbors in a face-centered cubic metal.

Next, we have to consider the correlation factor in the face-centered cubic lattice. The physical meaning of the correlation factor is the probability of reverse jump; after the atom has exchanged position with a vacancy, it has a high probability to return to its original position before the activated configuration is relaxed. The factor, f, has a range between zero and unity. When $f = 0$, it means the probability of reverse jump is 100%, so the atom and the vacancy are exchanging position back and forth, which will not lead to any random walk and it is a correlated walk. When $f = 1$, it means that after the jump, the atom will not return back to its original position, and it is a random walk because the next jump will depend on the random probability of a vacancy coming to the neighborhood of this atom. Or the vacancy will exchange position randomly with one of its 12 nearest neighbor atoms. In a face-centered cubic metal, it has been calculated that $f = 0.78$, so about ~80% jumps are random walk, and about ~20% are correlated walk. Finally, the diffusivity is given as

$$D = f n_c \lambda^2 \nu_0 \exp\left(-\frac{\Delta G_m + \Delta G_f}{kT}\right)$$

$$D = f n_c \lambda^2 \nu_0 \exp\left(\frac{\Delta S_m + \Delta S_f}{k}\right) \exp\left(-\frac{\Delta H_m + \Delta H_f}{kT}\right) = D_0 \exp\left[-\frac{Q}{kT}\right] \tag{2.30}$$

Thus, we have

$$D_0 = f n_c \lambda^2 \nu_0 \exp\left(\frac{\Delta S_m + \Delta S_f}{k}\right) \tag{2.31}$$

and $Q = \Delta H_m + \Delta H_f$.

Table 2.1 gives the self-diffusion coefficient (in cm^2/s) for some elements. The activation enthalpies, ΔH, ΔH_m, and ΔH_f are given in units of eV/atom. Below, we calculate the self-diffusivity of Al ($D_0 = 0.047\,cm^2/s$, $\Delta H = 1.28\,eV/atom$) at room temperature.

$$D = D_0 \exp\left(-\frac{\Delta H}{kT}\right) = 0.047 e^{-(1.28\times 23{,}000)/(2\times 300)}$$
$$= 0.047 \times 10^{-(1.28\times 23{,}000)/(2.3\times 2\times 300)} = 0.047 \times 10^{-22}\,cm^2/s$$

With this diffusivity, if we take time to be a day or 10^5 s, we will have $(Dt)^{1/2}$ less than 10^{-8} cm. In other words, it is less than one atomic jump. Thus the metal Al is very stable at room temperature. Most metals and semiconductors have diffusivity smaller than $10^{-22}\,cm^2/s$ at room temperature, except the low-melting-point metals such as Pb and Sn.

We shall make a rough estimate of the upper bound and lower bound of diffusivity in solids. As shown in Table 2.1, for the upper bound we estimate the diffusivity of the face-centered cubic metals at their melting points, for example, Al near 660 °C, to be around $10^{-8}\,cm^2/s$. This value is smaller than the diffusivity found in liquid or molten metals, which is about $10^{-5}\,cm^2/s$. As a lower bound if we take the jump distance to be an interatomic distance of ~0.1 nm and the time to be 100 days, we have $x^2/t = 10^{-23}\,cm^2/s$. Any diffusivity of this order of magnitude is not of practical interest and is difficult to measure. Using superlattice structures or layer removal techniques for concentration profiling, we can measure diffusivity around 10^{-19} to $10^{-21}\,cm^2/s$. In an intermediate range of diffusivities, we find interdiffusion distances of 10^{-6} to 10^{-5} cm in thin film reactions, which can be finished in time of 1000 s, giving values of $x^2/t = 10^{-15}$ to $10^{-13}\,cm^2/s$. We note that the film thickness of 10^{-6} cm or 10 nm is in nanoscale. Knowing D at a given temperature, we can easily estimate the diffusion distance for a given time by using the relation of $x^2 = 4Dt$.

Substitutional dopants in Si have diffusivities close to that of self-diffusion of Si in Si. In Figure 2.6, we show a plot of diffusivity versus temperature for dopants in Si. To produce a p–n junction in Si by dopant diffusion, the diffusion temperature must be close to 1000 °C. Figure 2.6 shows that noble and near-noble metal elements such as Cu and Li have diffusivities several orders of magnitude higher than that of Si self-diffusion. They diffuse interstitially in Si.

It is worth mentioning that noble metals (Au, Ag, Cu) and near-noble metals (Pt, Pd, Ni) diffuse interstitially in group IV elements (Si, Ge, Sn, Pb). For this reason, they react at room temperature. For example, a thin film of Au deposited on a Si wafer will disappear at room temperature in a few days because the Au has dissolved into the Si. Also, Cu reacts with Sn to form Cu_6Sn_5 intermetallic compound at room temperature and can induce spontaneous Sn whisker growth.

TABLE 2.1 Lattice Self-Diffusion in Some Important Elements[a]

Element	D_0 (cm^2/s)	ΔH (eV)	ΔH_f (eV)	ΔH_m (eV)	$\Delta S/k$
FCC					
Al	0.047	1.28	0.67	0.62	2.2
Ag	0.04	1.76	1.13	0.66	—
Au	0.04	1.76	0.95	0.83	1.0
Cu	0.16	2.07	1.28	0.71	1.5
Ni	0.92	2.88	1.58	1.27	—
Pb	1.37	1.13	0.54	0.54	1.6
Pd	0.21	2.76	—	—	—
Pt	0.33	2.96	—	1.45	—
BCC					
Cr	970	4.51	—	—	—
α-Fe	0.49	2.95	—	0.68	—
Na	0.004	0.365	0.39/0.42	—	—
β-Ti	0.0036	1.35	—	—	—
V	0.014	2.93	—	—	—
W	1.88	6.08	3.6	1.8	—
β-Zr	0.000085	1.2	—	—	—
HCP					
Co	0.83	2.94	—	—	—
α-Hf	0.86/0.28	3.84/3.62	—	—	—
Mg	1.0/1.5	1.4/1.41	0.79/0.89	—	—
α-Ti	0.000066	1.75	—	—	—
Diamond lattice					
Ge	32	3.1	2.4	0.2	10
Si	1460	5.02	~3.9	~0.4	—

[a]From Chapter I of *Diffusion Phenomena in Thin Film and Microelectronic Materials* [5] (Courtesy of D. Gupta).

2.2.8 Experimental Measurement of the Parameters in Diffusivity

We have shown in Eq. (2.30) that the activation enthalpy of vacancy diffusion consists of two components,

$$\Delta H = \Delta H_f + \Delta H_m$$

As knowing the activation enthalpies is important in understanding the mechanism of diffusion and in identifying the type of defects that mediates the diffusion, their measurements have been the key activities in studying diffusion. The value of ΔH can be determined by measuring D at several temperatures and by plotting $\ln D$ versus $1/kT$; from the slope of the straight-line plot, we obtain ΔH. To decouple ΔH into ΔH_f and ΔH_m, experimental techniques of thermal expansion, quenching plus resistance measurement, and positron annihilation have been used. Quenching plus

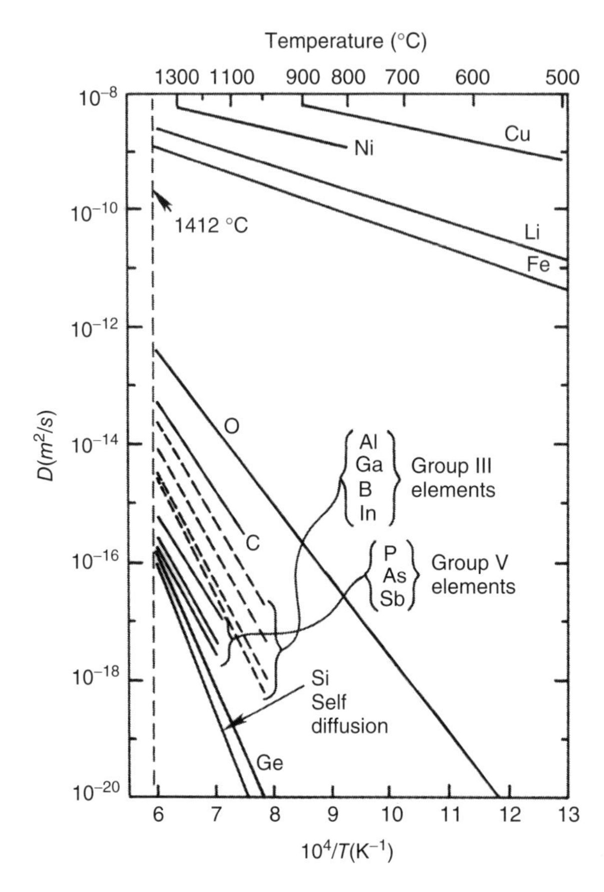

Figure 2.6 A plot of diffusivity (in logarithmic scale) versus inverse temperature for dopants in Si.

resistance technique has also been used to measure ΔH_m. These techniques are well covered in textbooks and reference books on diffusion. Here we discuss only briefly the measurement of ΔH_f by the Simmons–Balluffi method and the measurement of ΔH_m by resistance changes.

The concentration of vacancies in a solid is an equilibrium quantity. The concentration n_v/n increases with temperature as given in Eq. (2.29),

$$\frac{n_\nu}{n} = \exp\left(\frac{-\Delta G_f}{kT}\right) = \exp\left(\frac{\Delta S_f}{k}\right) \times \exp\left(\frac{-\Delta H_f}{kT}\right)$$

where n_v and n are respectively the number of vacancies and atoms in the solid. As temperature increases, more vacancies are formed by removing atoms from the interior to the surface of the solid. Consequently, the volume of the solid increases. Provided that we can decouple this volume increase from that because of thermal expansion, we can measure the vacancy concentration. Thermal expansion can be determined by measuring the lattice parameter change Δa as a function of temperature using X-ray diffraction, and we can express the fractional change by $\Delta a/a$. For the total volume change, we can use a wire of length L and measure the length change

ΔL as a function of temperature. We then have

$$\frac{\Delta n_\nu}{n} = 3\left(\frac{\Delta L}{L} - \frac{\Delta a}{a}\right) \tag{2.32}$$

The factor of 3 comes in because both ΔL and Δa are linear changes. Figure 2.7 shows $\Delta L/L$ and $\Delta a/a$ of Al wire as a function of temperature up to the melting point, as measured by Simmons and Balluffi. The results show that near the melting point, the vacancy concentration $n_\mathrm{v}/n = 10^{-4}$. Because $(\Delta L/L - \Delta a/a)$ is positive, the defect is predominantly vacancies. For interstitials, the difference is expected to be negative. From the two curves shown in Figure 2.7 for Al,

$$\frac{\Delta n_\nu}{n} = \exp(2.4)\exp\left(\frac{-0.76\,\mathrm{eV}}{kT}\right) \tag{2.33}$$

so we have $\Delta H_\mathrm{f} = 0.76\,\mathrm{eV}$ and $\Delta S_\mathrm{f}/k = 2.4$ for Al. These values are in good agreement with those listed in Table 2.1.

To measure ΔH_m, we can anneal an Al wire at a high temperature, say 600 °C, and quench the wire to room temperature. A supersaturation of vacancies is generated in the quenched wire, and then we anneal the wire at a higher temperature say at 200, 300, 400, and 500 °C. The supersaturated vacancies will diffuse to the surface of the wire and the resistance of the wire will decrease. In Figure 2.8, a ramping method is depicted. The solid curve from 0 K depicts the resistance of an Al wire, for example, as a function of temperature. In the method, we first anneal the wire at 600 °C, so the resistance of the wire is at point A. Second, we quench the wire

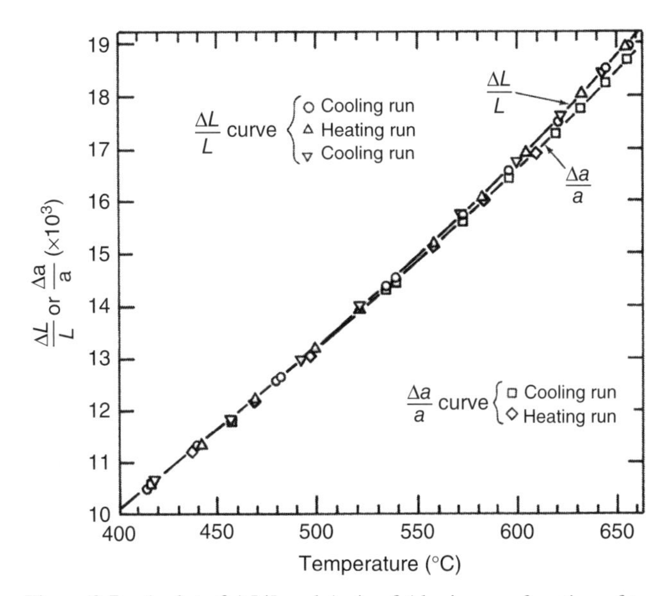

Figure 2.7 A plot of $\Delta L/L$ and $\Delta a/a$ of Al wire as a function of temperature up to the melting point, as measured by Simmons and Balluffi.

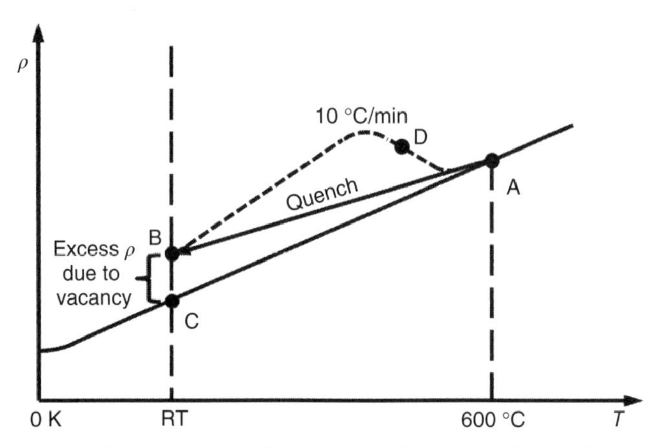

Figure 2.8 A schematic diagram to depict the ramping method of resistance change to measure the activation energy of motion.

to room temperature, from point A to point B. The resistance at point B is higher than the equilibrium resistance of Al at room temperature, which is represented by point C. The excess resistance between B and C is due to the excess vacancies from quenching. Third, we ramp the temperature of the wire up slowly at several rates of dT/dt, say 1, 5, and 10 °C/min. If we conduct *in situ* measurement of the resistance change of the wire during the ramping, the resistance will go up with the temperature following the broken line. Then at a certain temperature in the ramping, the resistance will decrease as vacancies begin to diffuse out, and, finally, the resistance returns to the solid curve when all the excess vacancies are gone. To calculate ΔH_m, we assume $D = D_0 \exp(-\Delta H_m/kT)$. Here, the activation enthalpy of diffusion is only the activation enthalpy of motion. We take x to be the radius of the Al wire, which is a constant, and

$$x^2 = \int_0^t 4D dt = \int_0^T 4D \frac{dt}{dT} dT = \int_0^T 4D_0 \frac{dt}{dT} \exp\left(-\frac{\Delta H_m}{kT}\right) dT$$

The integration is difficult because in the exponential, the variable T is in the denominator. To perform the integration, we follow Kissinger's approximation and obtain

$$x^2 = 4D \frac{dt}{dT} \frac{kT^2}{\Delta H_m} \exp\left(-\frac{\Delta H_m}{kT}\right) \tag{2.34}$$

Taking ln, we have

$$\ln\left(\frac{x^2}{4D} \frac{\Delta H_m}{k}\right) + \ln\left(\frac{1}{T^2} \frac{dT}{dt}\right) = -\frac{\Delta H_m}{kT} \tag{2.35}$$

It shows that by plotting ln $(1/T^2)(dT/dt)$ against $1/kT$, we can determine ΔH_m. To determine T, we can choose the midpoint of change of resistance, the point D, as shown in Figure 2.8. Because the midpoint of each ramping curve can be determined rather accurately, we can determine the corresponding temperature of the midpoint for each ramping rate at a given dT/dt.

By taking the attempt frequency or Debye frequency as $\nu_0 = 10^{13}$ Hz, we can estimate the entropy factor by measuring the preexponential factor of D_0, provided that we accept the theoretical value of the correlation factor f (=0.78 for face-centered cubic metals). We have

$$\exp\left(\frac{\Delta S}{k}\right) = \exp\left(\frac{\Delta S_{\mathrm{m}} + \Delta S_{\mathrm{f}}}{k}\right) = \frac{D_0}{f\nu_0\lambda^2 n_{\mathrm{c}}} \tag{2.36}$$

As we assume ΔS to be positive, we have $D_0 > f\nu_0\lambda^2 n_{\mathrm{c}}$. For Al,

$$D_0 > 0.78 \times 10^{13} \times (2.86 \times 10^{-8})^2 \times 12 = 0.076\,\mathrm{cm}^2/\mathrm{s}$$

In general, D_0 for self-diffusion in metals is of the order of $0.1 - 1$ cm^2/s, so the entropy change per atom in diffusion is of the order of unity times k.

2.3 NONLINEAR DIFFUSION

When a diffusion phenomenon is described by Fick's first law where the atomic flux or atomic jump frequency is directly proportional to the driving force of diffusion (concentration gradient or chemical potential gradient), it is a linear diffusion. It applies to the diffusion in a bulk sample over a long distance having a small concentration gradient. On the other hand, in the case of diffusion over a very short distance (of the order of 1 nm) with a very large concentration gradient, a kinetic nonlinear effect occurs where the atomic jump frequency is affected also by the high order terms of the driving force, the concentration gradient. It can occur in nanoscale materials.

In addition to the kinetic nonlinear effect, the thermodynamic nonlinear effect on diffusion has been treated by Cahn and Hilliard [6–9]. This thermodynamic nonlinear effect is taken from an energetic viewpoint that the free energy of an inhomogeneous alloy system is not only a function of concentration but also that of concentration gradient when the gradient is large. Specifically, the chemical potential gradient of an inhomogeneous solid solution has a square dependence on the concentration gradient, and hence it is positive. Cahn has presented a linearized fourth order diffusion equation, which has been used extensively to study spinodal decomposition and the interdiffusion in man-made layered structures or superlattices. Details of these published works are not reviewed here, except the removal of the nonlinear terms in Cahn's linearization derivation. The modified fourth order nonlinear diffusion equation that takes into account both the kinetic and thermodynamic nonlinear effect is presented. However, the topic of spinodal decomposition is covered in Chapter 5.

2.3.1 Nonlinear Effect due to Kinetic Consideration

We recall that when $\lambda F/kT \ll 1$, we can apply the "condition of linearization" of $\sinh x = x$ to Eq. (2.6), and we obtain

$$\nu_n = \nu\left(\frac{\lambda F}{kT}\right)$$

where the net jump frequency, ν_n, is proportional to the driving force, F. However, when the distance of diffusion is in nanoscale, the condition of linearization does not exist. We examine such a case as depicted in Figure 2.9 for a case of diffusion over a nanoscale distance of $\Delta x = N\lambda$, between plane 1 and plane 2 in a crystalline solid, where N is the number of atomic planes in between plane 1 and plane 2. We note that if Δx is of nanoscale, N is of the order of 10 because λ is typically about 0.3 nm. We assume an ideal dilute solution and take

$$\mu_2 = kT\ln(C_2\Omega)$$
$$\mu_1 = kT\ln(C_1\Omega)$$
$$\Delta\mu = \mu_1 - \mu_2 = kT\ln\frac{C_1}{C_2}$$

where μ_2 and μ_1 are chemical potentials and C_2 and C_1 are the concentration of plane 2 and plane 1, respectively. Then

$$F = \frac{\Delta\mu}{\Delta x} = \frac{kT}{N\lambda}\ln\frac{C_2}{C_1}$$

or

$$\frac{\lambda F}{kT} = \frac{\ln(C_2/C_1)}{N} \tag{2.37}$$

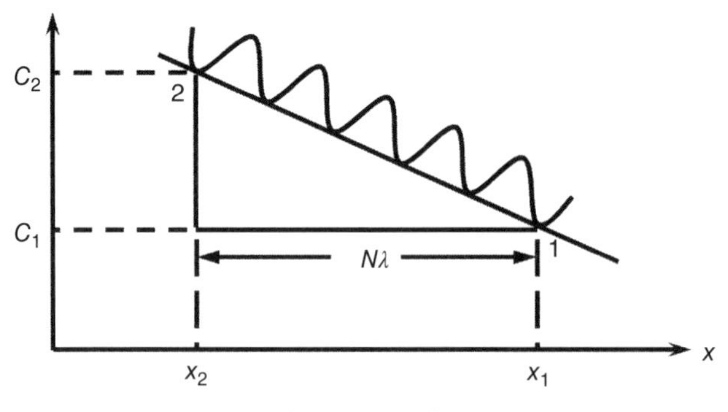

Figure 2.9 Schematic diagram of diffusion over a nanoscale distance of $\Delta x = N\lambda$, between plane 1 and plane 2 in a crystalline solid, where N is the number of atomic planes in between plane 1 and plane 2.

The last equation means that by knowing the ratio of C_2/C_1, we can select a value of N to evaluate the magnitude of $\lambda F/kT$ in order to see whether or not it is much less than unity, and in turn whether or not we can use the condition of linearization. For a high concentration gradient, the actual ratio of C_2/C_1 may vary from 10 to 1000, yet $\ln(C_2/C_1)$ varies only slightly, from 2.3 to 6.9. Therefore, if N is of 10 (if we take $\lambda = 0.3$ nm, then $\Delta x = N\lambda = 3$ nm), we see that $\lambda F/kT$ is of the order of unity, not much less than unity. Thus, in diffusion over a short distance of about a few nanometers, the condition of linearization breaks down, so we do not have linear diffusion. Then, we take [10]

$$\sinh x = x + \frac{x^3}{3!}$$

where $x = \lambda F/kT$, the jump frequency from Eq. (2.5) becomes

$$\nu = \frac{\lambda \nu_0}{kT} F + \frac{\nu_0}{24}\left(\frac{\lambda}{kT}\right)^3 F^3 \tag{2.38}$$

It is not linearly proportional to F. In turn, the diffusion flux will no longer be linearly proportional to the chemical potential gradient, but will contain a third order term. If we take $\mu = kT \ln C\Omega$, we have

$$J = -D\frac{\partial C}{\partial x}\left[1 + \frac{1}{24}\left(\frac{\lambda}{C}\right)^2 \left(\frac{\partial C}{\partial x}\right)^2\right]$$

So the equation of continuity will be a fourth order nonlinear differential equation. Explicitly, we obtain

$$\frac{\partial C}{\partial t} = D\frac{\partial^2 C}{\partial x^2}\left[1 + \frac{\lambda^2}{8C^2}\left(\frac{\partial C}{\partial x}\right)^2\right] \tag{2.39}$$

where $D = \nu\lambda^2$. The above nonlinear equation was derived from a kinetic consideration, and hence we may call it a kinetic nonlinear equation. It is worth pointing out that we have applied the ideal dilute solution chemical potential, $\mu = kT \ln C\Omega$, in the above derivation. The ideal dilute solution chemical potential is for a random and homogeneous solid solution, in which the enthalpy of mixing is zero and the entropy of mixing is ideal. However, for an inhomogeneous solid solution, when there is a steep local concentration variation or a steep concentration gradient, such as in the case of interdiffusion in a man-made superlattice, the free energy of the system is a function not only of concentration, but also of concentration gradient, to be discussed next. Then the chemical potential cannot be represented by that of the ideal dilute solution.

2.3.2 Nonlinear Effect due to Thermodynamic Consideration

Hillert and Cahn and Hilliard have considered the free energy of an inhomogeneous solid solution [6–9]. For an inhomogeneous solid solution, it may contain a large local concentration variations or a steep concentration gradient, and Hillert showed that its free energy is not only a function of concentration but also of concentration gradient. Cahn and Hilliard have treated the subject in detail, and they showed that the total free energy of a volume of an inhomogeneous system can be expressed as

$$\widetilde{F} = \int f\left(C, \frac{\partial C}{\partial x}\right) dV$$

where C and $\partial C/\partial x$ are independent variables, and f is the free energy per unit volume of the inhomogeneous solution. In Section 5.3.3, it is shown by using quasi chemical approach that we have

$$f = f_0 + K\left(\frac{\partial C}{\partial x}\right)^2 \tag{2.40}$$

where f_0 is the free energy of a homogeneous solid solution of the average concentration C_0 and it is a function of C only, and K is defined as the gradient energy coefficient and K can be a function of C and $\partial C/\partial x$. In practice, K has often been assumed to be a constant (as shown in Section 5.3.3) in order to get rid of nonlinear terms. As we consider nonlinear effects here, we first assume K to be a function of C only for simplicity. Equation (2.40) is a nonlinear equation. Because it is derived from an energy consideration, we call it a thermodynamic nonlinear equation.

On the basis of Eq. (2.40) and by assuming K to be a function of C only and by following the procedure of using Euler's equation as presented in Section 5.34, the chemical potential of the inhomogeneous solid solution is given as

$$\mu = \frac{\partial f_0}{\partial C} - \frac{\partial K}{\partial C}\left(\frac{\partial C}{\partial x}\right)^2 - 2K\frac{\partial^2 C}{\partial x^2} \tag{2.41}$$

where μ and $\partial f_0/\partial C$ are the chemical potential for the inhomogeneous and homogeneous solution, respectively. By substituting the last equation into the flux equation, $J = CMF = CM(-\partial\mu/\partial x)$

$$J = -CM\left[\frac{\partial^2 f_0}{\partial C^2}\frac{\partial C}{\partial x} - \frac{\partial^2 K}{\partial C^2}\left(\frac{\partial C}{\partial x}\right)^3 - 4\frac{\partial K}{\partial C}\frac{\partial C}{\partial x}\frac{\partial^2 C}{\partial x^2} - 2K\frac{\partial^3 C}{\partial x^3}\right] \tag{2.42}$$

In obtaining the last equation, as $\mu = \mu(C, C_x, C_{xx})$, where C, C_x $(=\partial C/\partial x)$, and C_{xx} $(=\partial^2 C/\partial x^2)$ are independent variables, we used the following expression,

$$\frac{d\mu}{dx} = \frac{\partial\mu}{\partial x} + \frac{\partial\mu}{\partial C}\frac{\partial C}{\partial x} + \frac{\partial\mu}{\partial C_x}\frac{\partial C_x}{\partial x} + \frac{\partial\mu}{\partial C_{xx}}\frac{\partial C_{xx}}{\partial x} \tag{2.43}$$

Then it follows that the diffusion equation is given as

$$\frac{\partial C}{\partial t} = \frac{\partial}{\partial x}\widetilde{D}_0\left[\frac{\partial C}{\partial x} - \frac{1}{f_0''}\frac{\partial^2 K}{\partial C^2}\left(\frac{\partial C}{\partial x}\right)^3 - \frac{4}{f_0''}\frac{\partial K}{\partial C}\frac{\partial C}{\partial x}\frac{\partial^2 C}{\partial x^2} - \frac{2K}{f_0''}\frac{\partial^3 C}{\partial x^3}\right] \tag{2.44}$$

where $\widetilde{D}_0 = CMf_0''$ and $f_0'' = \partial^2 f_0/\partial C^2$. We note that the above equation is Eq. (14) in Ref. No. 8. It is easy to see from the above equation that if K is assumed to be a constant, the two nonlinear terms drop out, and we have

$$\frac{\partial C}{\partial t} = \frac{\partial}{\partial x}\widetilde{D}_0\left[\frac{\partial C}{\partial x} - \frac{2K}{f_0''}\frac{\partial^3 C}{\partial x^3}\right] \tag{2.45}$$

Furthermore, if $\widetilde{D}_0$ is assumed to be independent of C, and of its derivatives including C_{xxx} $(=\partial^3 f_0/\partial x^3)$, the last equation becomes

$$\frac{\partial C}{\partial t} = \widetilde{D}_0\frac{\partial^2 C}{\partial x^2} + 2\widetilde{D}_0\frac{Kf_0'''}{(f_0'')^2}\frac{\partial C}{\partial x}\frac{\partial^3 C}{\partial x^3} - 2\widetilde{D}_0\frac{K}{f_0''}\frac{\partial^4 C}{\partial x^4} \tag{2.46}$$

where $f_0''' = \partial^3 f_0/\partial C^3$. Now if the nonlinear term in the above equation is removed, we obtain the linearization fourth order diffusion equation given by Cahn, as shown below.

$$\frac{\partial C}{\partial t} = \widetilde{D}_0\frac{\partial^2 C}{\partial x^2} - \frac{2K\widetilde{D}_0}{f_0''}\frac{\partial^4 C}{\partial x^4} \tag{2.47}$$

This linearized fourth order diffusion equation was used by Cahn to describe the kinetic behavior of spinodal decomposition, which is discussed in Chapter 5. A homogeneous solid solution inside spinodal is thermodynamically unstable; it decomposes spontaneously and gradually with compositional fluctuations by diffusion. Spinodal decomposition needs only diffusion and no nucleation. Compositional fluctuations begin and continue with diffusion in the direction against the concentration gradient; it is uphill diffusion. The driving force of diffusion is provided by the chemical potential gradient rather than the concentration gradient. The diffusion is described by the fourth order equation with a negative interdiffusion coefficient. The concentration gradient or the amplitude of the fluctuation increases with decomposition and finally two intermixed phased form. During the very early stage of spinodal decomposition, the kinetic nonlinear effect is unimportant because the concentration gradient is not large.

Chemical interdiffusion begins with two phases having an interface and ends with a single homogeneous phase, so the inhomogeneous state is thermodynamically unstable and proceeds toward homogenization by diffusion. Kinetically speaking, spinodal decomposition and chemical interdiffusion in man-made superlattice are quite similar except that diffusion fluxes are moving in opposite directions respectively between a homogeneous and an inhomogeneous state, so the same diffusion equation can be used to describe both, provided that the effect of the concentration gradient on free energy is taken into account. We present spinodal decomposition in Chapter 5. However, in the early stage of chemical interdiffusion over short distances and in the later stage of spinodal decomposition, the concentration gradient is very large, so the nonlinear effect must be considered. Cahn has addressed the nonlinear effect on spinodal decomposition, and so have Hilliard and his associates.

Because we emphasize nonlinear effects here, both thermodynamic and kinetic effects are considered together. In arriving at his linearized fourth order diffusion equation for spinodal decomposition, Cahn has dropped several nonlinear terms. Let us reexamine these terms. Because of the definition that the free energy of an inhomogeneous solid solution is a function of concentration and concentration gradient and the assumption that they are independent variables, nonlinear terms are introduced whenever a differentiation is taken in the process of the mathematical derivation of thc diffusion equation. This has been demonstrated by Eqs (2.41), (2.42), and (2.46). In the case of the early stage of spinodal decomposition, the nonlinear terms can be removed by a scaling procedure proposed by Cahn and Hilliard, as discussed below.

A scaling procedure with the expression of $C - C_0 = A\cos\beta x$ has been proposed by Cahn and Hilliard for evaluating the ranking of the terms. Specifically, if the concentration gradient is small, that is, A is small in $A\cos\beta x$, the magnitude of the nonlinear term in Eq. (2.46) is proportional to A^2 where the last term is proportional to A, and the nonlinear term can be dropped. However, this scaling procedure may not be applicable to cases where the concentration gradient is large because the amplitude of fluctuation is large. In this situation, we assume a different criterion to remove the nonlinear term. An alternative way to remove the nonlinear terms in Eqs (2.41) and (2.42) is to assume K to be a constant, and in Eq. (2.46) is to assume f_0'' to be a constant by approximating the free energy curve of a homogeneous binary solid solution with a parabolic function of concentration.

2.3.3 Combining Thermodynamic and Kinetic Nonlinear Effects

In the last section, the diffusion flux is assumed to be linearly proportional to $-\partial\mu/\partial x$. Strictly speaking, the kinetic nonlinear effect was not taken into account. If we combine the thermodynamic and kinetic nonlinear effects together and assume K to be a constant, we have

$$J = CMF\left[1 + \frac{1}{24}\left(\frac{\lambda F}{kT}\right)^2\right] \tag{2.48}$$

where

$$F = -f_0''\frac{\partial C}{\partial x}\left(1 - \frac{2K\left(\partial^3 C/\partial x^3\right)}{f_0''(\partial C/\partial x)}\right) \tag{2.49}$$

By substituting Eq. (2.49) into Eq. (2.48), we have

$$J = -CMf_0''\frac{\partial C}{\partial x}\left(1 - \frac{2K}{f_0''}\frac{C_{xxx}}{C_x}\right)\left[1 + \frac{1}{24}\left(\frac{\lambda f_0''}{kT}\frac{\partial C}{\partial x}\right)^2\left(1 - \frac{2K}{f_0''}\frac{C_{xxx}}{C_x}\right)^2\right] \tag{2.50}$$

In the right-hand side of both Eqs (2.48) and (2.49) the magnitude of the second terms is in general only a few percent of the first terms, they themselves are not negligible, but their product can be neglected. Therefore, we have reduced them and obtained the following flux and diffusion equations.

$$J = -\widetilde{D}_0 \left[\frac{\partial C}{\partial x} + \frac{1}{24}\left(\frac{\lambda f_0''}{kT}\right)^2 \left(\frac{\partial C}{\partial x}\right)^3 - \frac{2K}{f_0''}\frac{\partial^3 C}{\partial x^3} \right] \tag{2.51}$$

$$\frac{\partial C}{\partial t} = \frac{\partial}{\partial x}\widetilde{D}_0 \left[\frac{\partial C}{\partial x} + \frac{1}{24}\left(\frac{\lambda f_0''}{kT}\right)^2 \left(\frac{\partial C}{\partial x}\right)^3 - \frac{2K}{f_0''}\frac{\partial^3 C}{\partial x^3} \right] \tag{2.52}$$

where $\widetilde{D}_0 = CMf_0''$. We can use numerical analysis to solve Eq. (2.52) and to compare the relative contribution of the two higher order terms. In Eq. (2.52), if $\widetilde{D}_0$ is assumed to be independent of C and its derivative, we can bring it out of the differentiation and transform Eq. (2.52) into

$$\begin{aligned}\frac{\partial C}{\partial t} = \widetilde{D}_0\frac{\partial^2 C}{\partial x^2} + \frac{\widetilde{D}_0}{12}\left(\frac{\lambda}{kT}\right)^2 f_0'' f_0''' \left(\frac{\partial C}{\partial x}\right)^4 + \frac{\widetilde{D}_0}{8}\left(\frac{\lambda f_0''}{kT}\right)^2 \left(\frac{\partial C}{\partial x}\right)^2 \frac{\partial^2 C}{\partial x^2} \\ + \frac{2\widetilde{D}_0 K f_0'''}{(f_0'')^2}\frac{\partial C}{\partial x}\frac{\partial^3 C}{\partial x^3} - \frac{2K\widetilde{D}_0}{f_0''}\frac{\partial^4 C}{\partial x^4}\end{aligned} \tag{2.53}$$

Now if we further assume that the free energy curve of a homogeneous binary solid solution can be represented by a parabolic function of concentration, the two nonlinear terms containing f_0''' can be removed. Consequently, Eq. (2.53) is reduced to the following nonlinear fourth order diffusion equation,

$$\frac{\partial C}{\partial t} = \widetilde{D}_0\frac{\partial^2 C}{\partial x^2}\left[1 + \frac{1}{8}\left(\frac{\lambda f_0''}{kT}\right)^2 \left(\frac{\partial C}{\partial x}\right)^2\right] - \frac{2K\widetilde{D}_0}{f_0''}\frac{\partial^4 C}{\partial x^4} \tag{2.54}$$

Comparing the above equation to Cahn linearized fourth order diffusion equation in Eq. (2.47), we can conclude that the second term in the right-hand side of the above equation is due to the kinetic effect.

REFERENCES

1. Crank J. *Mathematics of Diffusion*. Fair Lawn, NJ: Oxford University Press; 1956.
2. Shewmon PG. *Diffusion in Solids*. 2nd ed. Warrendale, PA: The Minerals, Metals, and Materials Society; 1989.
3. Glicksman ME. *Diffusion in Solids*. New York: Wiley-Interscience; 2000.
4. Balluffi RW, Allen SM, Carter WC. *Kinetics of Materials*. New York: Wiley-Interscience; 2005.
5. Gupta D, Ho PS, editors. *Diffusion Phenomena in Thin Films and Micro-Electronic Materials*. Park Ridge, NJ: Noyes Publications; 1988.

6. (a) Cahn JW. On spinodal decomposition. Acta Metall 1961;9:795–801. (b) Hillert M, A solid-solution model for inhomogeneous systems. Acta Metall 1961; 6:525–535.
7. Cahn JW, Hilliard JE. Free energy of a nonuniform system. I: interfacial free energy. J Chem Phys 1958;28:258–267.
8. Cahn JW, Hilliard JE. Free energy of a nonuniform system. III: nucleation in a two-component incompressible fluid. J Chem Phys 1959;31:688–699.
9. Cahn JW, Hilliard JE. Spinodal decomposition: a reprise. Acta Metall 1971;19:151–161.
10. Tu KN. Interdiffusion in thin films. In: Huggins RA, editor. *Annual Review of Materials Science*. Vol. 15. Palo Alto: Annual Reviews Inc.; 1985. p 147.

PROBLEMS

2.1. What is correlation factor in the vacancy mechanism of diffusion?

2.2. If we take interdiffusion coefficient as $D = CMG''$, at inflection point of the curve of G, we have $G' = 0$. Does it means $D = 0$?

2.3. In a pure metal such as aluminum, show that $C_{Al}D_{Al} = C_v D_v$, where C_{Al} and D_{Al} are the concentration (number of atoms per unit volume) and atomic diffusivity of Al atoms, and C_v and D_v are the concentration of equilibrium vacancy and diffusivity of vacancy in Al, respectively.

2.4. Show that the following equation

$$C = \frac{A}{\sqrt{4\pi Dt}} \exp\left(-\frac{x^2}{4Dt}\right)$$

is a solution of diffusion equation of $\partial C/\partial t = D(\partial^2 C/\partial x^2)$.

2.5. Show that the following equation is a solution of the diffusion equation of $\partial C/\partial t = D(\partial^2 C/\partial x^2)$,

$$C = \overline{C} + \beta \sin\frac{\pi x}{l} \exp\left(-\frac{t}{\tau}\right)$$

And show that $\tau = l^2/D\pi^2$, where x and t are variables and the rest are constant parameters.

2.6. We anneal a pure aluminum wire at 650 °C and quench it to 100 °C. What is the concentration of the supersaturated vacancies in the wire after the quench? The equilibrium vacancy concentration in aluminum is given by

$$\frac{\Delta n}{n} = \exp(2.4)\exp\left(-\frac{0.76\,\mathrm{eV}}{kT}\right)$$

where n is the number of total lattice site per unit volume and Δn is the number of equilibrium vacancies per unit volume. The diameter of the wire is 2 mm. What is the time needed at 100 °C to remove all the supersaturate vacancies by diffusing them to the free surface of the wire? The diffusivity of vacancies in aluminum is given by

$$D_v = 0.1\exp\left(-\frac{0.68\,\mathrm{eV}}{kT}\right)\frac{\mathrm{cm}^2}{\mathrm{s}}$$

2.7. A block of copper was joined to a block of brass (Cu–Zn alloy) with inert markers placed at the interface. After the sample was held at 785 °C for a long time, analysis revealed that the marker had moved with a velocity of $v = 2.6 \times 10^{-9}$ cm/s in the direction of the brass. The composition at the marker was $X_{Zn} = 0.22$. The slope of the composition curve at the marker, $dX_{Zn}/dx = 0.89\ \text{cm}^{-1}$. The interdiffusion coefficient at the marker was $D = 4.5 \times 10^{-9}\ \text{cm}^2/\text{s}$. Calculate the diffusion coefficient of Zn (D_{Zn}) and Cu (D_{Cu}) in the brass at $X_{Zn} = 0.22$ at 785 °C.

2.8. The solution to diffusion equation of a thin layer of isotropic tracer into a rod of a pure metal is given by

$$C(x,t) = \frac{Q}{(\pi D t)^{1/2}} \exp\left(-\frac{x^2}{4Dt}\right)$$

where

$$\int_0^\infty C(x)dx = Q$$

Calculate the following,

$$\bar{x} = \frac{\int_0^\infty xC(x)dx}{\int_0^\infty C(x)dx} = \frac{1}{\sqrt{\pi}}\sqrt{4Dt}$$

2.9. Consider diffusion from the point source with diffusivity $D(t)$, which depends on time as

$$D(t) = D_0 \exp\left(-\frac{t}{\tau}\right)$$

The initial profile is proportional to Dirac's delta function; $C_0(x) = A\delta(x)$, where $A = C_0 h$ and C_0 is an initial concentration in the very thin layer of $-h/2 \leq x \leq h/2$. Find the profile of $C(t,x)$ at arbitrary time.

2.10. Consider the case of diffusion with sinks or traps. We let the vacancy supersaturation to be $\Delta C_V = C_V - C_V^{eq}$, which is maintained constant in the sample owing to the sources at the sample surface. Inside the sample, the vacancies tend to reach equilibrium by diffusing to sinks with the vacancy relaxation time, τ_V. We have

$$\frac{\partial C_V}{\partial t} = D_V \frac{\partial^2 C_V}{\partial x^2} - \frac{C_V - C_V^{eq}}{\tau_V}$$

Find the steady state solution.

2.11. We recall $J = Cv = CMF = C\frac{D}{kT}\left(-\frac{\partial \mu}{\partial x}\right)$. Now we consider diffusion on a curved surface, and we take the chemical potential to be

$$\mu = 2\gamma\Omega\left(\frac{1}{r} - \frac{1}{r_0}\right)$$

For a curved surface having a slowly varying curvature on which we assume $dz/dx \ll 1$, we have

$$\frac{1}{r} = \frac{d^2z/dx^2}{[1 + (dz/dx)^2]^{3/2}} = \frac{d^2z}{dx^2}$$

Derive a fourth order diffusion equation for the diffusion.

2.12. Calculate the entropy in

$$D_0 = f n_c \lambda^2 \nu_0 \exp\left(\frac{\Delta S_m + \Delta S_f}{k}\right)$$

2.13. It has been shown by using quasi chemical approach that the free energy of an inhomogeneous solid solution is given as

$$f = f_0 + K\left(\frac{\partial C}{\partial x}\right)^2$$

where f_0 is the free energy of a homogeneous solid solution of average concentration of C_0 and it is a function of C only. The parameter K is defined as the gradient energy coefficient and K has been assumed to be a positive constant. In this problem, we assume instead that K is a function C, rather than a constant. Now, by following the procedure of using Euler's equation as presented in Chapter 5 in the lecture notes, derive the chemical potential of the inhomogeneous solid solution to be

$$\mu = \frac{\partial f}{\partial C} = \frac{\partial f_0}{\partial C} - \frac{\partial K}{\partial C}\left(\frac{\partial C}{\partial x}\right)^2 - 2K\frac{\partial^2 C}{\partial x^2}$$

where μ and $\partial f_0/\partial C$ are the chemical potential for the inhomogeneous and homogeneous solution, respectively. Also explain why the last term has a negative sign.

CHAPTER 3

KIRKENDALL EFFECT AND INVERSE KIRKENDALL EFFECT

3.1 INTRODUCTION

The classic Kirkendall effect of interdiffusion in a diffusion couple of A and B showed that the flux of A diffusing into B is not equal to the flux of B diffusing into A. The effect was revealed by diffusion marker motion. Markers were placed between A and B before interdiffusion and they showed a displacement after interdiffusion. If the flux of A is equal to that of B, we expect no marker motion, when the atomic size of A and B are equal. The two fluxes could be equal if we assume a diffusion mechanism in which the interdiffusion of A and B would occur by a direct swap of A and B atoms; there will be no marker motion because one atom of A that moves to the right will have one atom of B moving to the left.

Kirkendall [1] performed the classic interdiffusion experiment between Cu and CuZn (brass) with Mo markers placed at the original interface. Using a sandwiched structure of Cu/CuZn/Cu with Mo markers in the two interfaces, he observed that

Kinetics in Nanoscale Materials, First Edition. King-Ning Tu and Andriy M. Gusak.

the Mo markers at the two interfaces had moved closer to each other with annealing, which can be measured quite easily. It indicated that more Zn had diffused out than Cu had diffused in. This means that the diffusion does not occur by a direct swap of positions between two neighboring atoms; otherwise there should be no marker motion. In order to balance the unbalanced fluxes of A and B, for substitutional diffusion, a flux of lattice defects or, specifically, a flux of vacancies will be needed. Kirkendall effect has led to the development of the basic concept of defect mediated mechanism of atomic diffusion, and that is why it is important. Atomic diffusion is a defect mediated kinetic process and the defect in a face-centered cubic metal is predominantly vacancy.

Recently, hollow nanoparticles of CoO or Co_3S_4 were formed when Co nanoparticles were annealed in oxygen or sulfur atmosphere, respectively, as discussed in Section 1.5. The formation of the hollow nanoparticles was explained on the basis of Kirkendall effect by assuming that the out-diffusion of Co is faster than the in-diffusion of oxygen or sulfur during the annealing. Thus, there is a renewed interest on Kirkendall effect in nanoscale materials.

To analyze the classic Kirkendall effect, Darken has provided a kinetic analysis by assuming that the diffusivity of Zn is faster than that of Cu in the CuZn alloy and a flux of vacancy is required to balance the flux of Cu against the flux of Zn [2]. Another key assumption in the analysis is that the vacancy distribution in the diffusion couple is equilibrium everywhere. In other words, the sources and sinks of vacancy are fully operative everywhere in the sample so that the vacancy concentration can be maintained at equilibrium everywhere. One mechanism of the source/sink operation is, for example, by dislocation climb to generate or to absorb vacancies.

On the basis of this assumption of equilibrium distribution of vacancies, there are three important implications. First, lattice shift or Kirkendall shift occurs, which can be measured by marker motion. Second, because of lattice shift, no stress is generated. Third, as equilibrium vacancy is assumed, there cannot be supersaturated vacancies, so no nucleation of void and in turn no void formation can take place. However, in actual diffusion couples, the vacancy sources and sinks are often only partially effective and voids are found. The voids are sometimes called *Frenkel voids* rather than Kirkendall voids. This is because theoretically speaking, under the assumption of Kirkendall shift, there should be no void formation. But experimentally, Kirkendall shift and Kirkendall voiding can coexist, so we can have these two competing Kirkendall effects [3, 4]. Both of them are caused by the divergence of vacancy flux that is generated owing to the difference in intrinsic fluxes of A and B components. It is known that by using hydrostatic compression, void formation can be suppressed and lattice shift can be recovered [3]. On the other hand, in nanosystems, where there is little space for dislocations, Kirkendall shift may be suppressed, and Kirkendall (Frenkel) voiding is observed in pure form [5–8]. On the stress issue, it is found that stress-induced bending occurs in the interdiffusion to form silicide between a metal thin film and Si wafer. The latter bent and it shows stress can exist in interdiffusion.

Inverse Kirkendall effect refers to the situation where a preexisting vacancy flux (generated by some external forces) affects the interdiffusion of A and B in an alloy of AB. A classic example is irradiation [9], under which segregation in a homogeneous

AB alloy occurs because the irradiation has produced excess vacancies and in turn a flux of diffusing vacancies in the alloy. The diffusion of the vacancies leads to the diffusion and segregation of A and B in the alloy. As the flux of A is different from that of B because the intrinsic diffusivity of A and B is different, segregation occurs.

Another example of inverse Kirkendall effect is a homogeneous AB alloy having a hollow shell structure in nanoscale, as shown in Figure 1.6. When such a hollow nanoshell of homogeneous alloy is annealed at a constant temperature, segregation of A and B occurs [10]. It is because of the Gibbs–Thomson effect that there will be a higher vacancy concentration near the inner shell surface than that near the outer shell surface. The vacancy concentration gradient induces a vacancy flux, and it leads to segregation in the hollow nanoshell. The faster diffusing species will segregate to the inner shell and create a gradient of chemical potential that will retard the vacancy flux. Provided that the vacancy potential is larger than the counterpotential of segregation, the hollow shell may transform to a solid sphere in order to reduce the total surface area.

Thus, when we consider the interdiffusion of A and B in a two-layer nanoshell structure, as shown in Figure 1.7b, the Kirkendall effect, the Gibbs–Thomson effect, and the inverse Kirkendall effect coexist, and how they interact with each other requires a detailed analysis. In this chapter, we use the term *Kirkendall effect* to include both Kirkendall shift and Kirkendall voiding (or Frenkel voiding). We present an analysis of the interaction of Kirkendall effect and inverse Kirkendall effect in nanoscale particles. The interaction is expressed in terms of the interaction of vacancy fluxes.

In the following, we first present a brief review of the classic Kirkendall effect. Next, a review of inverse Kirkendall effect is given. Then, the kinetics in a pure metal, an alloy, and a compound nanoshell is presented. Finally, the interaction between Kirkendall effect and inverse Kirkendall effect in the growth of a compound nanoshell is discussed briefly.

3.2 KIRKENDALL EFFECT

If two pieces of bulk metals are joined and heated, they generally interdiffuse to form alloy or intermetallic compounds. Here we assume that they have the same crystal structure, such as Cu and Ni, and when they interdiffuse, they form a continuous alloy or solid solution that has the same crystal structure as Cu and Ni. The effect of strain is ignored.

When A and B interdiffuse, the concentration profile in the region of interdiffusion broadens as shown in Figure 3.1. The mathematical solution of the profile is obtained by solving the one-dimension diffusion equation.

$$\frac{\partial C_A}{\partial t} = D\frac{\partial^2 C_A}{\partial x^2}$$

We first consider the solution of a thin layer of A diffusing into a semi-infinite long rod of B, similarly to the diffusion of a tracer layer, $\Delta\alpha$, of A into a semi-infinite long

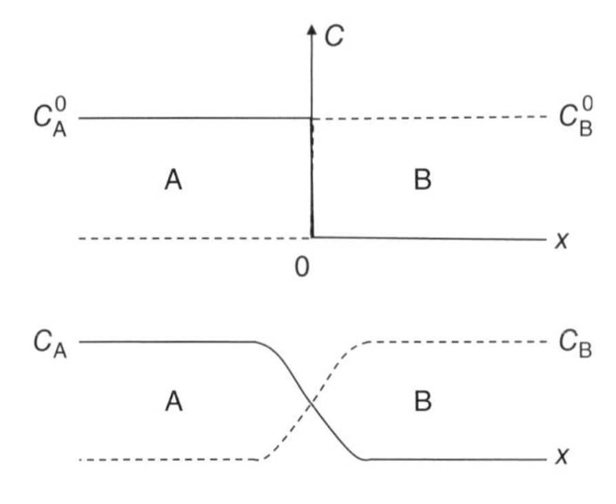

Figure 3.1 When A and B interdiffuse, the concentration profile broadens. The thin layer solution is extended to a semi-infinite couple by dividing A into many vertical slice of thickness of $\Delta\alpha$ and then an integration is taken to obtain the solution.

rod of A as discussed in Chapter 2.

$$C_A = \frac{\Delta\alpha C_A^0}{\sqrt{4\pi Dt}} \exp\left(-\frac{x^2}{4Dt}\right) \tag{3.1}$$

where C_A is the concentration of A atoms/cm^3 and C_A^0 is the initial concentration of A atoms/cm^3 in the thin layer of thickness "$\Delta\alpha$" before interdiffusion.

Then the thin layer solution is extended to a semi-infinite couple of A and B by dividing A into many vertical slices of thickness of $\Delta\alpha$, and integration is taken to obtain the following solution, as depicted in Figure 3.2.

$$C_A(x,t) = \frac{C_A^0}{\sqrt{4\pi Dt}} \int_{-\infty}^{0} \exp\left[-\frac{(x+\alpha)^2}{4Dt}\right] d\alpha \tag{3.2}$$

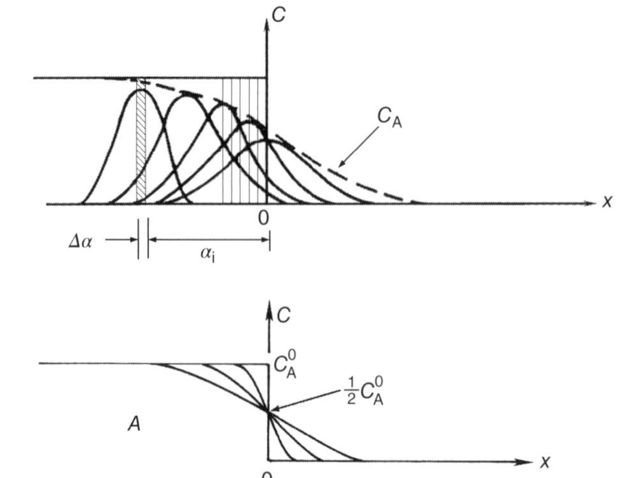

Figure 3.2 Mathematical solution of interdiffusion in the middle of composition remains at the original interface at $x = 0$.

Let a new dimensionless variable η to be

$$\eta = \frac{x+\alpha}{\sqrt{4Dt}}, \quad d\eta = \frac{d\alpha}{\sqrt{4Dt}}$$

then

$$\eta = \frac{x}{\sqrt{4Dt}} \quad \text{when } \alpha = 0$$

$$\eta = -\infty \quad \text{when } \alpha = -\infty$$

The last equation can be rewritten as,

$$C_{\mathrm{A}}(x,t) = \frac{C_{\mathrm{A}}^0}{\sqrt{\pi}} \int_{-\infty}^{x/\sqrt{4Dt}} \exp(-\eta^2) d\eta = \frac{C_{\mathrm{A}}^0}{2}\left[1 + \operatorname{erf}\left(\frac{x}{\sqrt{4Dt}}\right)\right] \tag{3.3}$$

The error function is defined as

$$\operatorname{erf}(z) = \frac{2}{\sqrt{\pi}} \int_0^z \exp(-\eta^2) d\eta \tag{3.4}$$

and we have $\operatorname{erf}(0) = 0$, $\operatorname{erf}(\infty) = 1$, and $\operatorname{erf}(-z) = -\operatorname{erf}(z)$. Thus at $x = 0$, we have

$$C_{\mathrm{A}} = \frac{C_{\mathrm{A}}^0}{2}$$

The composition profile changes as a function of time and temperature of interdiffusion. On the basis of the mathematical solution given in Eq. (3.3), the midpoint of composition is always at $x = 0$, that is, at the original interface of the couple. As shown in Figure 3.2, this mathematical midpoint is always located in the original interface. The original interface between A and B is defined as the Matano plane, and it does not move in interdiffusion. A simple physical meaning of the original interface (Matano plane) is that it can serve as the reference plane when we consider only the diffusion of A or B; the amount of A that has diffused across the original interface into the B side must equal to the loss of A from the A side. The reference plane applies to the diffusion of B too. However, the gain of A on the B side does not equal to the loss of B from the B side. Similarly, the gain of B on the A side does not equal to the loss of A from the A side. In other words, the flux of A does not equal to the opposite flux of B, and the difference is balanced by the flux of vacancies. The mathematical solution tends to miss this point because it does not consider the defect mechanism of diffusion as indicated by the Kirkendall effect.

How to find the Matano plane is discussed later. On interdiffusion in a bulk diffusion couple, the original interface disappears because of alloy formation and, furthermore, its position becomes unclear because of Kirkendall effect. Matano invented the method to find the original interface after interdiffusion. On the other hand, if we place marker in the original interface, the marker cannot be used to locate the original interface because it moves after interdiffusion. The marker plane is defined as the Kirkendall plane, and it moves in interdiffusion. Kirkendall effect showed that experimentally the diffusivity of A is not equal to that of B; therefore we cannot regard the mechanism of interdiffusion as a direct swap between A and B atoms. Kirkendall effect has changed our understanding of diffusion in solids, which must involve defects such as vacancies. To have a better understanding of the Kirkendall effect and the difference between Matano plane and Kirkendall plane, Darken's analysis is presented below, in which a moving reference frame and a fixed reference frame are coupled.

3.2.1 Darken's Analysis of Kirkendall Shift and Marker Motion

If more A atoms are assumed to diffuse into B, the lattice sites in the B must increase in order to accommodate the incoming excessive A atoms. Similarly, the lattice sites in A will decrease because more A atoms have left. The increase and decrease of lattice sites can be achieved respectively by the creation and elimination of vacant sites or vacancies in B and A, respectively. This is the reason we have to introduce a flux of vacancies in the interdiffusion between A and B. A very important assumption in Darken's analysis is that the vacancy concentration is at equilibrium everywhere in the diffusion couple. In other words, the vacancy sources in B are effective in creating vacant sites to accommodate the added A atoms, and the vacancy sinks in A are effective in eliminating vacant sites to accommodate the loss of A atoms. One of the atomic mechanisms of vacancy source and sink is dislocation climb as depicted in Figure 3.3. The climb of the edge dislocation loop in the right-hand side by taking up atoms at the end of the edge dislocation will lead to the addition or growth of an atomic plane. The climb of an edge dislocation loop in the left-hand side by taking up vacancies at the end of the edge dislocation will lead to the removal or shrinkage of a lattice plane. The combination of their operations, adding one atomic plane in

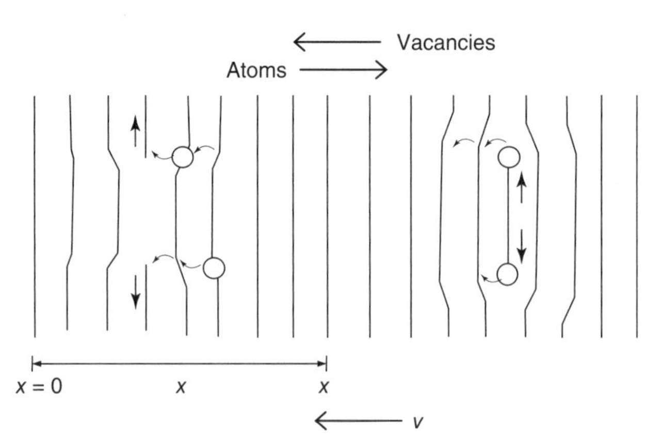

Figure 3.3 Schematic diagram of the atomic mechanisms of vacancy source and sink by dislocation climb.

the right and removing one atomic plane in the left, will cause all the lattice planes in between them to shift to the left by the distance of one atomic plane thickness. This is defined as *lattice shift* or *Kirkendall shift* in interdiffusion. If we place markers in the lattice having a lattice shift velocity of v, the markers will move with a velocity v, toward A (or against the faster diffusing species). We define the marker velocity v quantitatively later. The lattice shift is the atomic mechanism of marker motion in Kirkendall effect.

Because of the lattice shift, the vacancy is at equilibrium everywhere. Thus there is no vacancy supersaturation, and hence no void formation. Also, because of the lattice shift, the added lattice sites on the B side are balanced by the loss of lattice sites on the A side, so the net change in lattice sites is zero. If we further assume that the partial molar volume of A atom and B atom in the AB alloy is the same, it is a constant volume process, which means no strain. It also means that there is no stress in the interdiffusion. So in Darken's analysis, there is no void formation and no stress.

On the other hand, many experimental observations of interdiffusion in bulk diffusion couples have found void formation (Kirkendall or Frenkel void formation). Also, many studies of interdiffusion between a thin film and a Si wafer have found the bending of the wafer, which indicates that stress was produced in the interdiffusion. In these cases, we have to remove the assumption of a full lattice shift, so we do not have vacancy equilibrium everywhere in the sample. On the (001) single crystal surface of Si wafer, the density of vacancy source/sink is low if we assume that only the kinks on a surface step are the source/sink of vacancies.

It is worth mentioning that in an AB alloy, the molar volume of A may be different from that of B. Assuming A is larger than B, the volume in the B side will increase when A diffuses into B even if we assume that the diffusion occurs by a direct swap between A and B. Thus marker motion will move only to the side of A, the large molar volume side. Yet, experimentally, the smaller size atom tends to diffusion faster in interdiffusion, leading to marker motion to the smaller molar volume side. Furthermore, void formation due to Kirkendall effect must require excess vacancies; the effect of molar volume difference can cause stress but does not lead to void formation. No doubt, in the experimental measurement of marker motion, the consideration of molar volume change will make the measurement better, but it does not defeat the basic conclusion of Kirkendall effect that diffusion is defect mediated, which is conceptually correct.

Because of the motion of the lattice, Darken used two reference frames (coordinates) for analyzing the atomic fluxes; the laboratory frame (the fixed frame) and the marker frame (the moving frame). In the laboratory frame, the atomic fluxes are examined from a distance and the origin of the coordinate is typically selected to locate at one end of the diffusion couple where no interdiffusion occurs. In the marker frame, the atomic fluxes are examined by sitting on a marker, so the frame moves with the marker. In other words, the origin of the marker frame is located on a moving marker, so it is a moving frame.

If we examine the atomic fluxes, expressed by J_A and J_B, from the laboratory frame, the atomic fluxes consist of two terms as given below; the first term is due to diffusion and the second term is due to the motion of the lattice with a velocity v.

$$J_A = -D_A \frac{\partial C_A}{\partial x} + C_A v \tag{3.5}$$

$$J_B = -D_B \frac{\partial C_B}{\partial x} + C_B v \tag{3.6}$$

Thus

$$J_A + J_B = -D_A \frac{\partial C_A}{\partial x} - D_B \frac{\partial C_B}{\partial x} + Cv \tag{3.7}$$

where D_A and D_B are intrinsic diffusivities, and C_A and C_B are concentration of A and B in the AB alloy, respectively. Thus, $C_A + C_B = C$, where C is the total number of atoms per unit volume and it is a constant. Or if we consider one mole of alloy, we can take $C = N_A$, where N_A is the Avogadro's number and is the number of atoms per mole. In the equations, we use the lower case of x as the x axis of the coordinate of diffusion. We shall use the upper case X as atomic fraction of A and B as shown below. We define

$$X_A = \frac{C_A}{C_A + C_B} = \frac{C_A}{C} \tag{3.8}$$

$$X_B = \frac{C_B}{C_A + C_B} = \frac{C_B}{C} \tag{3.9}$$

where $X_A + X_B = 1$, and $CX_A + CX_B = C_A + C_B = C$. Furthermore, if we assume atomic volume is Ω, we have $C\Omega = 1$ or $C = 1/\Omega$ or $\Omega = 1/C$. Then we can express

$$X_A = \Omega C_A \text{ or } C_A = \frac{X_A}{\Omega}$$

$$X_B = \Omega C_B \text{ or } C_B = \frac{X_B}{\Omega}$$

Often it is convenient to use X instead of C in diffusion analysis, so Ω appears in the diffusion equation, as in the second part of this chapter on inverse Kirkendall effect.

As C is a constant, it implies that

$$\frac{\partial (C_A + C_B)}{\partial x} = 0$$

$$\frac{\partial C_A}{\partial x} = -\frac{\partial C_B}{\partial x}$$

$$\frac{\partial X_A}{\partial x} = -\frac{\partial X_B}{\partial x}$$

In the laboratory frame, the net flux is

$$J = J_A + J_B$$

and

$$\frac{\partial C}{\partial t} = -(\nabla \cdot J) = \frac{\partial}{\partial x}\left[D_A \frac{\partial C_A}{\partial x} + D_B \frac{\partial C_B}{\partial x} - Cv\right] \tag{3.10}$$

As C is constant and is independent of time, $\partial C/\partial t = 0$, and we have

$$D_A \frac{\partial C_A}{\partial x} + D_B \frac{\partial C_B}{\partial x} - Cv = \text{Const.} = K \tag{3.11}$$

To determine the constant K, we consider the origin at the end of the sample, $x = 0$, where no interdiffusion occurs, and the concentration of C_A and C_B are constant, and the concentration gradient of $\partial C_A/\partial x$ and $\partial C_B/\partial x$ are zero. Thus there is no marker motion, so $v = 0$. Therefore, the constant K in the last equation is zero. This means Eq. (3.7) = 0 or the net flux in the laboratory frame is zero, so $J_A + J_B = 0$, and $J_A = -J_B$. Therefore, we see that in laboratory frame, J_A and J_B are equal in magnitude but opposite in sign.

Now if we examine the atomic fluxes of A and B in the marker frame (or the moving frame), expressed respectively by j_A and j_B, we sit on a marker or on the origin of the frame. The atomic flux comes from diffusion alone as shown below, and because the origin is moving with the marker, there is no relative velocity difference between them.

$$j_A = -D_A \frac{\partial C_A}{\partial x} \tag{3.12}$$

$$j_B = -D_B \frac{\partial C_B}{\partial x} \tag{3.13}$$

So from Eqs (3.5) and (3.6) and $J_A + J_B = 0$, we have

$$j_A + j_B = -Cv = -j_V \tag{3.14}$$

We see that in the marker frame, the net flux of j_A and j_B is not zero, and $j_A \neq -j_B$, and where j_V is the vacancy flux that balances the difference between j_A and j_B. To continue, we rewrite the marker velocity as

$$v = \frac{1}{C}\left[D_A \frac{\partial C_A}{\partial x} + D_B \frac{\partial C_B}{\partial x}\right] = D_A \frac{\partial X_A}{\partial x} + D_B \frac{\partial X_B}{\partial x} = (D_B - D_A)\frac{\partial X_B}{\partial x} \tag{3.15}$$

In the following, we combine the above analyses in both frames and derive the atomic flux of B in the laboratory frame and obtain

$$J_B = -\widetilde{D} \frac{\partial C_B}{\partial x} \tag{3.16}$$

$$\widetilde{D} = X_A D_B + X_B D_A \tag{3.17}$$

If we use the laboratory frame to examine the atomic flux, we have, as given before,

$$J_B = j_B + C_B v = -D_B \frac{\partial C_B}{\partial x} + C_B v = -D_B \frac{\partial C_B}{\partial x} + C_B (D_B - D_A) \frac{\partial X_B}{\partial x}$$

$$= -\frac{(C_A + C_B)}{C} D_B \frac{\partial C_B}{\partial x} + \frac{C_B}{C}(D_B - D_A) \frac{\partial C_B}{\partial x} = -\frac{C_A D_B}{C} \frac{\partial C_B}{\partial x} - \frac{C_B D_A}{C} \frac{\partial C_B}{\partial x}$$

$$= -\frac{1}{C}(C_A D_B + C_B D_A) \frac{\partial C_B}{\partial x} = -(X_A D_B + X_B D_A) \frac{\partial C_B}{\partial x} = -\widetilde{D} \frac{\partial C_B}{\partial x}$$

where $\widetilde{D} = X_A D_B + X_B D_A$ is defined as the interdiffusion coefficient. Similarly, we have in the laboratory frame,

$$J_A = -\widetilde{D} \frac{\partial C_A}{\partial x}$$

This is because $\partial C_A / \partial x = -\partial C_B / \partial x$, we have $J_A = -J_B$ as shown before. At this point, we note the difference among the following three flux equations of B.

$$j_B = -D_B \frac{\partial C_B}{\partial x}$$

$$J_B = -D_B \frac{\partial C_B}{\partial x} + C_B v$$

$$J_B = -\widetilde{D} \frac{\partial C_B}{\partial x}$$

The same kind of equations can be written for the flux of A atoms too.

In the above analysis, we express the driving force of atomic flux in terms of concentration gradient, and we obtain $\widetilde{D} = X_A D_B + X_B D_A$. It can also be expressed in terms of chemical potential gradient (see Appendix B), and we can obtain an expression of the interdiffusion coefficient as $\widetilde{D} = C_B M G''$, where M is mobility and G'' is the second derivative of Gibbs free energy against concentration, so the interdiffusion coefficient takes the sign of G'', and it becomes negative inside a spinodal region, which is discussed in Chapter 4.

If we can measure experimentally both interdiffusion coefficient $\widetilde{D}$ and marker velocity v, we can solve the two equations for the two unknowns, D_A and D_B, which are the intrinsic diffusivities of A and B, respectively, in the alloy of AB. The measurement of marker velocity has been shown by Kirkendall. How to measure $\widetilde{D}$ by the Boltzmann and Matano analysis is presented below.

3.2.2 Boltzmann and Matano Analysis of Interdiffusion Coefficient

In Boltzmann and Matano analysis of interdiffusion coefficient, we first recall Fick's second law of diffusion,

$$\frac{\partial C}{\partial t} = \frac{\partial}{\partial x}\left(\widetilde{D} \frac{\partial C}{\partial x}\right)$$

To solve the above equation, we follow Boltzmann's conversion of variables,

$$C(x,t) = C(\eta) \text{ and } \eta = \frac{x}{t^{1/2}}$$

The advantage of the conversion is to change $C(x, t)$ of two variables into $C(\eta)$ of one variable, so that we can use the total differentiation of C with respect to η instead of partial differentiation of C with respect to x and t. Thus

$$\frac{\partial}{\partial t} = \frac{\partial \eta}{\partial t}\frac{d}{d\eta} = -\frac{x}{2t^{3/2}}\frac{d}{d\eta} = -\frac{1}{2t}\eta\frac{d}{d\eta}$$

$$\frac{\partial}{\partial x} = \frac{\partial \eta}{\partial x}\frac{d}{d\eta} = \frac{1}{t^{1/2}}\frac{d}{d\eta}$$

Getting back to Fick's second law, we have

$$-\frac{1}{2t}\eta\frac{dC}{d\eta} = \frac{1}{t^{1/2}}\frac{d}{d\eta}\left[\widetilde{D}(C)\frac{1}{t^{1/2}}\frac{dC}{d\eta}\right]$$

Dropping $1/t$, we have

$$-\frac{\eta}{2}\frac{dC}{d\eta} = \frac{d}{d\eta}\left(\widetilde{D}\frac{dC}{d\eta}\right)$$

Because these are total differentials, we can drop $1/d\eta$ and integrate both sides to give,

$$-\frac{1}{2}\int_0^{C'} \eta dC = \int_0^{C'} d\left(\widetilde{D}\frac{dC}{d\eta}\right) = \left[\widetilde{D}\frac{dC}{d\eta}\right]_0^{C'} \tag{3.18}$$

where C' is an arbitrary concentration, $0 < C' < C_0$, and C_0 is the concentration of A at $x = \infty$. Now we examine the physical picture of the interdiffusion. If we consider at a given time (i.e., t is fixed, so $d\eta = t^{-1/2}\,dx$), we have at both ends of the diffusion couple $dC/d\eta = 0$ at $C = 0$ and $C = C_0$. Therefore, in Eq. (3.18), if we integrate from $C = 0$ to $C = C_0$ along the "vertical" axis, it gives

$$-\frac{1}{2}\int_0^{C_0} \eta dC = \widetilde{D}\left.\frac{dC}{d\eta}\right|_{C=C_0} - \widetilde{D}\left.\frac{dC}{d\eta}\right|_{C=0} = 0 - 0 = 0$$

This means that at a fixed t,

$$\int_0^{C_0} xdC = 0 \tag{3.19}$$

This is because when the interdiffusion is considered at a fixed time, the variable η is the same as x. This relationship, Eq. (3.19), defines the Matano plane. The Matano plane is selected at where the quantity of A atoms $(1 - C_A)$ that have been removed from the left of the interface is equal to the quantity C_A that has been added to the

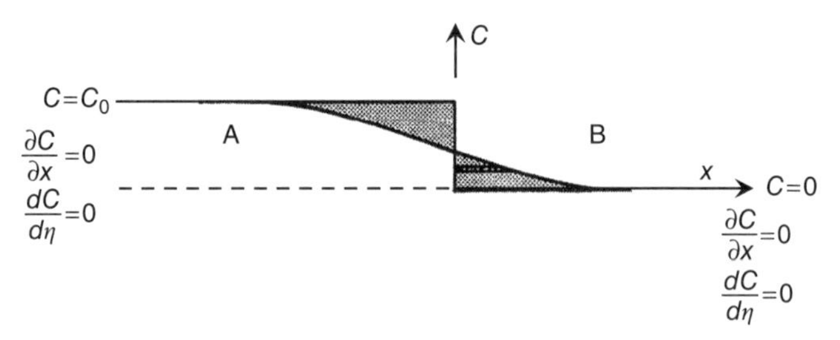

Figure 3.4 Matano interface divides the shaded area on the left of the interface to be equal to the shaded area on the right of the interface.

right of the interface. Graphically, it means in the sketch in Figure 3.4, the shaded area on the left of the interface is equal to the shaded area on the right of the interface. This reference interface is defined as Matano plane. It is at the same location as the original interface before interdiffusion. In other words, Matano plane identifies the original interface.

Why is Matano plane important? It defines the location of the origin of the x-axis, that is, $x = 0$, which is needed for carrying out the integration along the vertical axis in Eqs (3.18) and (3.19). Otherwise, the integral of

$$-\frac{1}{2}\int_0^{C'} \eta dC$$

is "undetermined" because it is integrated over C along the vertical axis, and the origin is arbitrary until it is defined by the Matano plane.

In Eq. (3.18), if we convert η back to x, and define $\widetilde{D}$ to be the interdiffusion coefficient, we have

$$\widetilde{D}(C') = -\frac{1}{2t}\left(\frac{dx}{dC}\right)_{C'}\int_0^{C'} xdC \tag{3.20}$$

This equation indicates that the measurement of C as a function of x (i.e., the concentration profile of A) can give $\widetilde{D}$ at a given concentration C' by a graphical method (when t is given), that is, by measuring the slope and the shaded area as shown in Figure 3.5. It is worth mentioning that if we plot the concentration profile of B, we obtain the same Matano plane. This is because the experimental concentration profiles are measured in laboratory frame, in which we have shown that $J_A = -J_B$.

On a measured interdiffusion concentration profile, we can choose an arbitrary concentration of C' and obtain both the slope and the shaded area, as shown in Figure 3.5. By using the Boltzmann–Matano analysis, we can measure the interdiffusion coefficient. When it is combined with the measurement of marker velocity in the diffusion couple, we can solve the intrinsic diffusion coefficient of D_A and D_B using the pair of equations of Eqs (3.15) and (3.16).

Figure 3.6 is a sketch of interdiffusion coefficient (the middle curve) of Cu–Ni alloys at 1000 °C. The intrinsic diffusion coefficient of Cu is the upper curve and that of Ni is the lower curve.

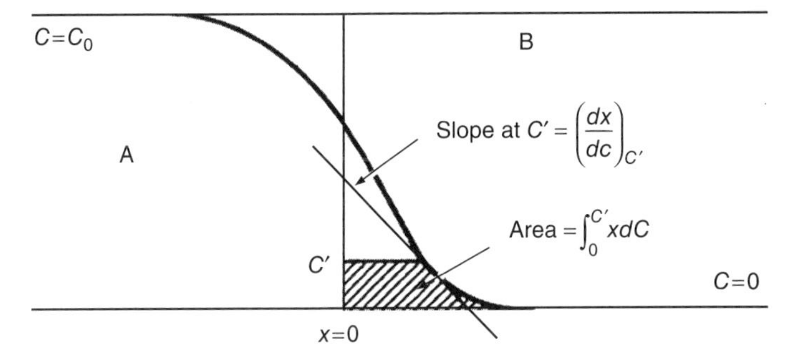

Figure 3.5 A graphical method to measure the interdiffusion coefficient $\widetilde{D}$ at a given concentration C' by measuring the slope and the shaded area as shown in the figure.

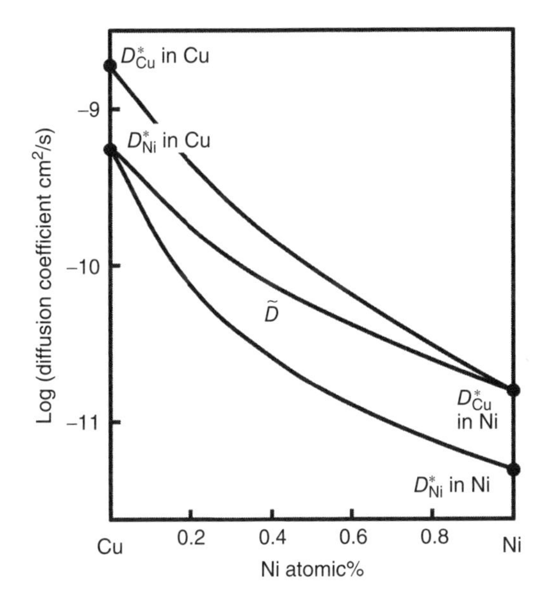

Figure 3.6 A sketch of interdiffusion coefficient (the middle curve) of Cu–Ni alloys at 1000 °C. The intrinsic diffusion coefficient of Cu is the upper curve and that of Ni is the lower curve.

To verify that Matano plane identifies the original interface, we consider a very simple case of the interdiffusion between A (Si) and B (Ni) to form a single layer intermetallic compound of $A_\beta B$ (Ni_2Si), as depicted in Figure 3.7. We assume that in the interdiffusion, Ni is the dominant diffusing species, so we can ignore the diffusion of Si. The growth of Ni_2Si occurs by dissolving Ni from the Ni layer at the Ni_2Si/Ni interface, the diffusion of Ni atoms across the Ni_2Si, and the reaction to form Ni_2Si at the Si/Ni_2Si interface. As Ni_2Si grows, the interface of Ni_2Si/Ni moves into Ni and the interface of Si/Ni_2Si moves into Si. The latter moves at half the rate of the former because the atomic ratio of Ni/Si is 2/1 in the compound, if we assume that the atomic volume as well as the molar volume of Ni and Si are the same. Furthermore, if we assume that the concentration in Ni_2Si is nearly constant and if we plot the concentration profile of Ni as shown in Figure 3.7, the Matano plane occurs at the location that is 2/3 from the Ni_2Si/Ni interface and 1/3 from the Si/Ni_2Si

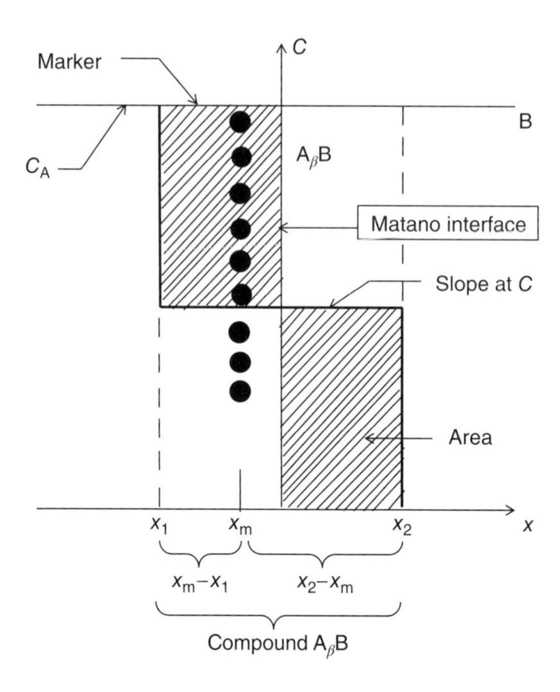

Figure 3.7 Schematic diagram of the interdiffusion between A (Si) and B (Ni) to form a single layer intermetallic compound of $A_\beta B(Ni_2Si)$.

interface, and it is the original interface between Ni and Si before the reaction or interdiffusion.

3.2.3 Activity and Intrinsic Diffusivity

We recall the interdiffusion coefficient as

$$\widetilde{D} = X_B D_A + X_A D_B$$

where D_A and D_B are defined as the intrinsic diffusivity of A atom and B atom in the AB alloy of a specific composition of C', for example. Now, what is the difference between the diffusion of a tracer A atom in the AB alloy and the diffusion of an A atom (nontracer A atom) in the same AB alloy? The difference is that the diffusion of the nontracer A atom has to be driven by a chemical potential gradient or a concentration gradient; otherwise, the A atom just does random walk in the homogeneous alloy and results in no flux of A atoms. However, in interdiffusion, we need to consider the flux of A atoms driven by the concentration gradient of A atoms. On the other hand, the diffusion of a tracer A atom in the AB alloy is driven by the concentration gradient of tracer A atoms. This is because there is no concentration gradient of the A atoms as the AB alloy is homogeneous.

If there is a strong chemical effect in forming the AB alloy, or in an intermetallic compound, we need to consider the chemical effect on diffusion, or chemical activity, besides the concentration gradient. Chemical activity means chemical strength. A simple picture of chemical activity can be obtained by considering whether the AB alloy is an ideal solid solution or a regular solid solution, or whether the enthalpy of mixing to form these solutions is zero or nonzero. Because the mixing occurs typically at ambient, the change of pV is usually negligible and we can take enthalpy as internal energy. On the basis of quasi chemical short-range interaction, the internal energy of mixing per mole is calculated in Chapter 5 as

$$\Delta E = N_a Z X_A X_B \varepsilon \tag{5.8}$$

where N_a is Avogadro number of atoms per mole, Z is the nearest neighbor number in the solid solution, X_A and X_B are the fraction of A and B atoms in the solution, and ε is defined as

$$\varepsilon = \varepsilon_{AB} - \frac{1}{2}(\varepsilon_{AA} + \varepsilon_{BB}) \tag{5.7}$$

where ε_{AB}, ε_{AA}, and ε_{BB} are interatomic bond energy between A and B, A and A, and B and B atoms, respectively, and they are all negative. To form AB solid solutions, the AB bond is stronger or more negative, so ε is negative. However, in an ideal solid solution where ε is zero, there is no net chemical effect. To envision the chemical interaction, we depict in Figure 3.8 the cross-section of the surface of an AB alloy that has a very low concentration of B. At equilibrium, the partial pressure of A and B above the alloy surface is given as p_A and p_B, respectively. For an ideal solution, when there is no chemical effect, the partial pressure is directly proportional to the concentration. This is Raoult's law, as depicted by the broken curve in Figure 3.9. For a regular solution, when there is a chemical effect to form alloy between A and B, the partial pressure of B is lowered in the alloy having a low concentration of B in the left-hand side of Figure 3.9. This is because the small number of B atoms will be tied down by the nearest neighboring A atoms due to the stronger chemical interaction between A and B, or, in other words, it is harder to break the AB bonds compared to AA bonds. This is Henry's law as depicted by the solid curve in Figure 3.9. However, as the concentration of B increases, by moving to the right-hand side of Figure 3.9, the chemical effect becomes less and less important for B atoms and the partial pressure of p_B returns to the broken curve.

The chemical effect on diffusion is similar. In an AB alloy of low concentration of B, to exchange a vacancy with a B atom is harder than that with an A atom. Again this is because to break the AB bonds is harder than to break the AA bonds. To use activity to represent the chemical effect, we recall chemical potentials for a B atom

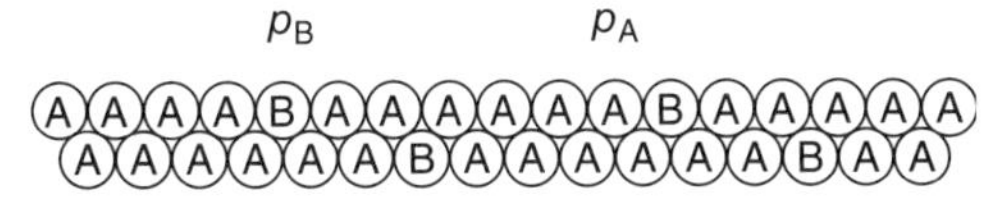

Figure 3.8 Schematic diagram of the cross-section of a surface of an AB alloy of very low concentration of B. At equilibrium, the partial pressure of A and B above the surface is given as p_A and p_B, respectively.

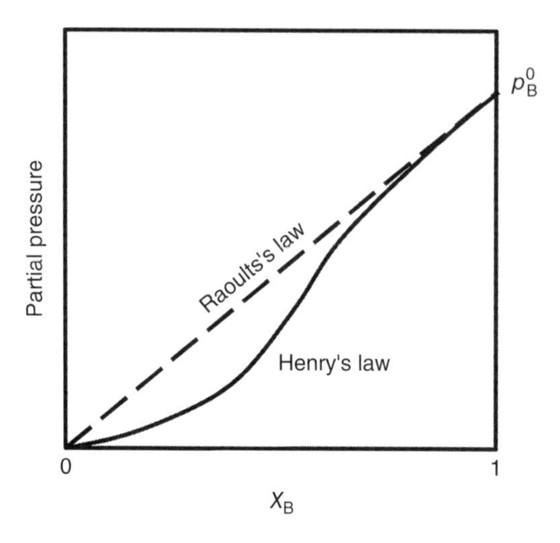

Figure 3.9 Schematic diagram of Raoult's law and Henry's law of partial pressure of solutions of AB as a function of alloy composition.

in a dilute ideal solution and a regular solution below.

$$\mu_B = kT \ln X_B$$
$$\mu_B = kT \ln a_B = kT \ln \gamma_B X_B$$

where X_B is atomic fraction of B atoms in the solution, a_B is defined as activity, and γ_B is defined as activity coefficient. For an ideal solution, $a_B = X_B$ and $\gamma_B = 1$.

In Section 2.2.2, we have obtained Fick's first law by taking

$$J_B = C_B \frac{D_B}{kT}\left(-\frac{\partial \mu_B}{\partial x}\right) = -D_B \frac{\partial C_B}{\partial x} \tag{3.21}$$

where D_B can be taken as the diffusivity of a tracer B atom in pure B and the diffusion is driven by the concentration gradient of tracer B atoms. This is because we can regard the mixture of a metal and a small amount of its tracer as an ideal dilute solution.

By following the same procedure for a regular solution, we substitute μ_B of the regular solution into the above equation, and we recall that $X_B/\Omega = C_B$, where Ω is atomic volume. Also, we have to change D_B into D_B^*, where D_B^* is the diffusivity of tracer B atoms in the AB alloy.

$$J_B = C_B \frac{D_B^*}{kT}\left(-\frac{\partial \mu_B}{\partial x}\right) \tag{3.22}$$

By substituting $\mu_B = kT \ln \gamma_B X_B$ into the last equation, we have

$$J_B = C_B \frac{D_B^*}{kT}\left(-\frac{\partial \mu_B}{\partial X_B}\frac{\partial X_B}{\partial x}\right) = -\frac{X_B}{\Omega}\frac{D_B^*}{kT}\frac{\partial X_B}{\partial x}\frac{\partial}{\partial X_B}(kT\ln X_B + kT\ln\gamma_B)$$

$$= -\frac{X_B}{\Omega}\frac{D_B^*}{kT}\frac{\partial X_B}{\partial x}\left[\frac{kT}{X_B}\left(1+\frac{\partial \ln\gamma_B}{\partial \ln X_B}\right)\right]$$

$$= -D_B^*\left(1+\frac{\partial \ln\gamma_B}{\partial \ln X_B}\right)\frac{\partial (X_B/\Omega)}{\partial x} = -D_B\frac{\partial C_B}{\partial x}$$

In the above equation, we have

$$D_B = D_B^*\left(1+\frac{\partial \ln\gamma_B}{\partial \ln X_B}\right) = D_B^*\varphi \tag{3.23}$$

where φ is a thermodynamic factor as defined below,

$$\varphi = \frac{X_B}{kT}\frac{\partial \mu_B}{\partial X_B} = \frac{X_B}{kT}\frac{\partial}{\partial X_B}(kT\ln X_B\gamma_B) = 1+\frac{\partial \ln\gamma_B}{\partial \ln X_B} \tag{3.24}$$

Similarly, the intrinsic diffusivity of A in the alloy of AB is

$$D_A = D_A^*\left(1+\frac{\partial \ln\gamma_A}{\partial \ln X_A}\right) = D_A^*\varphi$$

where D_A^* is the diffusivity of tracer A atoms in the AB alloy and we have

$$\widetilde{D} = X_B D_A + X_A D_B = (X_B D_A^* + X_A D_B^*)\varphi$$

In the above expression for $\widetilde{D}$, we have separated it into two parts, where φ is the thermodynamic part coming from the driving force and it includes the chemical effect, and the bracket term is the kinetic part, and both the diffusivities are functions of the atomic fraction. These diffusivities, in the bracket term, can be measured by tracer of A and B in the alloy. Because in an alloy of constant composition the chemical potential is constant, it offers no driving force for the tracer atoms; rather, the driving force for diffusion of tracer atoms depends on the concentration gradient of the tracer atoms. Thus the chemical effect of AB can be ignored. However, the tracer diffusivity is also a function of alloy composition because the chemical potential of the alloy changes with composition.

We recall that the intrinsic diffusivities of D_A and D_B can be measured by using Darken's analysis and Boltzmann and Matano analysis. The relationship between interdiffusivity and intrinsic diffusivities as a function of alloy composition is shown schematically in Figure 3.6 for Cu–Ni alloys. In the figure, what is the relationship between tracer diffusion of A in pure A and that in an AB alloy? Actually we can represent the diffusivity D_B in Eq. (3.21) by D_B^* as in Eq. (3.22), except that in Eq. (3.21) it is for pure A, rather than for an alloy. We see from Figure 3.6 that for dilute alloys, by a straight line extrapolation, we have $\ln D_B^*(\text{alloy}) = (1 + bC) \ln D_B^*(\text{pure})$, where b is an empirical number. If we plot the intrinsic diffusivities in linear scale, we have $D_B^*(\text{alloy}) = (1 + bC)D_B^*(\text{pure})$.

3.2.4 Kirkendall (Frenkel) Voiding Without Lattice Shift

In Darken's analysis, the marker velocity has been given as Eq. (3.14),

$$v = \frac{1}{C}\left[D_A \frac{\partial C_A}{\partial x} + D_B \frac{\partial C_B}{\partial x}\right] = D_A \frac{\partial X_A}{\partial x} + D_B \frac{\partial X_B}{\partial x} = (D_B - D_A)\frac{\partial X_B}{\partial x}$$

In obtaining the above equation, we have taken $j_A + j_B = -j_V$ as shown by Eq. (3.14). Thus, the vacancy flux is given as

$$j_V = Cv = C(D_B - D_A)\frac{\partial X_B}{\partial x} = (D_B - D_A)\frac{\partial C_B}{\partial x} \tag{3.25}$$

where we recall that X_B ($=C_B/C$) is the fraction of concentration of B, which is the faster diffusing species in the alloy. If this vacancy flux cannot be absorbed by lattice shift, it will lead to vacancy supersaturation and the nucleation of voids. Moreover, if we consider the growth of a single void, and assume the cross-sectional area of the interdiffusion couple is A', the number of vacancies transported in a period of dt will be $j_V A't$. The volume of the void will be $V = \Omega j_V A't$, where Ω is the atomic volume of a single vacancy. Furthermore, if we assume the void is spherical with a radius r, we have $4\pi r^3/3 = \Omega j_V A't$, and the growth rate of the void is

$$\frac{dr}{dt} = \frac{\Omega j_V A'}{4\pi r^2} = \frac{\Omega A'}{4\pi r^2}(D_B - D_A)\frac{\partial C_B}{\partial x} \tag{3.26}$$

In the above equation, the void growth rate changes with D_A, D_B, and $\partial C_B/\partial x$ besides r. In actual experiments, not one void but many voids were formed, and the total volume of all the voids will be given by $V = \Omega j_V A't$.

3.3 INVERSE KIRKENDALL EFFECT

In Section 1.5, we have considered the instability of a nanoshell of a pure element of Au. The nanoshell has a positive curvature at its outer surface and a negative curvature at its inner surface. Following Gibbs–Thomson potential difference, a vacancy gradient exists across the nanoshell, which can cause instability of the nanoshell by the out-diffusion of the vacancies and transforms the nanoshell into a solid nanosphere. We may regard these vacancies as nonequilibrium vacancies and we have had a discussion of them on the basis of Figure 1.5.

Now, if we consider a nanoshell of a homogeneous AB alloy, as depicted in Figure 1.6, again the Gibbs–Thomson potential difference will generate a flux of

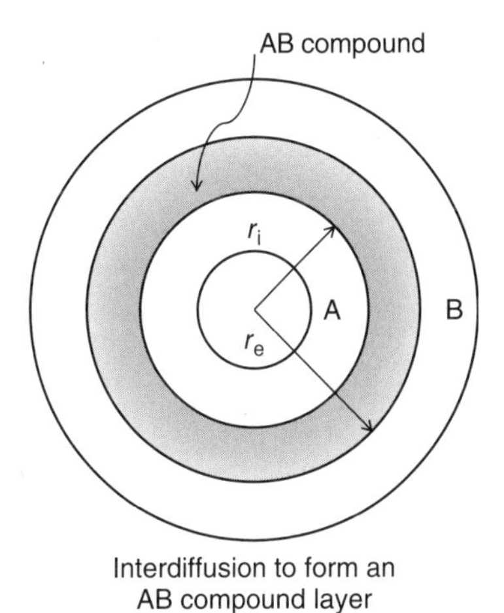

Figure 3.10 Inverse Kirkendall effect in nanoshell, where r_i and r_e are the radii of inner and outer (external) boundaries of an alloy nanoshell, respectively.

vacancies, and the diffusion of these vacancies will lead to segregation in the alloy. It is important to note that this segregation is a kinetic (or dynamic) one, contrary to the common thermodynamic segregation at interfaces. The thermodynamic segregation is due to thermodynamic or energetic reasons (dependence of interface energy on the local composition). Kinetic segregation exists only as long as the vacancy flux exists; if the vacancy flux stops, the kinetic segregation disappears.

This kinetic segregation has been called *inverse Kirkendall effect*. In the following, we discuss first the physical meaning of the inverse Kirkendall effect or the segregation. Then we analyze the kinetics of void shrinkage in a homogeneous alloy nanoshell of AB. Finally, we analyze the interaction between Kirkendall effect and inverse Kirkendall effect in the reaction of a concentric (spherical) or coaxial (cylindrical) nanoshell of a bilayer of A and B in forming an intermetallic compound of ApBq. Such a nanoshells is depicted in Figure 3.10.

A nanoshell of a pure element is depicted in Figure 1.5, a nanoshell of an alloy of AB, or a nanoshell of an intermetallic compound of ApBq is depicted in Figure 1.6, and a concentric nanoshell of A and B to form an intermetallic compound of ApBq in between them by interdiffusion is depicted in Figure 3.10.

It is worth mentioning that in the interdiffusion of A and B to form an alloy of AB or in the segregation of an AB alloy, we can assume an ideal solution of AB; thus we do not need to consider the chemical interaction between A and B. On the other hand, in the interdiffusion in an intermetallic compound of ApBq or in the formation of the intermetallic compound, we must consider the chemical interaction between A and B so that we need to include the thermodynamic factor φ of intrinsic diffusivity by assuming a regular solution, for example, and the analysis is more complicated.

In the kinetic analysis to be presented below, we focus on the kinetics of elimination of the central void in the nanoshell. We take the same approach by solving the diffusion problem in spherical coordinates under steady state. The boundary conditions are taken at the inner and outer nanoshell surfaces. The solution gives the concentration profile, especially that of the nonequilibrium vacancies. Then we obtain the flux of vacancies leaving the inner surface, so we can calculate the time needed to eliminate the central void. We recall that the simplest case of kinetics of instability of nanoshell of a pure element has been presented in Section 1.5.

We note that the formation of a nanoshell of Co_3S_4 during the annealing of Co nanospheres in sulfur ambient requires the out-diffusion of Co and in-diffusion of vacancies, which is opposite to what has been outlined in the above.

3.3.1 Physical Meaning of Inverse Kirkendall Effect

Inverse Kirkendall effect was discovered first in alloys under irradiation [9]. Such alloys often demonstrated segregation near the external surfaces. Actually, inverse Kirkendall effect appears under a vacancy flux generated by some external reasons and it involves two interrelated "subeffects." First, the mentioned vacancy flux causes counterfluxes of both components. Because of the inequality of intrinsic diffusivities of A and B in the homogeneous alloy, the counterfluxes have different velocities, leading to the segregation of the faster component at the side from which vacancies flow. Second, the segregation means the building-up of concentration gradient of the A and B components (uphill diffusion). Owing to the cross-effect as in irreversible processes (discussed in Appendix C on nonequilibrium vacancies), this concentration gradient (either A or B) creates a feedback on the vacancy flux. Namely, this feedback always makes the total vacancy flux less; it is one more example of the **Le Chatelier's** principle that if some external reasons attempt to change the state of the system, this system will create some feedback to counteract the change, and as a result to smooth and slow down the change.

Now let us concentrate on the first subeffect of inverse Kirkendall effect on segregation. We can imagine that some force "pushes" vacancy to jump in a certain direction and to exchange positions with a certain neighboring atom. Vacancy jump leads to opposite jumps of atoms, either A or B, in the alloy. So we ask whether A or B will most probably exchange places with the jumping vacancy. The probability per unit time for a given vacancy to exchange with atom A within a small time interval dt is a product of atomic fraction of A, X_A, and the frequency of exchange between the vacancy and the atom A, ν_{VA}, and dt; a product of $X_A \nu_{VA} dt$. Similar for exchange with atom B, we have $X_B \nu_{VB} dt$. Thus, the probability to choose A is

$$\frac{X_A \nu_{VA} dt}{X_A \nu_{VA} dt + X_B \nu_{VB} dt} = \frac{X_A D_A^*}{X_A D_A^* + X_B D_B^*}$$

Here we take into account the fact that the tracer diffusivity of a component is proportional to its jump frequency. Therefore, in the beginning, when a vacancy flux, j_V, is starting to flow in the initially homogeneous alloy, the fluxes of the A and the B components are

$$\Omega j_B = -\frac{X_B D_B^*}{X_A D_A^* + X_B D_B^*} \Omega j_V = X_B \nu_B$$

$$\Omega j_A = -\frac{X_A D_A^*}{X_A D_A^* + X_B D_B^*} \Omega j_V = X_A \nu_A \tag{3.27}$$

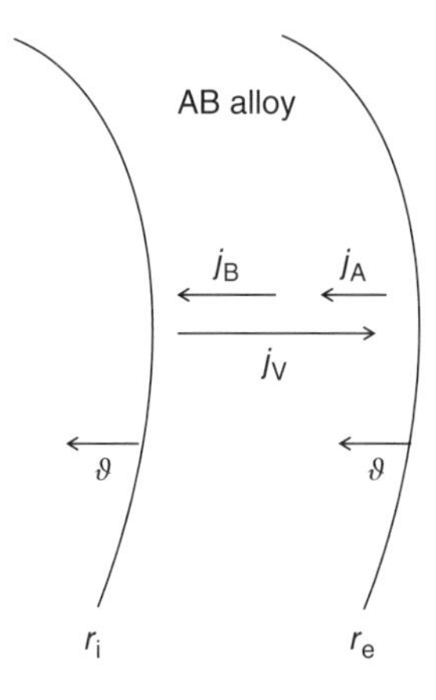

Figure 3.11 A thin film of homogeneous alloy of AB. Under the influence of some external force, it provides the fixed constant vacancy flux j_V directed from the left to the right as depicted in the figure. In the induced fluxes of A and B, we assume B to be the faster diffusing species.

In the above equations, we have replaced C by X/Ω in the general equation of $j = C<v>$, so $\Omega j = X<v>$, and v_A and v_B are the drift velocity of A and B atoms, respectively. The sum of the above two equation gives

$$\Omega(j_A + j_B) = -\Omega j_V$$

Or $j_A + j_B = -j_V$. However, the difference of drift velocities of A and B under the vacancy flux is, in fact, the core of inverse Kirkendall effect, which is shown by Eq. (3.28). In the pair of equations in Eq. (3.27), we divide the first by X_B and the second by X_A and take the difference and we obtain

$$v_B - v_A = \frac{D_A^* - D_B^*}{X_A D_A^* + X_B D_B^*} \Omega j_V \tag{3.28}$$

After the starting of segregation of components under inverse Kirkendall effect, the concentration gradients of A and B will appear, and typically the atomic diffusion will counteract the uphill diffusion so that the system will come to a steady-state separation. We can illustrate it by a simple example. We imagine a thin film of binary alloy of AB, and we assume B to be the faster diffusing species. Under the influence of some external force, it provides the fixed constant vacancy flux j_V directed from the left to the right, as depicted in Figure 3.11. As we have just discussed, this flux should lead to the accumulation of the faster component of B closer to the left boundary and the slower component of A closer to the right boundary, while both of them are in the same direction, opposing the vacancy flux. In turn, the atomic fluxes should lead to the motion of both boundaries of the thin film in the opposite direction of the vacancy flux (to the left) with a constant velocity equal to $v = -\Omega j_V = \Omega j_A + \Omega j_B$, and each boundary moves opposite to vacancies drift. For the boundary velocity, see the explanation of Eq. (7.2). At the moving boundary of r_i and r_e, the net flux of B and the net flux of A should be zero (see Fig. 3.11).

To find the steady-state concentration gradient under the condition of zero flux of components A and B with respect to the moving boundaries, we have

$$\Omega j_{\mathrm{A}} - X_{\mathrm{A}} v = D_{\mathrm{A}}^* \varphi \frac{\partial X_{\mathrm{A}}}{\partial x} + \frac{D_{\mathrm{A}}^* X_{\mathrm{A}}}{X_{\mathrm{V}}} \frac{\partial X_{\mathrm{V}}}{\partial x} + X_{\mathrm{A}} \Omega j_{\mathrm{V}} = 0 \tag{3.29}$$

$$\Omega j_{\mathrm{B}} - X_{\mathrm{B}} v = -D_{\mathrm{B}}^* \varphi \frac{\partial X_{\mathrm{B}}}{\partial x} + \frac{D_{\mathrm{B}}^* X_{\mathrm{B}}}{X_{\mathrm{V}}} \frac{\partial X_{\mathrm{V}}}{\partial x} + X_{\mathrm{B}} \Omega j_{\mathrm{V}} = 0 \tag{3.30}$$

We note that the expressions for Ωj_{A} and Ωj_{B} in Eqs (3.29) and (3.30), having the thermodynamic factor of φ, are given in Appendix C on nonequilibrium vacancies.

We can treat the last two equations as a set of two linear algebraic equations with two unknowns, and the two unknowns are the gradients of $\partial X_{\mathrm{B}}/\partial x, \partial X_{\mathrm{V}}/\partial x$. As discussed later in Section 3.3.3, we can treat the system as a pseudo ternary system of A, B, and vacancy. Nevertheless, in the pseudo ternary system, only two of the three are independent variables. To eliminate $\partial X_{\mathrm{V}}/\partial x$ from them, we multiple Eq. (3.29) by $D_{\mathrm{B}}^* X_{\mathrm{B}}$ and multiple Eq. (3.30) by $D_{\mathrm{A}}^* X_{\mathrm{A}}$ and subtract them and we have

$$\frac{\partial X_{\mathrm{B}}}{\partial x} = -\frac{X_{\mathrm{A}} X_{\mathrm{B}} (D_{\mathrm{B}}^* - D_{\mathrm{A}}^*)}{D_{\mathrm{A}}^* D_{\mathrm{B}}^* \phi} \Omega j_{\mathrm{V}} = -\frac{X_{\mathrm{A}} X_{\mathrm{B}} (1 - D_{\mathrm{A}}^*/D_{\mathrm{B}}^*)}{D_{\mathrm{A}}^* \phi} \Omega j_{\mathrm{V}} \tag{3.31}$$

The last equation demonstrates that the magnitude of segregation in inverse Kirkendall effect is determined mainly by the diffusivity of slow component, D_{A}^*, provided that the vacancy flux is fixed. If $D_{\mathrm{A}}^* \ll D_{\mathrm{B}}^*$, we can drop the term of their ratio in the last equation.

We examine a couple of examples of inverse Kirkendall effect on the shrinking kinetics of a binary alloy nanoshell and a compound nanoshell below.

3.3.2 Inverse Kirkendall Effect on the Instability of an Alloy Nanoshell

The analysis of inverse Kirkendall effect presented in the above can be applied to the instability of an alloy nanoshell or nanotube. If we take a thin film of a binary alloy and roll it into a nanotube, Gibbs–Thomson effect will generate a flux of vacancies in the nanotube and will lead to segregation and instability as presented above. Furthermore, the segregation will produce a cross-effect on the vacancy flux and may tend to reduce it and slow down the segregation as well as the shrinking of the central void in the nanotube or nanoshell. In this section, a very simple model of analysis is presented in order to show the methodology and procedure of the analysis so that the physical picture is clear. In the next section, a more detailed analysis of segregation in an alloy nanoshell assuming regular solution is given.

We begin by analyzing the vacancy flux in this case, and it is given below for a nanoshell. For a nanotube, it is very similar. Let r_{i} and r_{e} be the radii of inner and outer (external) boundaries of an alloy nanoshell, respectively, as shown in Figure 1.6

and Eq. (1.18).

$$C_V(r_e) = C_{V0}\exp\left(-\frac{2\gamma\Omega}{kTr_e}\right) \cong C_{V0}\left(1 - \frac{2\gamma\Omega}{kTr_e}\right)$$

$$C_V(r_i) = C_{V0}\exp\left(+\frac{2\gamma\Omega}{kTr_i}\right) \cong C_{V0}\left(1 + \frac{2\gamma\Omega}{kTr_i}\right)$$

Thus, as a vacancy gradient exists in the alloy nanoshell, a vacancy flux moves from the inner boundary to the outer boundary. By assuming a constant vacancy gradient because the thickness of the nanoshell is very small, the vacancy flux can be expressed as

$$j_V = -D_V\frac{\Delta C_V}{\Delta r} = -D_V\frac{C_V(r_e) - C_V(r_i)}{r_e - r_i} = D_V C_{V0}\frac{2\gamma\Omega}{kT}\left(\frac{1}{r_i} + \frac{1}{r_e}\right)\frac{1}{\Delta r} \tag{3.32}$$

Where $\Delta r = r_e - r_i$. The out-diffusion of this vacancy flux will lead to the segregation of the alloy. As we have discussed in Appendix C on nonequilibrium vacancies, the diffusion of the A and B atoms will influence the vacancy diffusion by the cross-effect and tend to reduce it. For simplicity, if we ignore the cross-effect as well as the chemical effect (ideal solution) and we substitute the above Eq. (3.32) into Eq. (3.31), we obtain the segregation effect. The slower diffusing species will move in the same direction as the vacancy and will segregate to the outer surface. The faster diffusing species will segregate to the inner surface.

To analyze the kinetic of shrinkage of the hollow alloy nanoshell, when we ignore the cross-effect as well as the chemical effect, it will be the same as the shrinking of a pure metal hollow nanoshell as shown in Section 1.5. We begin with the diffusion equation in spherical coordinate and assume spherical symmetry and steady-state process. The Laplacian is

$$\nabla^2 C_V = \frac{1}{r^2}\frac{\partial}{\partial r}\left(r^2\frac{\partial}{\partial r}C_V\right) = \frac{1}{r^2}\frac{1}{D_V}\frac{\partial}{\partial r}(r^2 j_V) = 0$$

Instead of following the standard procedure as presented in Section 1.5, we consider

$$\frac{\partial}{\partial r}(r^2 j_V) \approx 0$$

We solve this diffusion equation by taking $r^2 j_V = \text{const}$. We obtain the flux equation of vacancy as

$$\Omega j_V \approx \frac{K_v}{r^2}$$

Assuming this is the average j_V leaving the inner radius at $r = r_i$, we can estimate the time it takes to remove the total number of vacancies in the void, and roughly, it will obey the $r_i^3 \approx (\gamma\Omega D^*/kT)t$ relationship as given in Section 1.5.

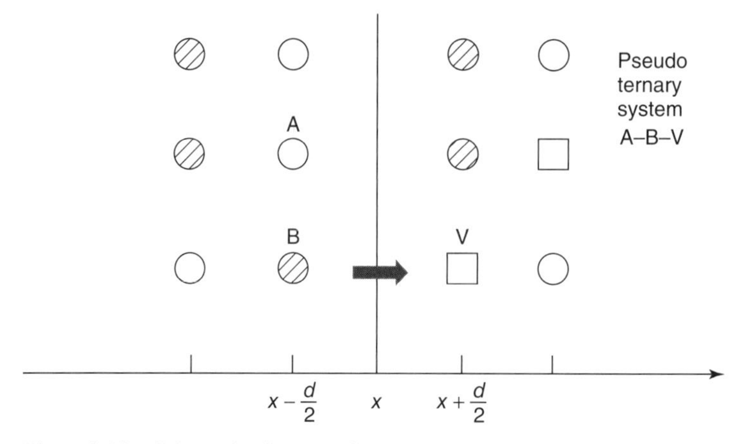

Figure 3.12 Schematic diagram of a B atom in alloy of AB making an attempt to exchange positions with a neighboring vacancy.

3.3.3 Inverse Kirkendall Effect on Segregation in a Regular Solution Nanoshell

When we consider the kinetics in a nanoshell of an AB alloy where the alloy is a regular solution, besides the chemical effect, we have the following effects. First, Gibbs–Thomson effect introduces a vacancy flux, which leads to the segregation of A and B by the inverse Kirkendall effect due to the uneven diffusion of A and B atoms in the compound. Because the vacancy flux is dominant, we should consider the alloy as a pseudo ternary system of A, B, and vacancy, in which two of them as well as their fluxes are independent. For example, we can select the flux of A atoms to be dependent on those of vacancies and the B atoms. When vacancy and B are independent, there is a cross-effect between the vacancy flux and the B atomic flux and the cross-effect tends to influence and reduce the vacancy flux. Figure 3.12 is a schematic diagram of a B atom in the AB alloy making an attempt to exchange position with a neighboring vacancy. Kinetic analysis of the cross-effect is given in Appendix C. Furthermore, we need to consider the chemical effect on the intrinsic diffusivities in the alloy due to the chemical interaction of A and B atoms.

We begin by Fick's second law of diffusion of the vacancies and the B atoms, and we assume that there is no source/sink of vacancies in the nanoshell. In the spherical coordinates, we write the diffusion equation for vacancies as

$$\frac{\partial X_V}{\partial t} = -\frac{1}{r^2}\frac{\partial}{\partial r}(r^2 \Omega j_V) + 0, \quad \text{where } r_i(t) < r < r_e(t)$$

In the above equation, we have taken the concentration of vacancy as $C_V = X_V/\Omega$ and the Fick's first law of diffusion of vacancy flux as $j_V = -D_V(\partial C_V/\partial r)$. The term zero at the end of the last equation indicates no source/sink of vacancies.

The diffusion equation for B atoms with concentration $C_B = X_B/\Omega$ is

$$\frac{\partial X_B}{\partial t} = -\frac{1}{r^2}\frac{\partial}{\partial r}(r^2 \Omega j_B), \quad \text{where } r_i(t) < r < r_e(t)$$

Here, the expression for both j_V and j_B should include the contributions from the vacancy flux due to the Gibbs–Thomson effect as well as from the cross-effect as shown in Eqs (3.33) and (3.34).

$$\Omega j_V(r) = (D_B^* - D_A^*)\varphi \frac{\partial X_B}{\partial r} - D_V \frac{\partial X_V}{\partial r} \tag{3.33}$$

where the first term on the right side of the equation is due to the cross-effect and the thermodynamic factor φ is included to express the chemical interaction between A and B (see Appendix C), and in the second term,

$$D_V = \frac{D_A^* X_A + D_B^* X_B}{X_V}$$

For the flux of component B, we have

$$\Omega j_B(r) = -D_B^* \varphi \frac{\partial X_B}{\partial r} + D_B^* X_B \frac{1}{X_V}\frac{\partial X_V}{\partial r} \tag{3.34}$$

The second term on the right side of the last equation is due to cross-effect.

In Eqs (3.33) and (3.34), we need to find out what $\partial X_B/\partial r$ and $\partial X_V/\partial r$ are. Let us go back to Fick's second law of diffusion and assume steady state so that we have

$$\frac{\partial}{\partial r}(r^2 \Omega j_V) \approx 0 \text{ and } \frac{\partial}{\partial r}(r^2 \Omega j_B) \approx 0$$

that means

$$\Omega j_V = \frac{K_V}{r^2}(> 0) \text{ and } \Omega j_B = \frac{K_B}{r^2}(< 0) \tag{3.35}$$

By using Fick's first law and assuming constant diffusivities, we have

$$\frac{\partial X_V}{\partial r} \approx -\frac{L_V}{r^2} \text{and } \frac{\partial X_B}{\partial r} \approx -\frac{L_B}{r^2} \tag{3.36}$$

The parameters of L_V and L_B are determined below.

For L_V, we consider vacancies at the inner and outer surfaces of nanoshell due to Gibbs–Thomson effect as the boundary conditions for the steady-state diffusion. As shown before, we have

$$X_V(r_i) = X_{V0}\left(1 + \frac{2\gamma\Omega}{kTr_i}\right) \text{ and } X_V(r_e) = X_{V0}\left(1 - \frac{2\gamma\Omega}{kTr_e}\right)$$

Because we have $\partial X_V/\partial r \approx -L_V/r^2 \Rightarrow$ by integration, we obtain $X_V(r_e) - X_V(r_i) = L_V((1/r_e) - (1/r_i))$.

In turn, we have

$$L_V = \frac{X_V(r_e) - X_V(r_i)}{(1/r_e) - (1/r_i)} = X_{V0}\frac{2\gamma\Omega}{kT}\frac{(1/r_e) + (1/r_i)}{(1/r_i) - (1/r_e)} = X_{V0}\frac{2\gamma\Omega}{kT}\frac{r_e + r_i}{r_e - r_i} \tag{3.37}$$

Next, to determine the parameter of L_B, we consider the flux of B atoms and vacancies leaving the inner surface. As the fluxes leave, the inner surface will move and the velocity of the boundary motion can be given as

$$\Omega j_B(r_i) - X_B(r_i)\frac{dr_i}{dt} = 0 \tag{3.38}$$

Equation (3.38) means that the flux of B with respect to the moving boundary during shrinking is zero because of mass conservation. We recall that in last section, we have obtained

$$\begin{aligned}
\Omega j_B &= -D_B^*\varphi\left(-\frac{L_B}{r^2}\right) + \frac{D_B^* X_B}{X_{V0}}\left(-\frac{L_V}{r^2}\right)\\
\frac{dr_i}{dt} &= -\Omega j_V(r_i) = -(D_B^* - D_A^*)\varphi\frac{\partial X_B}{\partial r}\Big|_{r=r_i} + D_V\frac{\partial X_V}{\partial r}\Big|_{r=r_i}
\end{aligned} \tag{3.39}$$

By substituting them into Eq. (3.38), we have

$$\begin{aligned}
&- D_B^*\varphi\left(-\frac{L_B}{r^2}\right) + \frac{D_B^* X_B}{X_{V0}}\left(-\frac{L_V}{r^2}\right)\\
&- X_B\left[-\left(D_B^* - D_A^*\right)\varphi\left(-\frac{L_B}{r^2}\right) + \frac{D_A^* X_A + D_B^*}{X_{V0}}\left(-\frac{L_V}{r^2}\right)\right] = 0
\end{aligned}$$

Rearranging the terms of L_B and L_V and applying the condition of $X_A + X_B = 1$, we have

$$L_B(X_A D_B^* + X_B D_A^*)\varphi = L_V\frac{X_A X_B(D_B^* - D_A^*)}{X_{V0}}$$

By substituting L_V from Eq. (3.37) into the last equation, we have

$$L_B = X_{V0}\frac{2\gamma\Omega}{kT}\frac{r_e + r_i}{r_e - r_i}\frac{X_A X_B(D_B^* - D_A^*)}{X_{V0}(X_A D_B^* + X_B D_A^*)\varphi} \tag{3.40}$$

Now let us go back to the equation of the velocity of the inner surface, Eq. (3.39),

$$\frac{dr_{\mathrm{i}}}{dt} = -\Omega j_{\mathrm{V}}(r_{\mathrm{i}}) = -(D_{\mathrm{B}}^* - D_{\mathrm{A}}^*)\varphi \frac{\partial X_{\mathrm{B}}}{\partial r}\Big|_{r=r_{\mathrm{i}}} + D_{\mathrm{V}} \frac{\partial X_{\mathrm{V}}}{\partial r}\Big|_{r=r_{\mathrm{i}}}$$

$$= -(D_{\mathrm{B}}^* - D_{\mathrm{A}}^*)\varphi \left(-\frac{L_{\mathrm{B}}}{r_{\mathrm{i}}^2}\right) + D_{\mathrm{V}}\left(-\frac{L_{\mathrm{V}}}{r_{\mathrm{i}}^2}\right)$$

$$= \frac{1}{r_{\mathrm{i}}^2}\left[\left(D_{\mathrm{B}}^* - D_{\mathrm{A}}^*\right)\varphi X_{\mathrm{V0}} \frac{2\gamma\Omega}{kT}\frac{r_{\mathrm{e}} + r_{\mathrm{i}}}{r_{\mathrm{e}} - r_{\mathrm{i}}} \frac{X_{\mathrm{A}} X_{\mathrm{B}}(D_{\mathrm{B}}^* - D_{\mathrm{A}}^*)}{X_{V0}(X_{\mathrm{A}} D_{\mathrm{B}}^* + X_{\mathrm{B}} D_{\mathrm{A}}^*)\varphi} - D_{\mathrm{V}} X_{\mathrm{V0}} \frac{2\gamma\Omega}{kT}\frac{r_{\mathrm{e}} + r_{\mathrm{i}}}{r_{\mathrm{e}} - r_{\mathrm{i}}}\right]$$

$$= \frac{1}{r_{\mathrm{i}}^2}\frac{2\gamma\Omega}{kT}\frac{r_{\mathrm{e}} + r_{\mathrm{i}}}{r_{\mathrm{e}} - r_{\mathrm{i}}}\left[\left(D_{\mathrm{B}}^* - D_{\mathrm{A}}^*\right)\frac{X_{\mathrm{A}} X_{\mathrm{B}}(D_{\mathrm{B}}^* - D_{\mathrm{A}}^*)}{(X_{\mathrm{A}} D_{\mathrm{B}}^* + X_{\mathrm{B}} D_{\mathrm{A}}^*)} - (X_{\mathrm{A}} D_{\mathrm{A}}^* + X_{\mathrm{B}} D_{\mathrm{B}}^*)\right]$$

$$= -\frac{1}{r_{\mathrm{i}}^2}\frac{2\gamma\Omega}{kT}\frac{r_{\mathrm{e}} + r_{\mathrm{i}}}{r_{\mathrm{e}} - r_{\mathrm{i}}}\left[\frac{D_{\mathrm{B}}^* D_{\mathrm{A}}^* \left(X_{\mathrm{A}} + X_{\mathrm{B}}\right)^2}{(X_{\mathrm{A}} D_{\mathrm{B}}^* + X_{\mathrm{B}} D_{\mathrm{A}}^*)}\right]$$

$$\frac{dr_{\mathrm{i}}}{dt} = -\frac{1}{r_{\mathrm{i}}^2}\frac{2\gamma\Omega}{kT}\frac{r_{\mathrm{e}} + r_{\mathrm{i}}}{r_{\mathrm{e}} - r_{\mathrm{i}}}\left[\frac{D_{\mathrm{B}}^* D_{\mathrm{A}}^*}{\left(X_{\mathrm{A}} D_{\mathrm{B}}^* + X_{\mathrm{B}} D_{\mathrm{A}}^*\right)}\right] \approx -\frac{1}{r_{\mathrm{i}}^2}\frac{2\gamma\Omega}{kT}\frac{r_{\mathrm{e}} + r_{\mathrm{i}}}{r_{\mathrm{e}} - r_{\mathrm{i}}}\frac{D_{\mathrm{A}}^*}{X_{\mathrm{A}}} \tag{3.41}$$

The last term in the above equation is obtained by assuming that $D_{\mathrm{B}}^* \gg D_{\mathrm{A}}^*$ and $X_{\mathrm{A}} D_{\mathrm{B}}^* \gg X_{\mathrm{B}} D_{\mathrm{A}}^*$. Thus the shrinking of the central void is limited by the slower diffusing species. By integration, we obtain that the time needed to eliminate the void is approximately proportional to initial value of r_{i}^3.

3.4 INTERACTION BETWEEN KIRKENDALL EFFECT AND GIBBS–THOMSON EFFECT IN THE FORMATION OF A SPHERICAL COMPOUND NANOSHELL

In the reaction of a concentric (spherical) or coaxial (cylindrical) nanoshell of a bilayer of A and B in forming an intermetallic compound of ApBq, we need to consider the interaction between Kirkendall effect and Gibbs–Thomson effect. Experimentally, the formation of nanoshells of CoO and Co_3S_4 by annealing nanospheres of Co in oxygen and sulfur ambient, respectively, has been reported. Kinetic analysis of the formation is presented here.

We analyze the interdiffusion in a pair of concentric spherical nanoshells of A and B to form an intermetallic compound layer between them, where A is the inside layer and B is the outside layer, as shown in Figure 3.10. We assume spherical symmetry and steady-state process in analyzing the formation. Experimentally, we take the example of Co_3S_4 compound formation during the annealing of nanospheres of Co in sulfur ambient, in which A is the metal Co and B is the sulfur. To simplify the analysis, we assume that the total fluxes of A (diffuses out) and of B (diffuses in) through the compound layer do not depend on the distance from the center:

$$\Omega J_B^{tot} = 4\pi R^2 \Omega j_B(R) = \text{const over } R$$
$$\Omega J_A^{tot} = 4\pi R^2 \Omega j_A(R) = \text{const over } R \tag{3.42}$$

Here each flux density of j_B, j_A can be expressed in terms of concentration times velocity ($j = Cv$), and, in turn, the velocity ($v = MF$) can be expressed in terms of mobility times of chemical driving force (which is equal to minus gradient of the corresponding chemical potential).

$$\Omega j_B(R) = -X_B v_B = -X_B \frac{D_B^*}{kT} \frac{\partial \mu_B}{\partial R} \tag{3.43a}$$

$$\Omega j_A(R) = -X_A v_A = -X_A \frac{D_A^*}{kT} \frac{\partial \mu_A}{\partial R} \tag{3.43b}$$

For all stoichiometric compounds, we can take atomic fractions X_A, X_B of B and A practically to be constant through the layer.

Now we make a mathematical rearrangement by using the constancy of the total fluxes over radius; we can multiply and divide simultaneously by the same integral $\int_{r_i}^{r_e} dR/R^2 X_V(R)$ and use the possibility of taking a constant factor from outside into the integrand in Eq. (3.42):

$$\Omega J_B^{tot} = \frac{\int_{r_i}^{r_e} (\Omega J_B^{tot} dR)/R^2}{\int_{r_i}^{r_e} dR/R^2} = \frac{-4\pi X_B D_B^*}{kT \int_{r_i}^{r_e} dR/R^2} \int_{r_i}^{r_e} (\partial(\mu_B)/\partial R) dR$$
$$= -4\pi X_B D_B^* \frac{\mu_B(r_e) - \mu_B(r_i)}{kT((1/r_i) - (1/r_e))} \tag{3.44}$$

Thus, the total B-flux through the growing layer of compound (or through the outside interface of the compound layer) is

$$\Omega J_B^{tot} = -4\pi X_B D_B^* r_i r_e \frac{\mu_B(r_e) - \mu_B(r_i)}{kT(r_e - r_i)} \tag{3.45a}$$

Similarly, the total A-flux through the growing layer of compound (or through the inside interface of the compound layer) is

$$\Omega J_A^{tot} = -4\pi X_A D_A^* r_i r_e \frac{\mu_A(r_e) - \mu_A(r_i)}{kT(r_e - r_i)} \tag{3.45b}$$

Now we should write the condition of flux balance for B or the metal (Co) at the outer boundary of the compound layer (because metal does not go outside of the nanoshell) and the flux balance for A or sulfur at the inner boundary of the compound layer (as sulfur does not go into the void).

$$4\pi r_e^2 \frac{dr_e}{dt}(X_B - 0) = \Omega J_B^{tot} - 0 \tag{3.46}$$

$$4\pi r_i^2 \frac{dr_i}{dt}(X_A - 0) = \Omega J_A^{tot} - 0 \tag{3.47}$$

Thus,

$$\frac{dr_e}{dt} = D_B^* \frac{r_i}{r_e} \frac{\mu_B(r_i) - \mu_B(r_e)}{kT(r_e - r_i)} \tag{3.48}$$

$$\frac{dr_i}{dt} = D_A^* \frac{r_e}{r_i} \frac{\mu_A(r_i) - \mu_A(r_e)}{kT(r_e - r_i)} \tag{3.49}$$

From standard thermodynamics, we have from Section 7.2.4 and also from Appendix C that

$$\mu_B = g + X_A \frac{\partial g}{\partial X_B}, \quad \mu_A = g - X_B \frac{\partial g}{\partial X_B} \tag{3.50}$$

So,

$$\begin{aligned} \mu_B(r_i) - \mu_B(r_e) &= g(r_i) + X_A \left.\frac{\partial g}{\partial X_B}\right|_{ri} - g(r_e) - X_A \left.\frac{\partial g}{\partial X_B}\right|_{re} \\ &\cong X_A \left(\left.\frac{\partial g}{\partial X_B}\right|_{ri} - \left.\frac{\partial g}{\partial X_B}\right|_{re} \right) + (g(r_i) - g(r_e)) \end{aligned} \tag{3.51}$$

In Eqs (3.48) and (3.49), we should use Gibbs–Thomson potential for calculating chemical potentials at the curved boundaries. Gibbs energy per atom at the **internal** boundary is **lower** by Gibbs–Thomson term, and at **external** boundary is **higher** by Gibbs–Thomson term:

$$g(r_i) = g^{compound} - \frac{2\gamma\Omega}{r_i}, \quad g(r_e) = g^{compound} + \frac{2\gamma\Omega}{r_e}, \quad \text{so that}$$

$$g(r_i) - g(r_e) = -2\gamma\Omega \left(\frac{1}{r_i} + \frac{1}{r_e} \right) \tag{3.52}$$

Here we take into account the fact that for an almost stoichiometric compound its composition is practically the same at both sides ($X_A(r_i) \cong X_A(r_e) = X_A^{compound}$), but the first derivatives of Gibbs energy are very different. They are determined by the common tangent rule (at inner boundary due to equilibrium between compound and remaining metallic core B; at external boundary due to equilibrium between compound and surrounding medium (in our case, sulfur A):

$$\left.\frac{\partial g}{\partial X_B}\right|_{ri} - \left.\frac{\partial g}{\partial X_B}\right|_{re} = \frac{g_B - (g^{\text{compound}} - (2\gamma\Omega/r_i))}{1 - X_B^{\text{compound}}} - \frac{(g^{\text{compound}} + (2\gamma\Omega/r_e)) - g_A}{X_B^{\text{compound}} - 0} =$$

$$= \frac{\Delta g}{X_B X_A} + 2\gamma\Omega\left(\frac{1}{X_A r_i} - \frac{1}{X_B r_e}\right) \tag{3.53}$$

Here Δg is a Gibbs free energy (per atom) of compound formation from the mixture of pure A and B in necessary proportion (see Fig. 7.11). In other words, it is the thermodynamic driving force of solid state reaction to form the compound phase. Combining Eqs (3.52) and (3.53), we obtain for B component,

$$\mu_B(r_i) - \mu_B(r_e) = \frac{1}{X_B}\left(\Delta g - \frac{2\gamma\Omega}{r_e}\right) \tag{3.54a}$$

Similarly, we obtain for A component,

$$\mu_A(r_i) - \mu_A(r_e) = \frac{1}{X_A}\left(-\Delta g - \frac{2\gamma\Omega}{r_i}\right) \tag{3.54b}$$

Physically, we can understand Eqs (3.54a) and (3.54b) by referring to interdiffusion in a bulk sample of A/B to form an intermetallic compound between them. The chemical potential difference across the compound layer is Δg. In the case of nanospherical shells, the chemical potential is modified by $2\gamma\Omega/r$ at the inner as well as the outer surface of the shell. Now we can go back to the growth rate in Eqs (3.48) and (3.49) and substitute the calculated driving forces in Eqs (3.54a) and (3.54b),

$$\frac{dr_e}{dt} = D_B^* \frac{r_i}{r_e} \frac{\Delta g - (2\gamma\Omega/r_e)}{X_B kT(r_e - r_i)} \tag{3.55}$$

$$\frac{dr_i}{dt} = -D_A^* \frac{r_e}{r_i} \frac{\Delta g + (2\gamma\Omega/r_i)}{X_A kT(r_e - r_i)} \tag{3.56}$$

Dividing Eq. (3.55) by Eq. (3.56), we have the equation for the relative change of sizes

$$\frac{dr_e}{dr_i} = -\frac{D_B^* X_A r_i^2}{D_A^* X_B r_e^2} \frac{\Delta g^{\text{compound}} - (2\gamma\Omega/r_e)}{\Delta g^{\text{compound}} + (2\gamma\Omega/r_i)} \tag{3.57}$$

With further simplification of the boundary conditions and using the following nondimensional parameters:

$$x = \frac{r_i}{r_o}, \quad y = \frac{r_e}{r_o}, \quad G = \frac{2\gamma\Omega}{\Delta g r_o}$$

where r_o is the initial radius of nanoparticle and Δg is the driving force of reaction and is equal to the formation energy of the sulfide. We have the main solution as

$$\frac{dy}{dx} = -\frac{x^3}{y^3}\frac{D_{\mathrm{m}}^*}{D_{\mathrm{o}}^*}\frac{1-c_{\mathrm{m}}}{c_{\mathrm{m}}}\frac{y-G}{x+G} \tag{3.58}$$

This equation differs from a similar equation by Alivisatos et al. [8] with the last term containing $G \neq 0$ (responsible for Laplace pressure and Gibbs–Thomson effect both for vacancies and for main components). When both radii of r_{e} and r_{i} are large ($r_{\mathrm{e}}, r_{\mathrm{i}} \gg 2\gamma\Omega/\Delta g$, $G \ll 1$), the equations become similar.

It is evident from Eq. (3.58) that in the case $G > 1$, the nanoshell formation is impossible (taking into account the fact that x and y start from almost 1). It means that in very small particles, $r_o < 2\gamma\Omega/\Delta g$, the reaction with void formation is impossible. It is interesting that this critical condition coincides with the critical radius for the nucleation of a new phase.

REFERENCES

1. Smigelkas AD, Kirkendall EO. Zinc diffusion in alpha brass. Trans AIME 1947;171:130–142.
2. Darken LS. Diffusion, mobility and their interrelation through free energy in binary metallic systems. Trans AIME 1948;175:184–201.
3. Hoglund L, Agren J. Analysis of the Kirkendall effect, marker migration and pore formation. Acta Mater 2001;49:1311–1317.
4. Strandlund H, Larsson H. Prediction of Kirkendall shift and porosity in binary and ternary diffusion couples. Acta Mater 2004;52:4695–4703.
5. Yin Y, Rioux RM, Erdonmez CK, Hughes S, Somorjai GA, Alivisatos AP. Formation of hollow nanocrystals through the nanoscale Kirkendall effect. Science 2004;304:711–714.
6. Tu KN, Gösele U. Hollow nanostructures based on the Kirkendall effect: design and stability considerations. Appl Phys Lett 2005;86:093111–093111-3.
7. Gusak AM, Zaporozhets TV, Tu KN, Goesele U. Kinetic analysis of the instability of hollow nanoparticles. Philos Mag 2005;85:4445–4464.
8. Yin Y, Erdonmez CK, Cabot A, Hughes S, Alivisatos AP. Colloidal synthesis of hollow cobalt sulfide nanocrystals. Adv Funct Mater 2006;16:1389–1399.
9. Marwick AD. Segregation in irradiated alloys: the inverse Kirkendall effect and the effect of constitution on void swelling. J Phys F 1978;8:1849–1861.
10. Gusak AM, Tu KN. Interaction between Kirkendall effect and inverse Kirkendall effect in nanoscale particles. Acta Mater 2009;57:3367–3373.

PROBLEMS

3.1. Assume interdiffusion proceeds in a binary diffusion couple according to parabolic law. Let $X_{\mathrm{K}}(t)$ be a coordinate of Kirkendall plane at time t, with respect to Matano plane (the initial interface) in the laboratory reference frame. Let $v_{\mathrm{K}}(t)$ be a velocity of markers of Kirkendall plane, and with respect to Matano plane at the same moment of t. Find the relation between X_{K}, v_{K}, and t.

3.2. Consider a diffusion couple with markers inserted throughout the entire interdiffusion zone, so that we can measure the velocities of markers (lattice drift velocity) not only at the original interface but also everywhere. Let $D_B > D_A$ at all concentrations. Plot qualitatively the curve of velocity of marker v_{marker} to show the dependence of velocity of marker on position.

3.3. Assume that we know the dependence of marker velocity (lattice shift velocity) on position or coordinate, $v_{marker}(x)$, at a given time t of annealing, as the solution of Problem 3.2. Find a graphical method to determine the position of Kirkendall plane, at this moment.

3.4. When a nanowire of Zn is annealed in oxygen ambient, a hollow ZnO tube is formed following Kirkendall effect, in which Zn diffuses out is much faster than oxygen diffuses in. Besides Kirkendall effect, inverse Kirkendall effect exists. Consider the transformation of a nanowire of Zn of 100 nm into a hollow circular tube of ZnO. What is the inner and outer radius of the ZnO tube? What is the expression of the vacancy flux in the inverse Kirkendall effect?

3.5. In inverse Kirkendall effect, we need a preexisting vacancy flux and we have given two examples. One is by irradiation and another is a nano hollow shell. Now if we bend a bilayer of alloy of AB to have a radius of R, so that the outer surface is under tension and the inner surface is under compression, do we have inverse Kirkendall effect? We may assume that we can use Stoney's equation to calculate the stress near the inner and the outer surface layer by giving a thickness of "h" to the alloy layer and a Young's modulus of the alloy. How do we define and calculate the preexisting vacancy flux?

CHAPTER 4

RIPENING AMONG NANOPRECIPITATES

4.1 INTRODUCTION

In phase transformations, Ostwald ripening follows nucleation and growth. This is especially true in the later stage of precipitation from a dilute alloy, where the ripening occurs among a large number of various sizes of small precipitates. Figure 4.1 is a schematic diagram depicting the dispersion of spherical precipitates. When precipitation is near complete, there is no driving force from the reduction of supersaturation in concentration; yet the precipitates have surface and surface energy. The basic driving force of ripening is the reduction of the total surface area and surface energy of all the precipitates, and the basic kinetic process is the dissolution of a small precipitate and the growth of a neighboring larger precipitate. However, it is a many-body problem as all the precipitates are participating in the ripening process. Thus if we consider a precipitate that has both a larger and a smaller neighbors, it is hard to say whether it will grow or shrink. To overcome the many-body issue, typically, a mean-field approach is taken to simplify the kinetic consideration, and the mean-field concentration serves as the reference of kinetic change of every precipitate, whether it grows or shrinks. This is analyzed in detail in this chapter.

Kinetics in Nanoscale Materials, First Edition. King-Ning Tu and Andriy M. Gusak.

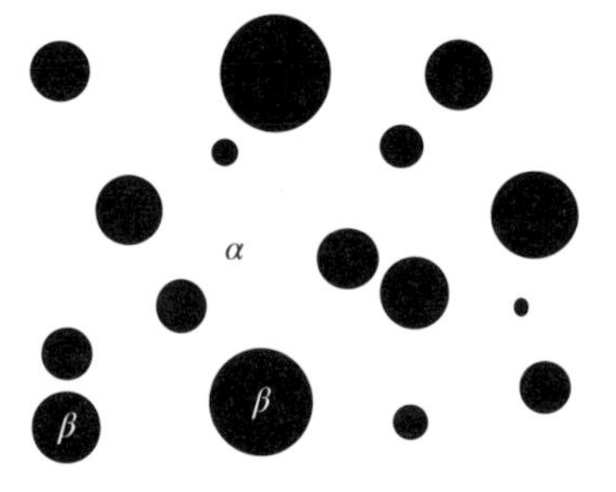

Figure 4.1 A schematic diagram depicting the dispersion of spherical precipitates.

In the derivation of Fick's second law or the continuity equation in Chapter 2, we consider flux divergence in a small cubic box and we assume that concentration can change in the small cubic box. This is true when we consider the diffusion in a gas phase or the diffusion of solute atoms in a solid phase. But in many solid state kinetic problems, when we consider the growth of a pure solid phase or a precipitate, we assume that there is negligible concentration change in the pure solid phase or in the precipitate. When a flux of atoms comes to the precipitate, it will grow in size rather than change in concentration. In other words, we do not consider dC/dt or dC/dx; rather, we consider dx/dt or dr/dt. It becomes a growth or a dissolution problem, and typically the following procedures are taken to solve the problem.

(1) Use Fick's second law to set up the diffusion equation and choose the coordinates and the initial and boundary conditions.

(2) Solve the diffusion equation to obtain the concentration profile. Often a steady state is assumed so we obtain C as a function of x or r.

(3) Use Fick's first law to obtain the atomic flux arriving at the growth front or leaving in the opposite direction at the dissolution front.

(4) Obtain the growth or dissolution equation by considering the conservation of mass or volume.

(5) Check the dimension or unit of the solution to see if it is correct.

We illustrate step (4) below and steps (1)–(3) and (5) in the next section. Let us consider the growth of a spherical particle by the diffusion in the spherical coordinate. The flux arriving at the particle surface per unit area and unit time is J and the radius of the particle is r. As the total number of atoms arriving at the particle surface in a period of Δt is $N = JA\Delta t$, where $A = 4\pi r^2$ is the surface area of the particle, the flux of atoms has added the following volume to the precipitate,

$$V = \Omega \Delta N = \Omega J 4\pi r^2 \Delta t$$

where Ω is atomic volume. On the other hand, the volume of the sphere is

$$V' = \frac{4}{3}\pi r^3 \text{ and } dV' = 4\pi r^2 dr$$

The volume change, dV', should equal to the added volume, V, in Δt. It will thicken the precipitate by dr or add a shell of volume of $4\pi r^2 dr$ to the precipitate in time dt; thus we have

$$4\pi r^2 dr = \Omega J 4\pi r^2 dt$$

so the growth rate of the particle is $dr/dt = \Omega J$. A more detailed presentation of the classic growth model of Ham is given in the next section, where the particle radius is of micron size or larger, and the Gibbs–Thomson effect is ignored in Ham's model. If we reverse the flux direction by considering a flux of atoms leaving the particle, we have the dissolution case.

In those cases of ripening (growth and shrinkage) among nanoscale particles, we must consider the Gibbs–Thomson effect and we present the Lifshiz–Slezov–Wagner (LSW) theory of ripening later [1, 2]. One of the key assumptions in LSW theory is that the volume fraction of total precipitates is negligibly small with respect to the total volume of the sample. Thus the radius of precipitates is of nanoscale. While it is a diffusion-controlled process, LSW theory showed that the rate of ripening obeys $t^{1/3}$ rather than $t^{1/2}$ dependence.

4.2 HAM'S MODEL OF GROWTH OF A SPHERICAL PRECIPITATE (C_r IS CONSTANT)

This is the classic model of growth of a precipitate under diffusion-controlled kinetics [3]. In Figure 4.2 we depict a spherical particle of radius r. The solute in the large sphere of radius r_0 takes part in the growth of the precipitate. Let R be the coordinate variable, and the diffusion equation in spherical coordination that has spherical symmetry, assuming a steady state, is

$$\frac{\partial^2 C}{\partial R^2} + \frac{2}{R}\frac{\partial C}{\partial R} = 0$$

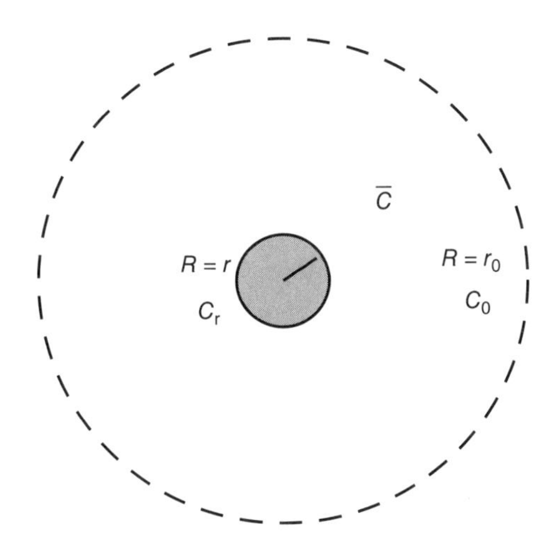

Figure 4.2 A schematic diagram depicting the growth of a spherical particle of radius r. The solute in the large sphere of diffusion of radius r_0 will contribute to the growth of the precipitate.

The solution as presented in Eq. (1.20) is

$$C = \frac{b}{R} + d \tag{4.1}$$

The boundary conditions are
At $R = r_0$, $C = C_0$, we have

$$C_0 = \frac{b}{r_0} + d \tag{4.2}$$

At $R = r$, $C = C_r$, we have

$$C_r = \frac{b}{r} + d \tag{4.3}$$

Now noting the difference between the last two equations, we have

$$C_r - C_0 = b\left(\frac{1}{r} - \frac{1}{r_0}\right) = b\frac{r_0 - r}{rr_0} \cong \frac{b}{r} \quad \text{where } r_0 \gg r \tag{4.4}$$

The approximation of $r_0 \gg r$ is an important assumption. It means that precipitates are small and are far apart, so the volume fraction of the precipitates is very small. Note that if we take the ratio of the volume of the precipitate to the volume of the diffusion field to be "*f*,"

$$f = \frac{(4\pi/3)\, r^3}{(4\pi/3)\, r_0^3} = \frac{r^3}{r_0^3} \to 0 \tag{4.5}$$

It is a very small value if $r_0 \gg r$; $f \to 0$. Furthermore, we can assume that the ratio of the total volume of all precipitates to the volume of the sample is very small too. We note that this is a very important assumption in LSW theory of ripening, which is discussed in the next section.

We have $b = r(C_r - C_0)$ from Eq. (4.4). Substituting b into Eq. (4.3), we have

$$C_r = \frac{r\left(C_r - C_0\right)}{r} + d \tag{4.6}$$

We have $d = C_0$, and Eq. (4.1) becomes

$$C(R) = \frac{\left(C_r - C_0\right) r}{R} + C_0 \tag{4.7}$$

This is the solution of the diffusion equation. Therefore,

$$\frac{dC}{dR} = -\frac{\left(C_r - C_0\right) r}{R^2}$$

At the particle/matrix interface for a particle of radius r, or $R = r$, we have

$$\frac{dC}{dR} = -\frac{C_r - C_0}{r} \tag{4.8}$$

Then the flux density of atoms arriving at the interface on the basis of Fick's first law is

$$J = +D\frac{\partial C}{\partial R} = \frac{D\left(C_0 - C_r\right)}{r}, \quad \text{at } R = r \tag{4.9}$$

Note that when $C_r > C_0$, $J < 0$, the net flux leaves the particle and it dissolves. When $C_r < C_0$, $J > 0$, the flux goes to the particle so the particle grows.

In the case of growth, if Ω is atomic volume, a volume is added to the spherical particle in time dt,

$$\Omega JAdt = \Omega J4\pi r^2 dt = 4\pi r^2 dr$$

where the last term is the increment of a spherical shell due to the growth. Hence

$$\frac{dr}{dt} = \Omega J = \frac{\Omega D\left(C_0 - C_r\right)}{r} \tag{4.10}$$

By integration, assume that $r=0$ when $t=0$,

$$r^2 = 2\Omega D\left(C_0 - C_r\right)t \tag{4.11}$$

We note that as we follow Ham's approach and have taken C_r as a constant, it means C_r is not a function of r (But, if the precipitate is very small, C_r will be a function of r as required by Gibbs–Thomson equation, which is discussed later.). From the above equation, we see that $r \cong t^{1/2}$. The rate dependence is $t^{1/2}$ as in diffusion-controlled kinetic processes. Or we have

$$r^3 = \left[2\Omega D\left(C_0 - C_r\right)t\right]^{3/2} \tag{4.12}$$

4.3 MEAN-FIELD CONSIDERATION

In this section, we take another approach to analyze the above precipitation problem by considering the reduction of average concentration in the matrix, $\Delta\overline{C} = C_0 - \overline{C}$, due to the formation of the precipitates, where the average concentration in the matrix is $\overline{C}$, which can be regarded as the "mean-field" concentration (this is the starting point of mean-field theory). Actually, mean-field approximation is valid also for the case of small volume fractions [5–8], but for the time being let us neglect this limitation. In the beginning, the average concentration is C_0, but it changes to $\overline{C}$ when the precipitates grow.

Let $1/\Omega = C_p$ to be the concentration in the solid precipitate. We have by mass balance,

$$\frac{4\pi}{3}r_0^3\left(C_0 - \overline{C}\right) = \frac{4\pi}{3}r^3\frac{1}{\Omega} = \frac{4\pi}{3\Omega}\left[2\Omega D\left(C_0 - C_r\right)t\right]^{3/2} \tag{4.13}$$

$$\overline{C} = C_0 - \left[\frac{2D\left(C_0 - C_r\right)\Omega^{1/3}}{r_0^2}t\right]^{3/2} = C_0 - \left[\frac{2Bt}{3}\right]^{3/2} \tag{4.14}$$

where $B \equiv 3D\left(C_0 - C_r\right)/C_p^{1/3}r_0^2$.

We note that the above equation is the same as Eq. (1.36) in Chapter 1 in Shewmon's book on *Diffusion in Solids* [4].

We can derive the last equation in a slightly different way. The growth of the precipitate reduces the concentration in the matrix. The amount of solute atoms that diffuses to the precipitate in time Δt is $J(r)4\pi r^2\Delta t$. The number of atoms should be equal to the reduction of the average concentration in the volume of the sphere of diffusion of r_0. Hence, if we take the average concentration in the matrix to be $\overline{C}$,

$$\frac{4\pi r_0^3}{3}\Delta\overline{C} = J(r)4\pi r^2\Delta t$$

Or, we have

$$\frac{\Delta\overline{C}}{\Delta t} = \frac{3}{4\pi r_0^3}4\pi r^2 J(r) = -\frac{3D}{r_0^3}(C_0 - C_r)\, r \tag{4.15}$$

The conservation of mass requires that

$$\frac{4\pi}{3}r_0^3\left(C_0 - \overline{C}\right) = \frac{4\pi}{3}r^3 C_p \tag{4.16}$$

where C_p is the concentration of solute in the solid precipitate and $C_p = 1/\Omega$. Hence,

$$r = r_0\left(\frac{C_0 - \overline{C}}{C_p}\right)^{1/3} \tag{4.17}$$

By substituting r into the rate equation in the above, we have

$$\frac{\Delta\overline{C}}{\Delta t} = -\frac{3D}{r_0^2}(C_0 - C_r)\frac{1}{C_p^{1/3}}\left(C_0 - \overline{C}\right)^{1/3} \tag{4.18}$$

Let

$$B \equiv \frac{3D(C_0 - C_r)}{C_p^{1/3} r_0^2}$$

We have

$$\frac{d\overline{C}}{dt} = -B\left(C_0 - \overline{C}\right)^{1/3}$$

By integration, we obtain

$$-\frac{3}{2}\left(C_0 - \overline{C}\right)^{2/3} = -Bt + \beta$$

at $t = 0$, $C_0 = \overline{C}$, so $\beta = 0$.

Thus we have the solution,

$$\overline{C} = C_0 - \left(\frac{2Bt}{3}\right)^{3/2} \tag{4.19}$$

which is the same as what we have obtained. Hence, we have

$$C_0 - \overline{C} \cong t^{3/2} \text{ for a three-dimentional growth}$$

Let

$$\overline{C} = C_0\left[1 - \left(\frac{2Bt}{3C_0^{2/3}}\right)^{3/2}\right] = C_0\left[1 - \left(\frac{t}{\tau}\right)^{3/2}\right] = C_0 \exp\left[-\left(\frac{t}{\tau}\right)^{3/2}\right] \tag{4.20}$$

if we assume $t \ll \tau$, where

$$\tau = \frac{C_p^{1/3} r_0^2 C_0^{2/3}}{2D\left(C_0 - C_r\right)} \cong \frac{r_0^2}{2D}\left(\frac{C_p}{C_0}\right)^{1/3} \tag{4.21}$$

Usually, if D, C_p, C_0 are known, we can design the experiment to control the growth of the precipitate.

4.4 GIBBS–THOMSON POTENTIAL

In the above analysis of the growth of a large-size precipitate, we have assumed the equilibrium concentration C_r to be constant, independent of the size of the precipitate. This assumption is not true when the particle is small, in the nanoscale. We must consider Gibbs–Thomson potential and allow C_r to be a function of r.

When we have a large number of particles, there is a distribution of size. There will be ripening action among them. To analyze the ripening among particles having a size distribution, we can use the concept of mean-field concentration, which can be regarded as the average concentration or a reference concentration for all the particles. We assume that in the distribution, there is a critical size r^* of particle, which is in equilibrium with the mean-field concentration. Then the ripening of any particle can be analyzed against the critical size particle. For those that are larger than the critical size, they will grow. For those that are smaller, they will shrink. Below, we first develop the Gibbs–Thomson potential and its link to the equilibrium concentration.

Consider a small sphere with radius r and the surface energy per unit area γ. The surface energy exerts a compressive pressure on the sphere because it tends to shrink to reduce the surface energy. In Chapter 1, we have shown that the pressure equals

$$p = \frac{F}{A} = \frac{(dE/dr)}{A} = \frac{\left(d4\pi r^2\gamma/dr\right)}{4\pi r^2} = \frac{8\pi r\gamma}{4\pi r^2} = \frac{2\gamma}{r}$$

If we multiply p by atomic volume Ω, we have the chemical potential

$$\mu_r = \frac{2\gamma\Omega}{r} \tag{4.22}$$

This is called *Gibbs–Thomson potential*. We see that for a flat surface, $r=\infty$, $\mu_\infty = 0$, so we have

$$\mu_r - \mu_\infty = \frac{2\gamma\Omega}{r} \tag{4.23}$$

We can apply this potential to determine the effect of curvature on solubility and then ripening among a set of particles of varying size.

4.5 GROWTH AND DISSOLUTION OF A SPHERICAL NANOPRECIPITATE IN A MEAN FIELD

We consider an alloy of $\alpha = A(B)$, where B is solute in solvent A. At a given low temperature, B will precipitate out. We consider a precipitate of B with radius r. The solubility of B in the alloy surrounding the precipitate is taken to be $X_{B,r}$. To relate the solubility to Gibbs–Thomson potential, we have the chemical potential of B as a function of its radius as

$$\mu_{B,r} - \mu_{B,\infty} = \frac{2\gamma\Omega}{r} \tag{4.24}$$

where γ is the interfacial energy between the precipitate and the matrix. If we define the standard state of B as pure B with $r=\infty$, we have

$$\mu_{B,r} = \mu_{B,\infty} + RT \ln a_B \tag{4.25}$$

where a_B is the activity. According to Henry's law

$$a_B = KX_{B,r}$$

where $X_{B,r}$ is the solubility of B surrounding a precipitate of radius r. At $r=\infty$,

$$\mu_{B,\infty} = \mu_{B,\infty} + RT \ln a_B$$

It implies that $RT \ln a_B = 0$, or $a_B = 1$. So $K = 1/X_{B,\infty}$.

Therefore

$$\mu_{B,r} = \mu_{B,\infty} + RT \ln \frac{X_{B,r}}{X_{B,\infty}} \tag{4.26}$$

Hence,

$$\ln \frac{X_{B,r}}{X_{B,\infty}} = \frac{\mu_{B,r} - \mu_{B,\infty}}{RT} = \frac{2\gamma\Omega}{rRT}$$

Or, if we consider kT per atom instead of RT per mole, we have

$$X_{\mathrm{B,r}} = X_{\mathrm{B},\infty} \exp\left(\frac{2\gamma\Omega}{rkT}\right) \tag{4.27}$$

The equilibrium solubility of B around a spherical particle of B of radius r is given by the above equation. When $r=\infty$, the exponential equals unity. So $X_{\mathrm{B,r}}$ goes up when r goes down. Now we replace $X_{\mathrm{B,r}}$ by C_{r} and $X_{\mathrm{B},\infty}$ by C_∞, which is the equilibrium concentration on a flat surface, and we have

$$C_{\mathrm{r}} = C_\infty \exp\left(\frac{2\gamma\Omega}{rkT}\right) \tag{4.28}$$

If $2\gamma\Omega \ll rkT$, we have

$$C_{\mathrm{r}} = C_\infty \left(1 + \frac{2\gamma\Omega}{rkT}\right)$$

$$C_{\mathrm{r}} - C_\infty = \frac{2\gamma\Omega C_\infty}{rkT} = \frac{\alpha}{r} \tag{4.29}$$

where $\alpha = (2\gamma\Omega/kT)\, C_\infty$

$$C_{\mathrm{r}} = C_\infty + \frac{\alpha}{r} \tag{4.30}$$

Thus, we obtain the very important result in Eq. (4.30) that C_{r} is not a constant but a function of r. Now we substitute C_{r} to the growth equation of

$$\frac{dr}{dt} = \Omega J = \frac{\Omega D\left(C_0 - C_{\mathrm{r}}\right)}{r}$$

We have

$$\frac{dr}{dt} = \frac{\Omega D}{r}\left(C_0 - C_\infty - \frac{\alpha}{r}\right) \tag{4.31}$$

Note that $C_0 - C_\infty > 0$ always. We can define a critical radius r^* such that

$$C_0 - C_\infty = \frac{\alpha}{r^*} \tag{4.32}$$

We can regard the concentration that is in equilibrium with r^* as the "mean-field" concentration. In considering the ripening of any particle of radius r, large or small, we just consider the ripening of this particle against the critical particle of r^*, or against the mean-field concentration. Then we have

$$\frac{dr}{dt} = \frac{\alpha\Omega D}{r}\left(\frac{1}{r^*} - \frac{1}{r}\right) \tag{4.33}$$

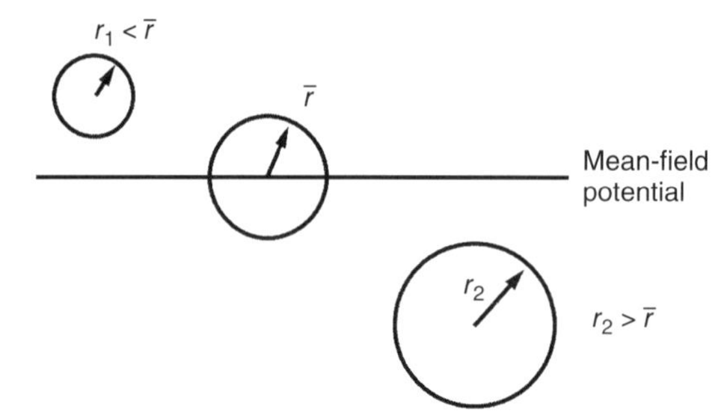

Figure 4.3 Three kinds of particle that behave with reference to the mean-field concentration.

The parameter r^* is defined such that

$$r > r^*, \frac{dr}{dt} > 0, \quad \text{the particle is growing}$$

$$r < r^*, \frac{dr}{dt} < 0, \quad \text{the particle is dissolving}$$

$$r = r^*, \frac{dr}{dt} = 0, \quad \text{the particle is in a state of unstable equilibrium}$$

It has a mean-field concentration $\overline{C}$ at the interface, or $C_{r^*} = \overline{C}$. How these three kinds of particles behave is depicted in Figure 4.3.

In ripening, the larger particles will grow at the expense of the mean-field and the smaller particles will shrink with respect to the mean-field. It will approach a dynamic equilibrium distribution of size of the particles. With time, the mean particle size will increase, and correspondingly the mean-field concentration will decrease. These are the key kinetic behaviors of ripening, and how to measure the particle distribution and the corresponding mean-field concentration experimentally has been challenging. The details are discussed in the next section.

Theoretically, the particle distribution function can be obtained by solving the continuity equation in size space as given by LSW theory of ripening. Knowing dr/dt is the beginning of the LSW theory. To obtain the continuity equation in size space requires knowing the velocity of dr/dt.

4.6 LSW THEORY OF KINETICS OF PARTICLE RIPENING

We rewrite Eq. (4.33) as below,

$$r^2 \frac{dr}{dt} = \alpha \Omega D \left(\frac{r}{r^*} - 1 \right) \tag{4.34}$$

$$\frac{dr^3}{dt} = 3\alpha \Omega D \left(\frac{r}{r^*} - 1 \right) = 3\alpha \Omega D (u - 1) \tag{4.35}$$

where we have $u = r/r^*$. We note that

$$\frac{du^3}{dt} = \frac{d}{dt}\left(\frac{r}{r^*}\right)^3 = \frac{1}{r^*}\left(\frac{dr^3}{dt}\right) - \frac{r^3}{(r^*)^6}\frac{d(r^*)^3}{dt} = \frac{1}{(r^*)^3}\left(\frac{dr^3}{dt} - u^3\frac{d(r^*)^3}{dt}\right)$$

By substituting Eq. (4.35) into the last equation, we obtain

$$\frac{du^3}{dt} = \frac{1}{(r^*)^3}\left\{3\alpha\Omega D(u-1) - u^3\frac{d(r^*)^3}{dt}\right\}$$

If we multiply both sides by $(r^*)^3 dt/d(r^*)^3$, we have

$$\frac{du^3}{\left(d(r^*)^3/(r^*)^3\right)} = \frac{dt}{d(r^*)^3}\left\{3\alpha\Omega D(u-1) - u^3\left(d(r^*)^3/dt\right)\right\}$$

Then we have

$$\frac{du^3}{d\ln(r^*)^3} = \frac{dt}{d(r^*)^3}\{3\alpha\Omega D(u-1)\} - u^3$$

Let

$$\tau = \ln\left(r^*\right)^3 \text{ and } \upsilon = 3\alpha\Omega D\left(dt/d\left(r^*\right)^3\right)$$

We convert the last equation into

$$\frac{du^3}{d\tau} = (u-1)\,\upsilon - u^3 \tag{4.36}$$

This is the kinetic equation in the LSW theory of ripening. Before we solve this equation, let us discuss the physical meaning of the parameters we introduced in the derivation of the equation. First, τ is a parameter serving as a measurement of time-dependent behavior of particle size. This is because $r^* \to \infty$ as $t \to \infty$. To understand the meaning of υ, we consider a constant value of $\upsilon = \upsilon_0 =$ constant. Let us take $\upsilon_0 = 27/4$, and the reason why we take this specific value is made clear later. In Figure 4.4, we plot three curves of du^3/dt against u of the following equation by taking $\upsilon_0 > 27/4$, $\upsilon_0 = 27/4$, $\upsilon_0 < 27/4$.

$$\frac{du^3}{d\tau} = (u-1)\,\upsilon_0 - u^3 \tag{4.37}$$

For the curve of $\upsilon_0 > 27/4$, it crosses the u-axis twice at u_1 and u_2, as shown. Those particles with $u > u_1$ will grow to u_2, and those particles with $u > u_2$ will also grow to u_2. Those particles smaller than u_1 will go to zero and disappear. This means the outcome of ripening is a mono size of particles, which is physically unreasonable. For the curve of $\upsilon_0 < 27/4$, all particles will decrease and disappear. Again it is unreasonable. For the curve of $\upsilon_0 = 27/4$, the curve shows a maximum that touches the u-axis at a single point at u_m. We analyze the solution below.

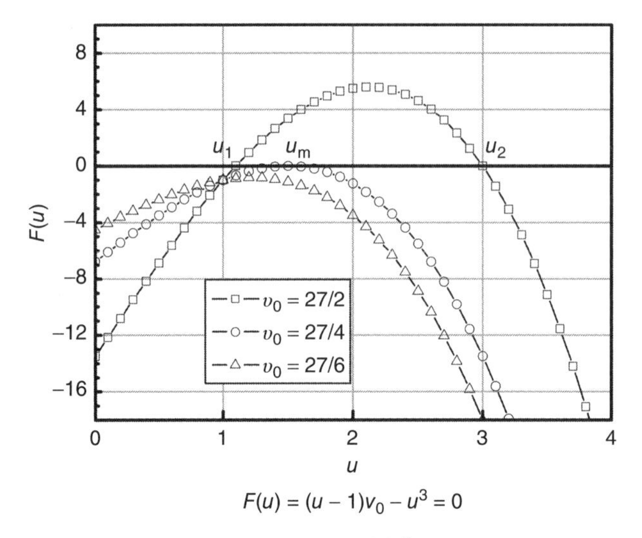

Figure 4.4 A plot of three curves of du^3/dt against u in Eq. (4.37) by taking $\upsilon_0 > 27/4$, $\upsilon_0 = 27/4$, and $\upsilon_0 < 27/4$.

At u_m, we have both the curve and its first derivative to be zero. It means

$$\frac{du^3}{d\tau} = (u - 1)\,\upsilon_0 - u^3 = 0$$

$$\frac{d}{du}\left(\frac{du^3}{d\tau}\right) = \frac{d}{du}\left[(u-1)\,\upsilon_0 - u^3\right] = \upsilon_0 - 3u^2 = 0 \tag{4.38}$$

From the last equation, we have

$$u = \left(\frac{\upsilon_0}{3}\right)^{1/2}$$

Substituting it back into Eq. (4.38), we have

$$\upsilon_0\left[\frac{2}{3}\left(\frac{\upsilon_0}{3}\right)^{1/2} - 1\right] = 0$$

The solution gives $\upsilon_0 = 0$ and $\upsilon_0 = 27/4$. Knowing $\upsilon_0 = 27/4$, we can calculate u at u_m.We have $u_m = (\upsilon_0/3)^{1/2} = [27/(3 \times 4)]^{1/2} = 3/2$. Another way to look at u_m is that we have the following pair of equations,

$$\left(u_m - 1\right)\upsilon_0 - u_m^3 = 0$$
$$\upsilon_0 - 3u_m^2 = 0$$

The solution of these two equations together is $u_m = 3/2$. As we define $u_m = r/r^*$, in ripening, the maximum distribution occurs at 3/2 of r^*.

As v_o is constant, we have

$$v_0 = 3\alpha\Omega D \frac{dt}{d(r^*)^3}$$

By integration, we obtain

$$\left(r^*\right)^3 - \left(r_0^*\right)^3 = \frac{3\alpha\Omega D}{v_0} t \tag{4.39}$$

where r_0^* is the critical radius of particle at the onset of ripening, taken at $t = 0$. $r^* = <r>$, which is the mean radius. As we obtained $v_0 = 27/4$ and $\alpha = 2\gamma\Omega C_\infty/RT$, we have

$$\left(r^*\right)^3 - \left(r_0^*\right)^3 = \frac{8\gamma\Omega^2 DC_\infty}{9RT} t = Bt \tag{4.40}$$

where $B = 8\gamma\Omega^2 DC_\infty/9RT$. We note that Eq. (4.40) shows the rate change of r^* is dependent on $t^{1/3}$ not $t^{1/2}$, while ripening is a diffusion-controlled process.

Using transmission electron microscopy, we can measure the size distribution of particles and in turn determine r^*. The slope of $(r^*)^3$ versus t gives B. We note that in B, there are two kinetic parameters that are the controlling parameters in the ripening process. They are γ and D, where γ is surface energy as given in the Gibbs–Thomson potential and D is lattice diffusivity. To determine them, we need one more equation or, in other words, we need to do one more independent experimental measurement besides the TEM images of particle distribution [7, 8]. Let us derive the other equation below from the measurement of the mean-field composition.

About mean radius $<r>$, we write

$$-\frac{dC_0}{dt} = \int_0^\infty V(r,t) f(r,t)\, dr \tag{4.41}$$

where C_0 is the average matrix concentration or the mean-field composition and

$$V(r,t) = \frac{dn}{dt} = \frac{d}{dt}\left(\frac{4}{3}\pi r^3 \frac{1}{\Omega}\right) = 4\pi r^2 \frac{1}{\Omega}\frac{dr}{dt} \tag{4.42}$$

where $n = (4\pi r^3/3)(1/\Omega)$. Equation (4.42) means that the rate at which a particle gains mass is equal to the negative rate at which the matrix loses mass or the negative rate of decreasing of the mean-field concentration. The function $f(r, t)$ in Eq. (4.41) is the distribution function of particles of various radius r at time t. As we have dr/dt given by Eq. (4.33), we have

$$\frac{dn}{dt} = 4\pi r^2 \frac{1}{\Omega}\frac{dr}{dt} = 4\pi Dr\left(C_0 - C_\infty - \frac{\alpha}{r}\right) \tag{4.43}$$

Hence,

$$-\frac{dC_0}{dt} = \int_0^\infty 4\pi Dr\left(C_0 - C_\infty - \frac{\alpha}{r}\right) f(r,t)\, dr$$

For a conservative ripening and under a steady-state condition, $dC_0/dt = 0$, so we have

$$\left(C_0 - C_\infty\right)\int_0^\infty rf(r,t)\,dr = \alpha\int_0^\infty f(r,t)\,dr$$

Thus

$$\frac{1}{r^*} = \frac{C_0 - C_\infty}{\alpha} = \frac{\displaystyle\int_0^\infty f(r,t)\,dr}{\displaystyle\int_0^\infty rf(r,t)\,dr} = \frac{1}{<r>} \tag{4.44}$$

Hence $r^* = <r>$ = the mean particle radius. We recall that in Eq. (4.32) we have

$$C_0 - C_\infty = \frac{\alpha}{r^*}$$

where C_0 is the mean-field concentration in the very beginning of ripening, $t = 0$. We can take $r^* = r_0^*$. It implies that

$$\left(r_0^*\right)^3 = \frac{\alpha^3}{\left(C_0 - C_\infty\right)^3}$$

Now, we define

$$C - C_\infty = \frac{\alpha}{<r>}$$

where C is precisely the value of concentration in equilibrium with the particle for which $r = <r>$ at time t. We can take $<r> = r^*$. Thus we have

$$\left(r^*\right)^3 = \frac{\alpha^3}{\left(C - C_\infty\right)^3} \tag{4.45}$$

Then we have

$$\frac{1}{\left(C - C_\infty\right)^3} - \frac{1}{\left(C_0 - C_\infty\right)^3} = \frac{1}{\alpha^3}\left[\left(r_0^*\right)^3 - \left(r^*\right)^3\right] \tag{4.46}$$

Upon a longtime annealing in ripening, the mean-field concentration C will decrease as $<r>$ increases, thus $1/(C - C_\infty)^3 \gg 1/(C_0 - C_\infty)^3$. We have from Eqs (4.40) and (4.29),

$$\frac{1}{\left(C - C_\infty\right)^3} = \left(\frac{8\gamma\Omega^2 DC_\infty}{9RT}t\right)\left(\frac{RT}{2\gamma\Omega C_\infty}\right)^3 = \frac{D(RT)^2}{9\Omega\left(\gamma C_\infty\right)^2}t = At \tag{4.47}$$

where $A = D(RT)^2/9\Omega\left(\gamma C_\infty\right)^2$. We can rearrange Eq. (4.47) as

$$C - C_\infty = (At)^{-1/3}$$

$$C = A^{-1/3}t^{-1/3} + C_\infty \tag{4.48}$$

The last equation implies that a linear plot of C against $t^{-1/3}$ will give the slope $A^{-1/3}$ with an intercept of C_∞ at $t^{-1/3}=0$. Such a plot can yield, in principle, the value of equilibrium concentration (or solubility), which can be measured by Curie temperature in Ni-based dilute alloys, for example. By knowing the size distribution and the mean-field concentration and by using the parameters of B and A, as indicated respectively in Eqs (4.40) and (4.48), we can determine the two most important kinetic parameters in ripening; they are D (diffusivity) and γ (surface energy). As diffusivity and surface energy can be measured independently by other methods, we have a comparison and can judge the agreement. It has served as an experimental check of the kinetic analysis of ripening.

4.7 CONTINUITY EQUATION IN SIZE SPACE

To describe the evolution of particle size distribution during ripening, we need the time-dependent equation of size distribution to be presented below. Let $f(r, t)dr$ be the number of particles per unit volume between r and $r+dr$ at time t. The total number of particles per unit volume is

$$N_P = \int_0^\infty f(r,t)\,dr$$

Next, we consider an interval of width dr in the size space at time t. During an amount of time δt, the particles in the interval will have grown by an amount of $\delta r = v(r, t)\delta t$, where $v(r, t)$ is growth velocity and assumed to be constant over the interval. A fraction $\delta r/dr$ of the particles in the interval will therefore be lost to the next interval owing to growth (or dissolution) during δt. The number of particles leaving the interval dr will thus be the number of particles originally in it multiplied by the fraction lost, that is,

$$f(r,t)\,dr\frac{\delta r}{dr} = f(r,t)\,\delta r = f(r,t)\,v(r,t)\,\delta t$$

However, this will be partly compensated by particles gained from the neighboring lower interval, which, by the same reasoning, is

$$f(r-dr,t)\,v(r-dr,t)\,\delta t$$

So the net change in the number of particles in the interval is

$$\left[-f(r,t)\,v(r,t) + f(r-dr,t)\,v(r-dr,t)\right]\delta t$$

In differentiation, we have

$$-\frac{\partial (fv)}{\partial r} = \frac{f(r-dr,t)\,v(r-dr,t) - f(r,t)\,v(r,t)}{dr}$$

Thus,

$$\delta t\left[f(r-dr,t)\,v(r-dr,t)-f(r,t)\,v(r,t)\right]=-\frac{\partial(fv)}{\partial r}dr\delta t$$

This must, by definition of continuity, equal to the change in the number of particles per unit volume per unit time, times the total time interval δt, which is denoted by the quantity

$$\frac{\partial f}{\partial t}dr\delta t$$

By equaling these two quantities, we have the continuity equation in size space,

$$\frac{\partial f}{\partial t}=-\frac{\partial(fv)}{\partial r}$$

Note the similarity between the last equation and the continuity equation in atomic diffusion below,

$$\frac{\partial C}{\partial t}=-\frac{\partial J_x}{\partial x}=-\frac{\partial(Cv)}{\partial x}$$

Instead of taking Fick's first law of $J_x=-D(dC/dx)$, we take $J_x=Cv$.

4.8 SIZE DISTRIBUTION FUNCTION IN CONSERVATIVE RIPENING

To determine the particle size distribution function, we solve the continuity equation in size space. We rewrite the equation as

$$\frac{\partial f}{\partial t}+\frac{\partial}{\partial r}\left(f\frac{dr}{dt}\right)=0$$

We recall Eq. (4.33) that

$$\frac{dr}{dt}=\frac{\alpha\Omega D}{r}\left(\frac{1}{r^*}-\frac{1}{r}\right)$$

We recall that in Section 4.6, we have converted the above equation into

$$\frac{du^3}{d\tau}=(u-1)\,\upsilon-u^3$$

We can replace the distribution function and its pair of variables $f(r,t)$ by $\varphi(u,\tau)$, and we have the continuity equation as

$$\frac{\partial\varphi(u,\tau)}{\partial\tau}+\frac{\partial}{\partial u}\left(\varphi\frac{du}{d\tau}\right)=0$$

The general solution of this equation is given as below,

$$g(u) = \frac{4u^2}{9}\left(\frac{3}{3-2u}\right)^{+11/3}\left(\frac{3}{3+u}\right)^{+7/3}\exp\left(-\frac{2u}{3-2u}\right), \quad \text{for } u < 3/2$$

$$g(u) = 0, \quad \text{for } u > 3/2$$

where $g(u)$ is defined as

$$g(u)\,du = \frac{\varphi du}{\int_0^{u_m} \varphi du}$$

4.9 FURTHER DEVELOPMENTS OF LSW THEORY

LSW theory is one of the most widely cited and beautiful theories in materials science. It predicts two universal laws: (i) linear time dependence of the cubed mean size and (ii) universal asymptotic size distribution for the relative sizes. Now we know that the first law (for mean size) is satisfied with most of the experimental data, but the second law (universal size distribution) is not satisfied with experiments [5–8]. Typically, the size distribution goes to zero not at relative size $u = r/r^* = 1.5$ (as predicted by LSW) but instead at about $u = 2$, so that experimental distribution is broader and lower than the LSW prediction. The main reason for the discrepancy is that the derivation of LSW model is based on three basic suggestions: steady-state concentration distribution, mean-field approximation, and stress-free approximation. Numerous investigations demonstrated that the steady-state approximation is OK, but the second and third approximations often fail. In particular, mean-field approximation is, in general, not OK: it works well only when the volume fraction of a new phase is about 1% or less. Detailed analysis of the role of finite volume fraction and of stress effects can be found in [5–8].

Another development of LSW theory is an application of its ideas to ripening in the open system. Typical example is a flux-driven ripening (FDR) of Cu6Sn5 scallops simultaneously with Cu-molten tin reaction [9]. In this case, one has nonconservative ripening, when the total volume is growing and total interface area remains approximately constant.

REFERENCES

1. Lifshitz IM, Slyozov VV. The kinetics of precipitation from supersaturated solid solutions. J Phys Chem Solids 1961;19:35–50.
2. Wagner C. Theorie der Alterung von Niederschlägen durch Umlösen (Ostwald-Reifung). Z Elektrochem 1961;65:581–591.
3. Ham FS. Theory of diffusion-limited precipitation. J Phys Chem Solids 1958;6:335–351.
4. Shewmon P. *Diffusion in Solids*. 2nd ed. Warrendale (PA): TMS; 1989.
5. Voorhees PW. The theory of Ostwald ripening. J Stat Phys 1985;38:231–252.
6. Voorhees PW. Ostwald ripening of two-phase mixtures. Annu Rev Mater Sci 1992;22:197–215.

7. Ardell AJ, Nicholson RB. Coarsening in Ni–Al systems. J Chem Phys Solids 1966;27:1793–1794.
8. Ardell AJ. In: Lorimer GW, editor. *Phase Transformations 87*. London: Institute of Metals; 1988. 485–494.
9. Gusak AM, Tu KN. Kinetic theory of flux-driven ripening. Phys Rev B 2002;66:115403-1–115403-14.

PROBLEMS

4.1. What is mean-field concentration in the theory of ripening, and does it increase or decrease with aging time?

4.2. In diffusion-limited ripening, why is the rate $r^3 = Kt$, rather than $r^2 = Kt$, where r and t are critical (or mean filed) radius and time, respectively.

4.3. Use computer to plot the function $F(u)$ against u,

$$F(u) = u^3 - (u-1)\,a = 0$$

where $a = 27/2$, $a = 27/4$, and $a = 27/6$. Discuss the results.

4.4. In ripening under a small volume fraction, it has been shown the critical radius "r^*" is equal to mean radius $<r>$, where $C_0 - C_\infty = a/r^*$ and $C - C_\infty = a/<r>$. Using the conservation relationship that

$$-\frac{dC_0}{dt} = \int_0^\infty V(r,t) f(r,t)\,dr$$

where $V(r,t) = dn/dt = (d/dt)\left((4\pi/3)\,r^3\,(1/\Omega)\right)$, and n is the number of atoms in a spherical particle of radius r, and Ω is atomic volume. And $f(r, t)$ is distribution function of particles.
Recall that $dr/dt = (D\Omega/r)\left(C_0 - C_\infty - (a/r)\right)$
Assuming a steady state, show that $r^* = <r>$, and $<r>$ is given by

$$<r> = \frac{\displaystyle\int_0^\infty r f(r,t)\,dr}{\displaystyle\int_0^\infty f(r,t)\,dr}$$

4.5. We modify Ham's model of precipitation growth when the growth rate is limited by interface-reaction-controlled kinetics. In linear approximation, we can write the flux across the interface as $J_R(r) = -k\cdot(C(r) - C^{eq}(r))$, where $C^{eq}(r) = C^{eq} + (a/r)$. At precipitation we can neglect the Gibbs–Thomson term a/r, so that $J_R(r) = -k\cdot(C(r) - C^{eq})$. Here rate constant k has a dimension of velocity. Find the growth equation.

4.6. Describe the growth of a spherical void in pure metal having supersaturated vacancies.

4.7. Describe the growth of a spherical void in a binary alloy having supersaturated vacancies.

4.8. Assume that at ripening stage the radii of all precipitates are still less than the characteristic length, $r \ll D/k \equiv l$.

4.9. **(1)** What will be the relation between critical size and mean size (in diffusion-controlled ripening they are just equal)?

(2) What will be the time law for mean size?

4.10. In solder reaction between molten eutectic SnPb and Cu at 200 °C, an array of hemispherical scallops is formed at the interface. Calculate the ripening rate among the scallops.

4.11. Plot the function $g(u)$ on p. 115 against u for $u < 3/2$.

CHAPTER 5

SPINODAL DECOMPOSITION

5.1 INTRODUCTION

Spinodal decomposition is a phase transformation by phase separation that occurs in bulk materials, but the scale of separation is in nanoscale [1–6]. Typically, it starts from a homogeneous solid solution and transforms into an inhomogeneous solid solution, and the scale of inhomogeneous domains is nanoscale. It requires no nucleation, and the initiation of the two new phases occurs by composition fluctuation, and the growth of the fluctuation occurs by uphill diffusion. Why a homogeneous alloy chooses this mode of phase change can be explained by examining the phase diagram of miscibility gap and the corresponding free energy curve as depicted in Figure 5.1. The free energy curve has a bow shape in which the middle part is concaved downward and two ends are concaved upward. The point where the curvature changes sign is defined as the inflection point. There are two inflection points in the free energy curve. If we plot the locus of the inflection points as a function of temperature and the two common tangent points in the free energy curve (which coincide with two minima in the case of symmetric curve) as a function of temperature, we obtain the

Kinetics in Nanoscale Materials, First Edition. King-Ning Tu and Andriy M. Gusak.

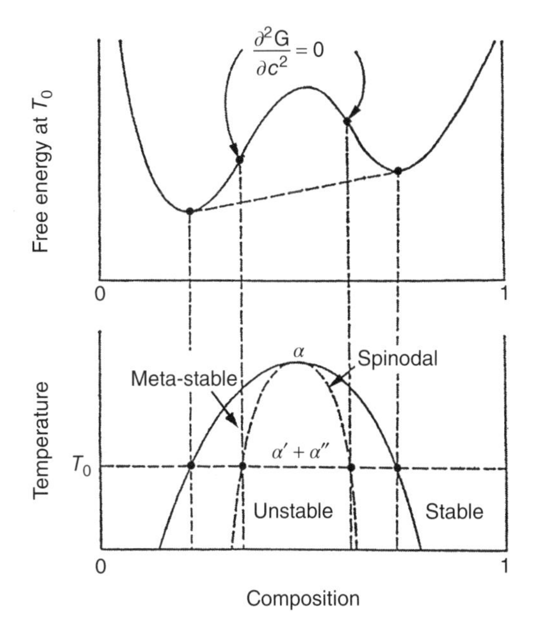

Figure 5.1 Schematic phase diagram of miscibility gap and the corresponding free energy curve.

broken curve and the solid curve in the phase diagram, respectively. The solid curve is the miscibility curve and the broken curve is the chemical spinodal curve. The zone or the area enclosed by the spinodal curve is an unstable zone in which spinodal decomposition occurs, as indicated in Figure 5.1. The two zones between the broken curve and the solid curve are metastable zones, in which the mode of nucleation and growth of phase change will occur. Because nucleation requires undercooling to overcome the nucleation barrier, it is metastable until nucleation barrier is overcome. The two zones outside the solid curve, on the two sides of the phase diagram, are stable zones, so no phase change will occur. The stability can be recognized by the fact that the free energy cannot be lowered anymore by any transformation. If we draw a common tangent line connecting the two valleys in the free energy curve, the tangent line represents the lowest free energy of the system at fixed total (average) concentration, corresponding to mechanical macroscopic mixture of two phases with concentrations of the tangency points and their volume fraction determined by the lever rule. The free energy difference between the free energy curve and the tangent is the driving force of phase change under fixed average concentration. However, the response to the driving force depends on the composition whether it is in the unstable zone or in the metastable zones; it occurs by nucleation and growth in the latter and by spinodal decomposition in the former.

In Figure 5.2, we enlarge the free energy curve to show that in the unstable zone, the free energy curve is concaved downward. Thus if there is a small fluctuation of composition everywhere at the composition of C_0 to $C_0 + \delta_C$ in some volume region V^{fluct} and $C_0 - \delta_C$ in another region of the same size (for conservation of matter), the free energy will decrease. Thus, we can have a spontaneous phase separation. On the other hand, in the metastable zone, where the free energy curve is concaved upward,

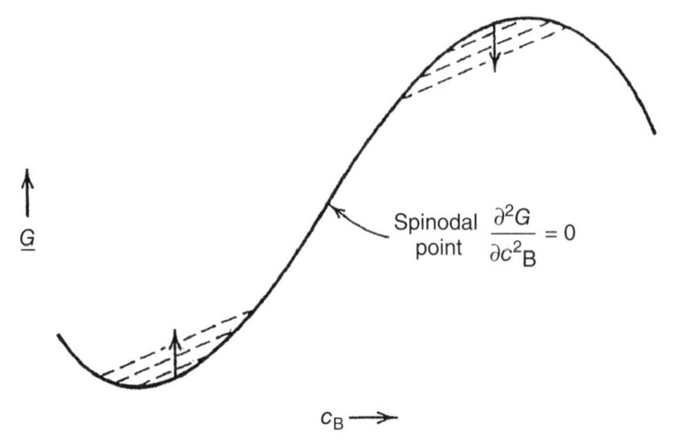

Figure 5.2 Enlarged free energy curve in Figure 5.1 to show that in the unstable region, the free energy curve is concaved downward. If there is a small fluctuation of composition at the composition of C_0 everywhere to $C_0 + \delta_C$ in some volume region V^{fluct} and $C_0 - \delta_C$ in another region of the same size (for conservation of matter), the free energy will decrease.

a similar small composition fluctuation will lead to free energy increase, which will not happen, so the zone is metastable. No doubt, in the metastable zone, there is a free energy gain in dropping to the tangent line, but it can occur only by nucleation and growth, which involves a very large composition change, not by a small composition fluctuation.

We calculate the free energy change in a composition fluctuation in the spinodal zone from C_0 to $C_0 \pm \delta_C$ as below. At composition C_0, the Gibbs free energy of an alloy with total volume V^{total} is G_{C_0} where $G_{C_0} = g(C_0) \cdot V^{\text{total}}$, where $g(C_0)$ is a density of Gibbs free energy (Gibbs free energy per unit volume). We use the upper case of G here to represent Gibbs free energy of the total volume of a sample and the lower case of g to represent Gibbs free energy per unit volume. By using a Taylor expansion series at C_0 for the volumes with deviated composition, we have

$$g(C_0 \pm \delta C) \cong g(C_0) + (\pm\delta C)\left.\frac{\partial g}{\partial C}\right|_{C_0} + \frac{1}{2}(\pm\delta C)^2 \left.\frac{\partial^2 g}{\partial C^2}\right|_{C_0} + \cdots \tag{5.1}$$

When we ignore the higher order terms, the change in Gibbs free energy is

$$\begin{aligned}\Delta G_{\text{SD}} &= G - G_0 = g(C_0) \cdot (V^{\text{total}} - 2V^{\text{fluct}}) + g(C_0 + \delta C)V^{\text{fluct}} \\ &\quad + g(C_0 - \delta C)V^{\text{fluct}} - g(C_0) \cdot V^{\text{total}} \\ &= (g(C_0 + \delta C) + g(C_0 - \delta C) - 2g(C_0))V^{\text{fluct}} \\ &= (\delta C)^2 \left.\frac{\partial^2 g}{\partial C^2}\right|_{C_0} \cdot V^{\text{fluct}}\end{aligned} \tag{5.2}$$

Abovementioned considerations may be easily generalized on the arbitrary small fluctuation of the concentration field all over the system (under the constraint of matter conservation). Provided that in the last term the second derivative of g is negative, the ΔG_{SD} is negative. In the next section, we discuss this.

In the beginning of composition fluctuation that leads to spinodal decomposition, there should be no change in the crystal structure before and after the decomposition. It means that the two decomposed phases should have the same crystal structure as the phase before decomposition. Otherwise, the nucleation of a new crystal structure is required. Therefore, we expect, for example, a face-centered-cubic homogeneous solid solution will decompose into two face-centered-cubic phases of slightly different composition. Thus, there could be a slight change in lattice parameter between the two phases because lattice parameter changes with the change of composition of an alloy phase. From the structure point of view, we expect that spinodal decomposition can occur in glassy phases that have amorphous structure; it means an amorphous phase can decompose into two amorphous phases without structure change.

5.2 IMPLICATION OF DIFFUSION EQUATION IN HOMOGENIZATION AND DECOMPOSITION

In Figure 5.3, we plot the function $y = \sin x$, its first derivative, $y' = \cos x$, and its second derivative, $y'' = -\sin x$. We see that when the curve of y is concave downward (from $y = 0$ to $y = \pi$), the curvature y'' is negative, but when the curve of y is concave upward (from $y = \pi$ to $y = 2\pi$), the curvature y'' is positive. We imagine the curve y to be a concentration profile or a free energy curve, and we can analyze its curvature similarly.

We can apply the above finding to one-dimensional diffusion equation, and we can show that the annealing of a periodic structure of composition, in which one of the periods is shown in Figure 5.4, will lead to homogenization. This is a unique property

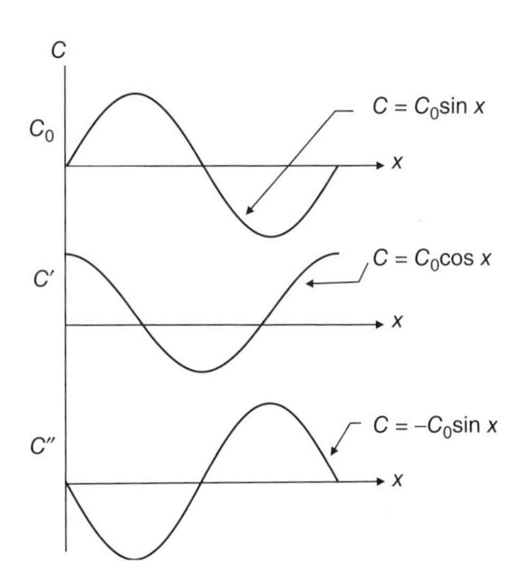

Figure 5.3 Plotting the function $y = \sin x$, its first derivative, $y' = \cos x$, and its second derivative, $y'' = -\sin x$. When the curve of y is concave downward (from $y = 0$ to $y = \pi$), the curvature y'' is negative, but when the curve of y is concave upward (from $y = \pi$ to $y = 2\pi$), the curvature y'' is positive.

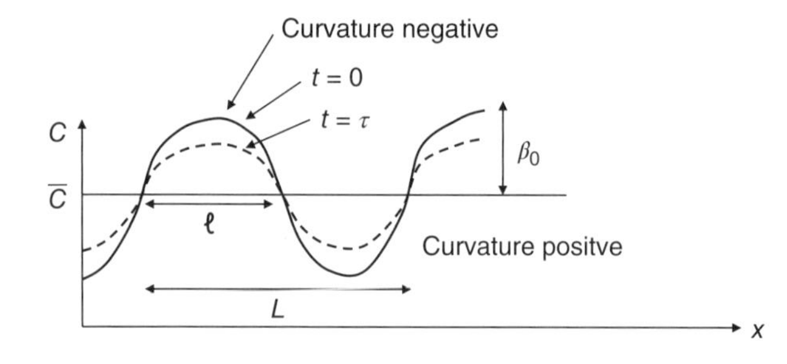

Figure 5.4 Schematic diagram showing that in the annealing of a periodic structure of composition, it will lead to homogenization.

of the diffusion equation. In the diffusion equation as shown in Eq. (5.3), provided that D and $\partial^2 C/\partial x^2$ are positive, $\partial C/\partial t$ is positive, which means that the composition will increase with time,

$$\frac{\partial C}{\partial t} = D\frac{\partial^2 C}{\partial x^2} = + \tag{5.3}$$

Thus if we imagine that the curve in Figure 5.3 is a composition profile, we apply the condition in Eq. (5.3) to the part of the profile ($y=\pi$ to $y=2\pi$) that is concave upward. The composition will increase with time, meaning that the valley of composition will be filled.

But when the composition profile is concave downward (from $y=0$ to $y=\pi$), the curvature y'' is negative, so $\partial^2 C/\partial x^2$ is negative, thus $\partial C/\partial t$ is negative (concentration decreases), provided that D is positive as shown in Eq. (5.4) below,

$$\frac{\partial C}{\partial t} = D\frac{\partial^2 C}{\partial x^2} = - \tag{5.4}$$

Thus, the hill in the concentration profile will be flattened. The above results mean that when D is positive, diffusion will lead to homogenization.

However, if D is negative, the opposite will occur; the hill will grow higher and the valley will dip deeper. We have mentioned that in spinodal region, D is negative (uphill diffusion), the outcome is phase separation. Of course, uphill diffusion means that it cannot be reduced to random walk of atoms on the grid of equivalent sites (entropic factor). In that case, diffusivity should be positive because it is proportional to the ratio of mean squared distance and time (both values are positive). Uphill diffusion always means that atoms migrate under a strong energy factor making difference between energetically favorable and unfavorable positions.

When the spacing of the period is large, linear diffusion occurs and it obeys Fick's second law. When the period is small as in man-made superlattice [7–9], we need to consider higher order diffusion, which is covered later in the chapter. The linear solution presented below for a periodic structure will serve as the starting point of spinodal decomposition or interdiffusion in superlattices.

At $t=0$, the concentration profile is depicted by the solid curve in Figure 5.3 as

$$C = \overline{C} + \beta_0 \sin\frac{\pi x}{l} \tag{5.5}$$

where C is the mean concentration, β_0 is the amplitude of the initial profile, and $l = L/2$ is half of the period, L, of the concentration profile. At $t > 0$, the concentration profile is given as

$$C = \overline{C} + \beta_0 \sin \frac{\pi x}{l} \exp\left(-\frac{t}{\tau}\right) \tag{5.6}$$

We can substitute it into the diffusion equation, and we find that it is a solution of the diffusion equation if

$$\tau = \frac{l^2}{D\pi^2}$$

We define τ as the relaxation time. The amplitude at $x = l/2$ is

$$C - \overline{C} = \beta = \beta_0 \exp\left(-\frac{t}{\tau}\right)$$

Thus, at $t = \tau$, $\beta = \beta_0/e$, and at $t = 2\tau$, $\beta = \beta_0/e^2$. It decays rapidly with time and with the short period of l.

5.3 SPINODAL DECOMPOSITION

5.3.1 Concentration Gradient in an Inhomogeneous Solid Solution

In a homogeneous solid solution, we can assume that there is no concentration gradient. In an inhomogeneous solid solution, we assume that there will be regions that have concentration gradient. These concentration gradients will affect the free energy of the solution. In turn, it will change the chemical potential of an atom in the inhomogeneous region and consequently the driving force of diffusion of the atom. In the following, we show first that the contribution of concentration gradient to free energy is positive. To do so, we need to calculate the energy of mixing to form a homogeneous solid solution as well as an inhomogeneous solid solution for comparison. Next, we should obtain the chemical potential of an atom in the inhomogeneous solid solution. Then by applying the principle of flux divergence, we can obtain the diffusion equation, which will be a fourth order diffusion equation. Finally, we can apply the diffusion equation to spinodal decomposition.

To consider the energy of mixing A and B to form an inhomogeneous solid solution, we first consider the energy of mixing to form a homogeneous solid solution. Then we compare them to show that the difference is positive. We use the quasi chemical approach or short-range interaction of nearest neighboring atoms to calculate the mixing energy [2]. If we let ε_{AB}, ε_{AA}, and ε_{BB} to be the bond energy of AB bond, AA bond, and BB bond, respectively, we note that they are negative. We define

$$\varepsilon = \varepsilon_{AB} - \frac{1}{2}(\varepsilon_{AA} + \varepsilon_{BB}) \tag{5.7}$$

In spinodal decomposition, we assume the AB bond energy to be higher (less negative) than half of the sum of AA bond and BB bond energy, thus ε is positive. Bond energy is negative, so higher energy means less negative or weaker bond. Thus, spinodal decomposition happens if "hetero-bonds" A-B are weaker than the average of "homo-bonds" A-A and B-B, and, of course, if temperature is low enough to provide bond energy victory because of the energetic favor over entropic factor (tendency to random mixing). On the other hand, in most alloy systems where the alloy prefers homogenization, that is, AB bond energy is lower than half of the sum of AA bond and BB bond energy (hetero-bond stronger), ε is negative.

In the following, we show first that in one mole of the homogeneous solution,

$$\Delta E_{\text{homomixing}} = P_{\text{AB}}\varepsilon = N_{\text{a}}ZX_{\text{A}}X_{\text{B}}\varepsilon = N_{\text{a}}ZX_{\text{A}}(1-X_{\text{B}})\varepsilon = ZC(1-C)N_{\text{a}}\varepsilon \quad (5.8)$$

where P_{AB} is the number of AB bonds in the solution, Z is the number of the nearest neighbors of an atom in the solid solution, and X_{A} and X_{B} are the fraction of A atoms and B atoms in the solid solution, respectively. If we define in one mole, there are n_{A} atoms of A and n_{B} atoms of B, and N_{a} to be Avogadro's number, we have

$$X_{\text{A}} = \frac{n_{\text{A}}}{n_{\text{A}} + n_{\text{B}}} = \frac{n_{\text{A}}}{N_{\text{a}}} \text{ and } X_{\text{B}} = \frac{n_{\text{B}}}{N_{\text{a}}}$$

In Eq. (5.8), for convenience, we let $C = X_{\text{A}}$ and $1 - C = X_{\text{B}}$. Note that here C is equal to atomic fraction or concentration fraction (unit-less), rather than the conventional definition of concentration that has the unit of the number of atoms per unit volume.

Therefore, the formation energy or the energy of mixing to form one mole of an inhomogeneous solid solution is

$$\Delta E_{\text{in homomixing}} = ZC(1 - C)N_{\text{a}}\varepsilon + K\left(\frac{\partial C}{\partial x}\right)^2 N_{\text{a}}\varepsilon \quad (5.9)$$

where K is the coefficient of concentration gradient energy and is positive. Comparing Eqs. (5.8) and (5.9), we see that in a homogeneous solid solution, where $dC/dx = 0$, the two equations are the same. However, in an inhomogeneous solid solution, $dC/dx \neq 0$, the mixing energy of the inhomogeneous solid solution is higher than that of the homogeneous solid solution because the second term is positive because K and ε are positive.

5.3.2 Energy of Mixing to Form a Homogeneous Solid Solution

Consider 1 mol of solid solution contains n_{A} atoms of A and n_{B} atoms of B. Let P_{AA}, P_{BB}, and P_{AB} be the number of AA bonds, BB bonds, and AB bonds in the solid solution, respectively. Hence the energy of the solid solution will be

$$E_{\text{solution}} = P_{\text{AA}}\varepsilon_{\text{AA}} + P_{\text{BB}}\varepsilon_{\text{BB}} + P_{\text{AB}}\varepsilon_{\text{AB}} \quad (5.10)$$

We note that the number of A atoms times the number of bonds per atom is equal to the number of AB bonds plus twice the number of AA bonds, and it can be expressed as

$$n_A Z = P_{AB} + 2P_{AA}$$

where Z is the number of nearest neighbors and in a face-centered-cubic lattice, $Z = 12$. The factor of 2 in front of P_{AA} is because each AA bond involves two A atoms and it has been counted twice. So we have

$$P_{AA} = \frac{n_A Z}{2} - \frac{P_{AB}}{2}$$

Similarly,

$$P_{BB} = \frac{n_B Z}{2} - \frac{P_{AB}}{2}$$

By substituting P_{AA} and P_{BB} into Eq. (5.10), we have

$$E_{\text{solution}} = \frac{1}{2} Z n_A \varepsilon_{AA} + \frac{1}{2} Z n_B \varepsilon_{BB} + P_{AB}\left[\varepsilon_{AB} - \frac{1}{2}\left(\varepsilon_{AA} + \varepsilon_{BB}\right)\right]$$

We note that the first two terms on the right-hand side of the last equation are just the energy of pure A and pure B before mixing. So the energy of mixing to form a homogeneous AB solution is equal to the energy of the solid solution minus the energy of the unmixed components;

$$\Delta E_{\text{homomixing}} = P_{AB}\varepsilon \tag{5.11}$$

Here, the question is what is P_{AB} in the homogeneous solid solution? Now we consider an arbitrary two neighboring sites, site 1 and site 2, in 1 mol of the solution. The probability of site 1 to be occupied by A atoms will be the total number of A atom in 1 mol of the solution divided by the total number of lattice sites in 1 mol of the solution, so it is

$$X_A = \frac{n_A}{N_a}$$

Similarly, the site 2 to be occupied by a B atom will be X_B. Thus the combined probability of site 1 occupied by A and site 2 occupied by B is $X_A X_B$. And, the combined probability of site 1 occupied by B and site 2 occupied by A is $X_B X_A$. Thus, the probability of a neighboring pair of sites having an AB pair is $2X_A X_B$. As 1 mol of solid solution contains $\frac{1}{2} Z N_a$ pair of lattice sites, the total number of AB pair is

$$P_{AB} = \left(\frac{1}{2} Z N_a\right)(2X_A X_B) = Z N_a X_A X_B = Z N_a C(1 - C) \tag{5.12}$$

Finally, for a solid solution having a face-centered-cubic lattice, $Z = 12$, as presented in Eq. (5.8), we have

$$\Delta E_{\text{homomixing}} = P_{AB}\varepsilon = 12C(1 - C)N_a \varepsilon$$

Note that in above considerations we neglected the short-range order or, in other words, correlation between occupancy of the neighboring sites. Indeed, generally speaking, if some site is occupied by, say, atom A, and atoms A and B "do not like each other" (energetic tendency to separation), then the pair probability of A-B configuration in the neighboring site should be lower than just the product of atomic fractions. Therefore, our considerations in the above have the qualitative character.

5.3.3 Energy of Mixing to Form an Inhomogeneous Solid Solution

We follow the quasi chemical approach to calculate the energy of mixing of an inhomogeneous solid solution in which there are concentration gradients [2]. Let the concentration gradient be directed along x-axis perpendicular to (111) planes of fcc lattice. The check point is that when we take $dC/dx = 0$, the energy of mixing returns to that of the homogeneous solid solution. We consider the mixing of 1 mol of atoms, in which there are n_A of A atoms and n_B of B atoms. Then X_A and X_B are the fraction of A and B, respectively. Again we let $C = X_A$ and $1 - C = X_B$. In Figure 5.5, we depict the side view of three vertical (111) atomic planes in the face-centered-cubic (fcc) solid solution. In the planes, we assume the solid dots to be A atoms of concentration of C and cross dots to be B atoms of concentration of $1 - C$.

In the previous section, we have shown that the mixing energy depends on P_{AB}, and not on P_{AA} or P_{BB}. Thus we calculate the AB interaction in the three atomic planes as shown in Figure 5.5a. In the middle plane, it has $C(x)$ of A atoms and

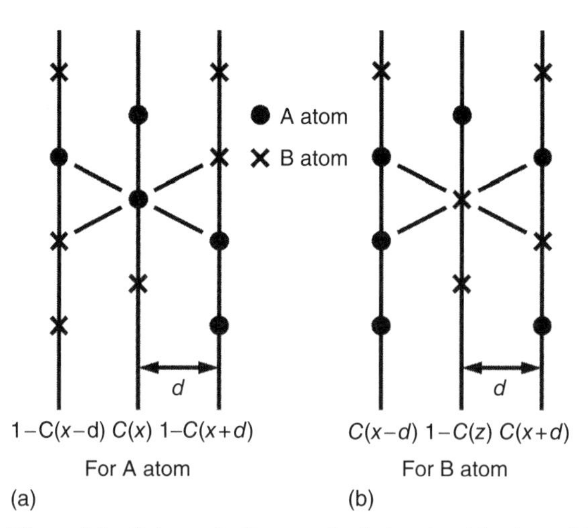

Figure 5.5 Schematic diagram depicting the side view of three vertical (111) atomic planes in the fcc solid solution. In the planes, we assume the solid dots to be A atoms of concentration of C and cross dots to be B atoms of concentration of $1 - C$. (a) We first consider the A atoms in the middle plane and their interaction with B atoms in the neighboring planes. (b) Similarly, we consider the B atoms in the middle plane and their interaction with A atoms in the neighboring planes.

$[1 - C(x)]$ of B atoms. As each atom has six nearest neighboring atoms in the plane and three nearest neighboring atoms in the left neighboring plane and three nearest neighboring atoms in the right neighboring plane, if we assume a closed-packed structure, say (111)-plane of fcc lattice, the AB pairs within the plane is as given below if we follow the same derivation as presented in the previous section for Eq. (5.12)

$$6C(1-C)N_a \tag{5.13}$$

Next we consider the AB interaction between the middle plane and its neighboring planes. Let us first consider the A atoms in the middle plane and their interaction with B atoms in the neighboring planes, as shown in Figure 5.5a. The middle plane has $C(x)$ of A atoms. The left-hand side neighboring plane has $1 - C(x - d)$ of B atoms, and the right-hand side neighboring plane has $1 - C(x + d)$ of B atoms, where "d" is the interplanar distance (in general, it is related to the lattice parameter "a" by the known equation $d = a/\sqrt{h^2 + k^2 + l^2} = a/\sqrt{3}$). The AB interaction or the AB pairs from the consideration of A atoms in the middle plane can be given as

$$3\frac{1}{2}\{C(x)[1 - C(x + d)] + C(x)[1 - C(x - d)]\}N_a \tag{5.14}$$

Similarly, we consider the B atoms in the middle plane and their interaction with A atoms in the neighboring planes, as shown in Figure 5.5b. The AB interaction or the AB pairs from the consideration of B atoms in the middle plane can be given as

$$3\frac{1}{2}\{[1 - C(x)]C(x + d)] + [1 - C(x)]C(x - d)\}N_a \tag{5.15}$$

We note that the factor ½ is because we consider first A and then B. The factor of 3 is because there are three nearest neighbors on either side for fcc structure. Now by substituting $C(x \pm d)$ from Taylor expansion, as shown below, into Eqs (5.14) and (5.15),

$$C(x \pm d) = C(x) \pm d\frac{\partial C(x)}{\partial x} + \frac{d^2}{2}\frac{\partial^2 C(x)}{\partial x^2}$$

we obtain the following equations for the consideration of A atoms and B atoms.

For A atoms and for simplicity, we take $C = C(x)$,

$$\frac{3}{2}\left\{C\left[1 - \left(C + d\frac{\partial C}{\partial x} + \frac{d^2}{2}\frac{\partial^2 C}{\partial x^2}\right)\right] + C\left[1 - \left(C - d\frac{\partial C}{\partial x} + \frac{d^2}{2}\frac{\partial^2 C}{\partial x^2}\right)\right]\right\}$$

$$= \frac{3}{2}\left\{C(1 - C) - Cd\frac{\partial C}{\partial x} - C\frac{d^2}{2}\frac{\partial^2 C}{\partial x^2} + C(1 - C) + Cd\frac{\partial C}{\partial x} - C\frac{d^2}{2}\frac{\partial^2 C}{\partial x^2}\right\}$$

$$= 3C(1 - C) - \frac{3}{2}d^2 C\frac{\partial^2 C}{\partial x^2} = 3C(1 - C) - \frac{1}{2}a^2 C\frac{\partial^2 C}{\partial x^2} \tag{5.16}$$

Similarly, for B atoms,

$$\frac{3}{2}\left\{(1-C)\left(C+d\frac{\partial C}{\partial x}+\frac{d^2}{2}\frac{\partial^2 C}{\partial x^2}\right)+(1-C)\left(C-d\frac{\partial C}{\partial x}+\frac{d^2}{2}\frac{\partial^2 C}{\partial x^2}\right)\right\}$$
$$=3C(1-C)-\frac{3}{2}\left[d^2(1-C)\frac{\partial^2 C}{\partial x^2}\right] \tag{5.17}$$

By summing up Eqs (5.13) + (5.16) + (5.17), we have

$$\mathrm{Sum}=12C(1-C)+\frac{3}{2}d^2(1-2C)\frac{\partial^2 C}{\partial x^2}$$

$$=12C(1-C)+\frac{1}{2}a^2(1-2C)\frac{\partial^2 C}{\partial x^2} \tag{5.18}$$

In the above sum, the first term is the contribution from homogeneous mixing, and the second term is from the contribution of inhomogeneous mixing. However, we need to convert the second derivative in the second term into the first derivative because we can more easily recognize the inhomogeneous contribution from the gradient or slope of concentration. To do so, we consider a product of uv, where

$$u=\frac{3}{2}d^2(1-2C)$$

$$v=\frac{\partial C}{\partial x}$$

Thus,

$$\frac{\partial u}{\partial C}=-3d^2 \text{ and } dv=\frac{\partial}{\partial x}\frac{\partial C}{\partial x}dx=\frac{\partial^2 C}{\partial x^2}dx$$

Because $d(uv)=udv+vdu$, we have

$$\int udv=uv|_{-\infty}^{\infty}-\int vdu$$

$$\int u\frac{\partial^2 C}{\partial x^2}dx=0-\int\frac{\partial C}{\partial x}\frac{\partial u}{\partial C}\frac{\partial C}{\partial x}dx=0-\int\frac{\partial u}{\partial C}\left(\frac{\partial C}{\partial x}\right)^2dx$$

From the above equation, we can replace the integrant in the left-hand side,

$$u\frac{\partial^2 C}{\partial x^2}=\frac{3}{2}d^2(1-2C)\frac{\partial^2 C}{\partial x^2}$$

by the integrant in the right-hand side,

$$-\frac{\partial u}{\partial C}\left(\frac{\partial C}{\partial x}\right)^2=3d^2\left(\frac{\partial C}{\partial x}\right)^2$$

Substituting the above result into the sum in Eq. (5.18), we have

$$\text{sum} = 12C(1-C) + 3d^2\left(\frac{\partial C}{\partial x}\right)^2 = 12C(1-C) + a^2\left(\frac{\partial C}{\partial x}\right)^2$$

The energy of mixing in an inhomogeneous solution is given by Eq. (5.9),

$$\Delta E_{\text{inhomo-mixing}} = 12C(1-C)N_a\varepsilon + a^2\left(\frac{\partial C}{\partial x}\right)^2 N_a\varepsilon \tag{5.19}$$

Within the spinodal composition region, ε is positive. Thus the energy of inhomogeneous mixing is higher than that of homogeneous mixing because the second term in the right-hand side of the above equation is positive. For a homogeneous solution, the second term goes to zero because $dC/dx = 0$.

However, when ε is negative, which occurs in a binary system that forms alloys over the entire composition range, it is different. On the basis of Eq. (5.19), the inhomogeneous mixing has a lower energy than the homogeneous mixing because the second term becomes negative. This seems misleading! Yet, this property has a very simple physical meaning; for a random alloy, when all atoms occupy fully randomly the lattice sites, it is not the most energetically favorable configuration for the alloy with a negative mixing energy, which means there is a tendency to ordering. Let us consider an alloy system at a composition of 50A50B. The most favorable configuration means formally the maximum local gradient from a site (plane) of pure A to a site (plane) of pure B. In the case of fcc lattice, it means Au–Cu type ordered phase, and in the case of bcc lattice, it means Cu–Zn type (B2) phase. However, if we want to exclude the gradient of the order of $1/d \sim 10^8\,\text{cm}^{-1}$ from a phenomenological description, it seems better just to change the reference level in Taylor expansion of the Gibbs free energy. The most energetically favorable homogeneous mixing (in a phenomenological sense) will have the chess board type ordering in which all the bonds are AB bonds. So it has the lowest energy of mixing, and any composition fluctuation to have concentration gradient will decrease the number of AB bonds and create more AA and BB bonds, resulting in the increase of the mixing energy. Thus, in this case, we should treat the $g(C_0)$ as the free energy of ordered phase or mechanical mixture of two ordered phase. Therefore, the concept that concentration gradient has a positive contribution to mixing energy is correct.

5.3.4 Chemical Potential in Inhomogeneous Solution

In the above, we have obtained the free energy change in an inhomogeneous solid solution. Let us convert it into a chemical potential so that the chemical potential gradient will be the driving force of atomic diffusion [3–6]. In other words, we need the chemical potential gradient of atoms in a concentration gradient.

Let us consider a cylindrical sample of inhomogeneous solution of the volume of Al, where A is the cross-section and l is the length. Let the concentration change only along the sample length (x-axis). The free energy can be given as

$$G_T = A\int_0^l \left[g_{\text{homo}}(C) + K\left(\frac{dC}{dx}\right)^2\right]dx \tag{5.20}$$

At equilibrium, we need to minimize G_T by varying the concentration C with respect to x. The minimization is subject to the condition that the average composition remains constant, that is, we average out the concentration fluctuation.

$$\int_0^{\infty} (C - C_0)dx = 0 \tag{5.21}$$

In applied mathematics, when we need to minimize an integral of

$$\int P(x, y, y')dx$$

that is subject to the condition (constraint) that

$$\int Q(x, y, y')dx = \text{const.}$$

(where $y' = dy/dx$), it must satisfy the Euler equation given below,

$$\frac{d}{dx}\left(\frac{\partial U}{\partial y'}\right) - \frac{\partial U}{\partial y} = 0$$

where $U = P - Q\alpha$, and α is a Lagrangian multiplier, determined from the boundary conditions.

In our problem here, we have $x = x$, $y = C$, $y' = dC/dx$. So

$$P = g_{\text{homo}}(C) + K\left(\frac{dC}{dx}\right)^2 = g(y) + K(y')^2$$

$$Q = C - C_0$$

$$U = g_{\text{homo}}(C) + K\left(\frac{dC}{dx}\right)^2 - \alpha(C - C_0)$$

$$\frac{d}{dx}\left(\frac{\partial U}{\partial y'}\right) = \frac{d}{dx}\left[2K\left(\frac{dC}{dx}\right)\right] = 2K\frac{\partial^2 C}{\partial x^2}$$

$$\frac{\partial U}{\partial y} = \frac{\partial U}{\partial C} = \frac{\partial g(C)}{\partial C} - \alpha$$

Then the Euler equation is

$$2K\frac{\partial^2 C}{\partial x^2} - \frac{\partial g_{\text{homo}}(C)}{\partial C} + \alpha = 0$$

$$\alpha = \frac{\partial g_{\text{homo}}(C)}{\partial C} - 2K\frac{\partial^2 C}{\partial x^2}$$

We note that α can be called a reduced chemical potential of an atom in the inhomogeneous solution per unit volume that we want. The analogous equation in a homogeneous solution is

$$\alpha = \frac{\partial g_{\text{homo}}}{\partial C} = \mu_{\text{B}} - \mu_{\text{A}}$$

where

$$\mu_{\text{B}} = \frac{1}{\Omega} \frac{\partial G(N_{\text{A}}, N_{\text{B}}, T, p)}{\partial N_{\text{B}}} \quad \text{and}$$

$$\mu_{\text{A}} = \frac{1}{\Omega} \frac{\partial G(N_{\text{A}}, N_{\text{B}}, T, p)}{\partial N_{\text{A}}}$$

are the usual chemical potentials, and N_{A} and N_{B}, respectively, are the total number of A and B atoms in the system. Physical sense of usual chemical potentials is the change of Gibbs free energy when we **add** the atoms of some sort (one atom, or mole). Physical sense of the **reduced** chemical potential is the change in Gibbs free energy when we reversibly **replace** A atoms by B atoms.

5.3.5 Coherent Strain Energy

We need to consider coherent strain energy in spinodal decomposition [3–5]. As there is a composition fluctuation or a composition modulation in spinodal decomposition, a lattice parameter modulation exists correspondingly. Coherent strain occurs between the two phases in modulation. This is because the lattice parameter of the phases after decomposition will be affected by the compositional change. The lattice parameter change or mismatch will induce coherent strain in the sample; hence, we must include the strain energy in the free energy of spinodal decomposition.

To calculate the coherent strain energy, we assume a one-dimensional model that the composition modulation occurs in x-direction. To maintain coherency, we require the atomic spacing to be normal to the x-axis, that is, the yz plane must be the same. Even though the composition modulation occurs between neighboring yz planes, in the model we must strain them so that they stick together without mismatch along the y- and z-directions. This means that the lattice spacing in the yz plane remains constant, but we allow expansion or shrinkage along the x-direction. To proceed, we assume "a_0" to be the lattice parameter of the average composition C_0 before fluctuation or modulation occurs. And we assume "a" to be the lattice parameter after the composition fluctuation of $C_0 \pm \delta C$. Then the strain in x-, y-, and z-direction is

$$\delta = \frac{a - a_0}{a_0}$$

The situation of keeping the strain in the y- and z-direction and relaxing the strain in x-direction is similar to that in the epitaxial growth of a thin film on a single crystal substrate under biaxial strain. If we assume that the thin film and the single crystal substrate have a cubic lattice, the thin film and the substrate will have the same lattice parameters across the yz epitaxial interface. Yet the thin film is strained in the

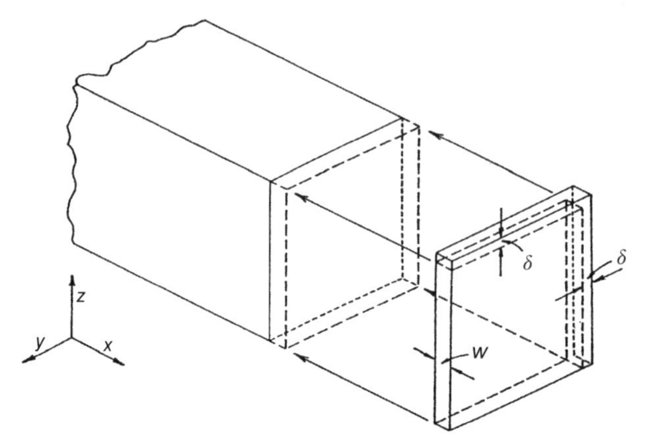

Figure 5.6 We assume that the x-axis is the [100] direction in a cubic crystal and calculate the strain energy or the reversible work done to deform a slab of crystal of strain δ by taking two steps.

x-direction, and thus the unit cell of the thin film undergoes a transition from cubic to tetragonal. And if we further assume the x-axis is the [100] direction in a cubic crystal, we calculate the strain energy or the reversible work done to deform a slab of crystal of strain δ by taking the following two steps, which is depicted in Figure 5.6.

Step 1: To compress or expand the slab hydrostatically of strain δ in x, y, z directions.

Step 2: To clamp the yz plane of the slab to the crystal and allow the x-direction to relax reversibly.

In Step 1, the stress required to produce a hydrostatic strain (no shear) of δ is

$$\sigma_x = \sigma_y = \sigma_z = \delta(C_{11} + 2C_{12})$$

This is obtained from the matrix of stress–strain relationship.

The elastic work done per unit volume is given as

$$W_{\mathrm{E}}^{(1)} = \frac{1}{2}\sum_{i}^{3} \sigma_i \varepsilon_i = \frac{1}{2}\sum_{i}^{3} \delta(C_{11} + 2C_{12})\delta = \frac{3}{2}(C_{11} + 2C_{12})\delta^2$$

In Step 2, we have relaxation in x-direction only, so $\varepsilon_y = \varepsilon_z = 0$. Then

$$\sigma_x = C_{11}\varepsilon_x + C_{12}(\varepsilon_y + \varepsilon_z) = C_{11}\varepsilon_x$$

The elastic work done in Step 2 per unit volume is

$$W_{\mathrm{E}}^{(2)} = \frac{1}{2}\sum_{i}^{3} \sigma_i \varepsilon_i = \frac{1}{2}\sigma_x \varepsilon_x = \frac{1}{2}\frac{\sigma_x^2}{C_{11}} = \frac{1}{2}\frac{(C_{11} + 2C_{12})^2\delta^2}{C_{11}}$$

The net work performed on the slab to achieve coherency on the crystal is

$$W_E = W_E^{(1)} - W_E^{(2)} = \frac{3}{2}(C_{11} + 2C_{12})\delta^2 - \frac{(C_{11} + 2C_{12})^2}{2C_{11}}\delta^2$$

$$= \left[C_{11} + C_{12} - \frac{2(C_{12})^2}{C_{11}}\right]\delta^2 = Y\delta^2 = \frac{E}{1-\nu}\delta^2$$

where E is Young's modulus and ν is Poisson's ratio, and

$$Y = C_{11} + C_{12} - \frac{2(C_{12})^2}{C_{11}}$$

$$C_{11} = \frac{E(1-\nu)}{(1-2\nu)(1+\nu)}$$

$$C_{12} = \frac{E\nu}{(1-2\nu)(1+\nu)}$$

The total strain energy of a sample of cross-section A and length l is

$$W_E = A\int_0^l Y\delta^2 dx$$

To relate the strain to the composition modulation, we let a_0 to be the lattice parameter of the unstrained solid, corresponding to the average composition C_0. By Taylor expansion around C_0, we have

$$a = a_0 + (C - C_0)\frac{da}{dC} + \cdots = a_0\left[1 + \frac{1}{a_0}\frac{da}{dC}(C - C_0) + \cdots\right]$$

$$= a_0[1 + \eta\Delta C + \cdots]$$

where $\eta = (1/a_0)(da/dC) = (a - a_0/a_0)(1/\Delta C) = \delta/\Delta C$, and $\delta = a - a_0/a_0$, and $\Delta C = C - C_0$.

Thus we have

$$W_E = A\int_0^l Y\eta^2(\delta C)^2 dx$$

Then the total energy for spinodal decomposition is

$$G_T = A\int_0^l [g_0(C) + K\left(\frac{\Delta C}{\Delta x}\right)^2 + Y\eta^2(\Delta C)^2]dx < 0$$

5.3.6 Solution of the Diffusion Equation

If we follow the procedure as given before by applying the Euler equation, assuming $Y\eta^2$ is independent of strain and gradient energy, we obtain the diffusion equation as

$$\frac{\partial C}{\partial t} = \widetilde{D}\left\{\left[1 + \frac{2Y\eta^2}{g''}\right]\frac{\partial^2 C}{\partial x^2} - \frac{2K}{g''}\frac{\partial^4 C}{\partial x^4}\right\}$$

The solution of the above equation has the general form of

$$C - C_0 = A(\beta, t)e^{i\beta x}$$

where $A(\beta, t)$ is an amplitude function that increases with time, and $\beta = 2\pi/\lambda$ and λ is the "period of concentration modulation" or the wavelength. By substituting the solution to the diffusion equation, we have

$$\frac{d^n C}{dx^n} = (i\beta)^n A(\beta, t)e^{i\beta x} \text{ and } \frac{dC}{dt} = \frac{dA(\beta, t)}{dt}e^{i\beta x}$$

$$\frac{\partial A(\beta, t)}{\partial t}e^{i\beta x} = \widetilde{D}\left\{\left[1 + \frac{2Y\eta^2}{g''}\right](-\beta^2 A(\beta, t)e^{i\beta x}) - \frac{2K}{g''}\beta^4 A(\beta, t)e^{i\beta x}\right\}$$

We can drop the term $e^{i\beta x}$ and let $A = A(\beta, t)$. After rearranging, we have

$$\frac{1}{A}\frac{\partial A}{\partial t} = \frac{\partial \ln A}{\partial t} = -\widetilde{D}\beta^2\left[1 + \frac{2Y\eta^2}{g''} + \frac{2K\beta^2}{g''}\right]$$

or $\partial \ln A = R(\beta)\partial t$, where

$$R(\beta) = -\widetilde{D}\beta^2\left[1 + \frac{2Y\eta^2}{g''} + \frac{2K\beta^2}{g''}\right]$$

The solution for $A(\beta, t)$ is

$$A(\beta, t) = A(\beta, 0)e^{R(\beta)t}$$

The quantity $R(\beta)$ is defined as the "amplitude factor." Finally, the solution of the diffusion equation is

$$C - C_0 = A(\beta, 0)e^{R(\beta)t}e^{i\beta x}$$

As $e^{i\beta x} = \cos\beta x + i\sin\beta x$, we take the real part and we have

$$C - C_0 = A(\beta, 0)e^{R(\beta)t}\cos\beta x$$

We recall that $\beta = 2\pi/\lambda$. To determine β or λ, we consider

$$R(\beta) = -\frac{\widetilde{D}}{g''}[g'' + 2Y\eta^2 + 2K\beta^2]\beta^2$$

If $R(\beta) > 0$, the amplitude or the wave will grow. We recall that in spinodal region, the interdiffusion coefficient and g'' are negative, which means that growth proceeds if

$$g'' + 2Y\eta^2 + 2K\beta^2 < 0$$

In the above inequality, g'' is negative, but the other two terms are positive. Then there exists a critical wave number β_C, which satisfies the equation of

$$g'' + 2Y\eta^2 + 2K\beta_C^2 = 0$$

$$\beta_C = \left(-\frac{g'' + 2Y\eta^2}{2K} \right)^{1/2}$$

$$\lambda_C = \frac{2\pi}{\beta_C}$$

For those waves having $\beta < \beta_C$, or $\lambda > \lambda_C$, they will grow. For those waves having $\beta > \beta_C$, or $\lambda < \lambda_C$, they will decay. But the fastest growing wave, β_m, is the one when $R(\beta_m)$ is a maximum; we have

$$\frac{dR(\beta)}{d\beta} = -[g'' + 2Y\eta^2 + 2K\beta^2](2\beta) - 4K\beta^3 = 0$$

So,

$$\beta_m^2 = -\frac{g'' + 2Y\eta^2}{4K} = \frac{\beta_C^2}{2}$$

Hence $\beta_m = \beta_C/\sqrt{2}$, or $\lambda_m = \sqrt{2}\lambda_C$, that is, the fastest growing wave has a wave length of square root of 2 of that of the critical wave.

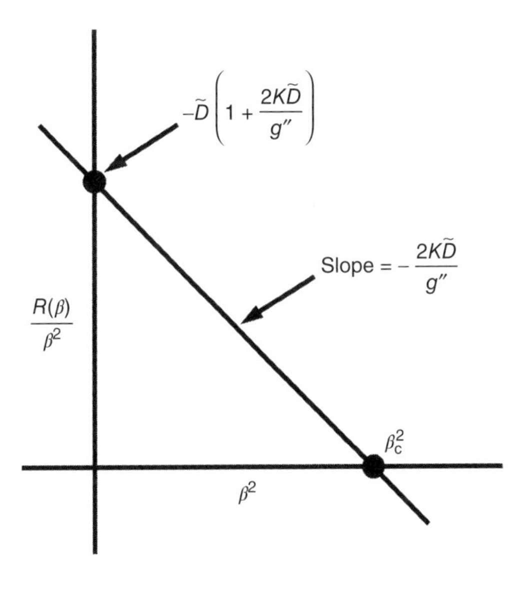

Figure 5.7 A plot of $R(\beta)/\beta^2$ versus β^2 should be a linear curve having a slope of $-2KD/g''$ and having an intercept of $-D(1 + 2Y\eta^2/g'')$.

We note that β_m depends on Y. Because $g'' < 0$, the largest possible value of β_m occurs when Y takes the smallest value it can have in the crystal.

From the equation of $R(\beta)$, we see that a plot of $R(\beta)/\beta^2$ versus β^2 should be a linear curve having a slope of $-2K\widetilde{D}/g''$ and having an intercept of $-\widetilde{D}(1 + 2Y\eta^2/g'')$, as shown in Figure 5.7. Hence, the slope and the intercept can be determined, and in both of them we have the four parameters of g'', $\widetilde{D}$, K, and $2Y\eta^2/g''$. This means that if we know two of them, we will be able to calculate the other two. Typically, we can estimate g'' and $Y\eta^2$, so we can calculate K and $\widetilde{D}$.

REFERENCES

1. Hilliard JEChapter 12 on. Spinodal decomposition. In: Aaronson HI, editor. *Phase Transformations*. Metals Park, Ohio: ASM; 1968. p 497–560.
2. Hillert M. A solid-solution model for inhomogeneous systems. Acta Metall 1961;9:525–535.
3. Cahn JW. On spinodal decomposition. Acta Metall 1961;9:795–801.
4. Cahn JW, Hilliard JE. Free energy of a nonuniform system I: interfacial free energy. J Chem Phys 1958;28:258–267.
5. Cahn JW, Hilliard JE. Free energy of a non-uniform system III: nucleation in a two-component incompressible fluid. J Chem Phys 1959;31:688–699.
6. Cahn JW, Hilliard JE. Spinodal decomposition: a reprise. Acta Metall 1971;19:151–161.
7. Cook HE, Hilliard JE. Effect of gradient energy on diffusion in gold–silver alloys. J Appl Phys 1969;20:2191–2198.
8. Cook HE, deFontaine D, Hilliard JE. A model for diffusion on cubic lattices and its application to the early stages of ordering. Acta Metall 1969;17:765–773.
9. Greer AL, Spaepen F. In: Chang LL, Giessen BC, editors. *Synthetic Modulated Structures*. Vol. 11. Orlando: Academic Press; 1985. p 419–486.

PROBLEMS

5.1. What is the difference between "spinodal decomposition" and "nucleation and growth" in phase transformations?

5.2. What is gradient energy in spinodal decomposition and why is it positive?

5.3. In spinodal decomposition, is there an interface between the two decomposed phases?

5.4. What is uphill diffusion and why can it occur against concentration gradient?

5.5. In spinodal decomposition, why is there a critical period of wave length?

5.6. Estimate roughly the minimal particle size under which spinodal decomposition becomes impossible. (Do not use elastic terms.)

5.7. Man-made superlattices of metals, such as Cu/Ni, Ag/Au, have been prepared to study the effect of gradient energy on interdiffusion. It has been found that the metallic superlattices tend to have poor epitaxial growth between neighboring layers. Because of the very short interdiffusion distance in nanometer range, diffusivity at low temperatures, relative to bulk samples, can be measured. However, it has an observable effect of rapid interdiffusion in the initial stage. Why?

5.8. In spinodal decomposition, we consider an initially homogeneous composition of

$$C(x) = C_0 + \sin\left(\frac{2\pi x}{\lambda}\right)$$

where $C_0 = 0.5$ and λ is the wave length of fluctuation. Show that the following expression of K is positive.

$$K = [1 - 2C(x)]\frac{\partial^2 C(x)}{\partial x^2}$$

5.9. The free energy of a system undergoing spinodal decomposition is given as

$$G_T = A\int_0^l \left[G_0(C) + \eta^2 Y(C - C_0)^2 + K\left(\frac{dC}{dx}\right)^2\right] dx$$

The parameters in the above equation have the same definition as given in the text of the chapter. Use Euler equation to find the chemical potential as well as the diffusion equation.

CHAPTER 6

NUCLEATION EVENTS IN BULK MATERIALS, THIN FILMS, AND NANOWIRES

6.1 INTRODUCTION

In the discussion on phase changes, Figure 5.1 showed a schematic diagram of a miscibility gap and the corresponding Gibbs free energy curve. It is the simplest binary phase diagram, yet it is fundamentally most informative on phase transformations. There are three regions in the phase diagram; the unstable, the metastable, and the stable. They are classified on the basis of how to minimize the free energy in phase changes. In the central part is the unstable region wherein the phase change of spinodal decomposition occurs by means of very small fluctuations in composition, as discussed in Chapter 5. In the metastable region on both sides of the unstable region, the free energy can be lowered only by the precipitation of a new phase that has a

Kinetics in Nanoscale Materials, First Edition. King-Ning Tu and Andriy M. Gusak.

composition very different from that of the matrix. The precipitation has to occur by nucleation because of the very large difference in composition between the nucleus and the matrix. It is metastable because the nucleation has an energy barrier. Without overcoming the barrier, the transformation cannot occur, so the system is metastable even though the tendency to transform exists. Beyond the metastable regions is the stable region, where the free energy is the lowest and no change will occur.

In this Chapter, We discuss the thermodynamics and kinetics of nucleation. Nucleation is required in most physical and chemical phenomena of phase changes. In essence it is a fluctuation of composition and structure in the matrix of a metastable phase [1–5]. If we consider the solidification of a pure phase of water, we have to use the concept of metastability, although there is no composition change, in order to induce the nucleation of ice in water we need the undercooling of water to do so.

Figure 6.1 is a replot of the two straight solid lines of bulk materials presented in Figure 1.1 of Chapter 1. The one having the larger slope represents the free energy of a liquid phase (water) and the other one having the smaller slope represents the free energy of a solid phase (ice). The two curves cross each other at the melting (solidification) point, T_m. But at the melting point, water will not transform to ice or vice versa. This is because nucleation has an energy barrier. A surface or interface between water and ice has to be created in the nucleation for solidification. As surface energy is positive, it becomes the energy barrier of nucleation. To overcome the barrier, an undercooling of water is required in solidification, as indicated by the temperature $T < T_m$. We have $T_m - T = \Delta T$ as the undercooling. The water is metastable in the state of undercooling. Thus, a fundamental concept of nucleation is that it occurs in a metastable state, not in an equilibrium state. Owing to the nucleation energy barrier, stable nucleation cannot occur without overcoming the energy barrier. Because of the nucleation energy barrier, the frequency of nucleation, that is, the number of successful nucleation events per unit time per unit volume will be greatly affected by how large the energy barrier is and, in turn, how large the undercooling is. Nature tends to choose the lowest energy barrier to proceed, and it leads to the consideration of homogeneous nucleation and heterogeneous nucleation. The latter has a much lower energy barrier than the former due to the help of heterogeneity, and in turn a much

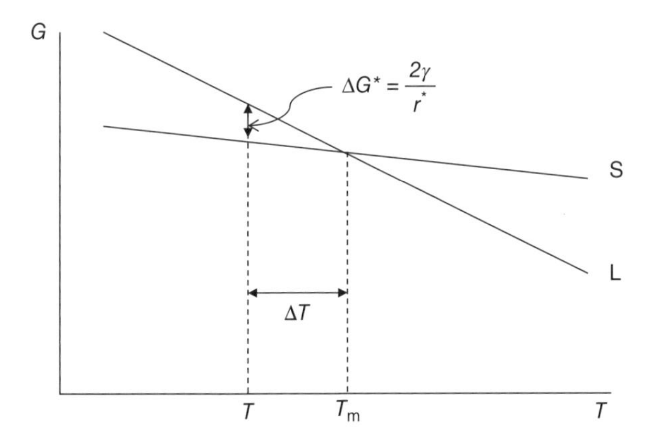

Figure 6.1 A schematic diagram of two straight free energy curves. The one having the larger slope represents the liquid phase (water) and the one having the smaller slope represents the solid phase (ice). r^* is a critical radius of spherical nucleus, ΔG^* is a thermodynamic driving force per unit volume of the nucleus.

higher frequency of occurrence. Hence, another important concept of nucleation is that homogeneous nucleation is rare in real world and nearly all actual nucleation events tend to be heterogeneous.

To promote homogeneous nucleation, we must remove those factors that can enhance heterogeneous nucleation. For example, it is known that tiny droplets of high purity water can show a large degree of undercooling below 0 °C in solidification due to the absence of heterogeneous nucleation. However, water in river will freeze very close to 0 °C because of heterogeneous nucleation. To induce artificial rain, powder is spread in the sky to enhance condensation of rain drops by heterogeneous nucleation. In Chapter 1, we have discussed the lowering of melting point in nanosize spheres, but it is a different event. On the other hand, if we can place pure water within a hollow nanotube, a large undercooling can be detected.

In solid state phase transformations, we have the precipitation of GP (Guenier-Preston) zones. It occurs in Al – 1% Cu alloy by homogeneous nucleation of a single atomic layer of Cu in the matrix of Al because the Cu atomic layer forms a very low energy coherent interface with the Al matrix. So the energy barrier of homogenous nucleation of GP zone is much lower than that of the nucleation of the stable θ-phase of Al_2Cu, which has an incoherent interface with the Al. From the point of view of free energy change in nucleation, the nucleation of θ-phase has much larger gain than the nucleation of GP zone; yet it is the kinetic reason rather than thermodynamic reason that GP zone occurs. The GP zone is metastable, and it will go through two more transitional phase formations, θ'' and θ', before the final stable θ-phase forms.

In cellular precipitation of lamellar Sn from Pb-rich PbSn alloy, it occurs by heterogeneous nucleation in the grain boundaries of PbSn alloy near room temperature. This is because lattice diffusion of Sn in the alloy is negligible near room temperature, so without lattice diffusion, homogeneous nucleation cannot occur. Yet grain boundary diffusion is fast enough in a Pb-rich PbSn alloy at room temperature, so nucleation occurs in grain boundaries and it is a kinetically controlled heterogeneous nucleation event. In Section 6.3, we present a detailed discussion on the event.

6.2 THERMODYNAMICS AND KINETICS OF NUCLEATION

6.2.1 Thermodynamics of Nucleation

We shall calculate the energy gain due to the undercooling as shown in Figure 6.1. Let the Gibbs free energy change per unit volume of the liquid phase be

$$G_V^l = H_V^l - TS_V^l$$

And the solid phase be

$$G_V^S = H_V^S - TS_V^S$$
$$\Delta G_V = G_V^l - G_V^S = \Delta H_V - T\Delta S_V$$

At the melting point, the system is at equilibrium, so

$$\Delta G_V = 0 \quad \Delta H_V = T_m \Delta S_V = L \quad \Delta S_V = \frac{\Delta H_V}{T_m}$$

where L is latent heat. Assume ΔS_V and ΔH_V are independent of the temperature near the melting point,

$$\Delta G_V = \Delta H_V - \frac{T}{T_m}\Delta H_V = \frac{\Delta H_V}{T_m}(T_m - T) = \frac{L}{T_m}\Delta T$$

We shall show that $\Delta G_V = 2\gamma/r^*$, where γ is the surface energy per unit area of the nucleus and r^* is the radius of the critical nucleus. Here we review quickly the classical model of homogeneous nucleation of a spherical nucleus of radius r. The energy of formation is

$$\Delta G = -\frac{4\pi}{3} r^3 \Delta G_V + 4\pi r^2 \gamma \tag{6.1}$$

where ΔG_V is positive and the negative sign is because it is a gain in energy in the phase change, and γ is positive because we need to pay for it. Let us recall Chapter 1, Section 1.2, where we showed that for a nanoparticle, the surface energy of $4\pi r^2 \gamma$ is equal to the Gibbs–Thomson potential energy. Thus, even though a nucleus is a nanoparticle, we do not add Gibbs–Thomson potential energy to its energy of formation. To calculate the energy change in forming the critical nucleus and the radius of critical nucleus, we set $d\Delta G/dr$ to zero and obtain

$$\Delta G^* = \frac{16\pi}{3} \frac{\gamma^3}{(\Delta G_V)^2} \tag{6.2}$$

$$r^* = +\frac{2\gamma}{\Delta G_V} \tag{6.3}$$

In Figure 6.2, a schematic plot of ΔG versus r is shown and ΔG^* and r^* are indicated. As $\Delta G_V = L\Delta T/T_m$, we have

$$\Delta G^* = \frac{16\pi\gamma^3 T_m^2}{3L^2(\Delta T)^2}$$

If we assume the rate of nucleation is given as

$$I^* = K \exp\left(-\frac{\Delta G^*}{kT}\right) = K \exp\left[-\frac{16\pi\gamma^3 T_m^2}{3L^2(\Delta T)^2 kT}\right] \tag{6.4}$$

A plot of ln (I^*) versus $1/T(\Delta T)^2$ should be nearly a straight line. It is a line of abrupt rise, because it has a square dependence on ΔT. This means that physically homogeneous nucleation has a strong dependence on undercooling. No nucleation will take place without sufficient undercooling, yet nucleation occurs rapidly when the undercooling is sufficient. At a low undercooling, no homogeneous nucleation will

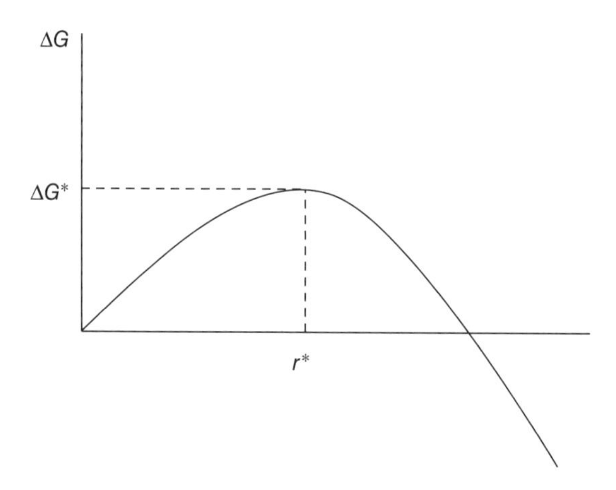

Figure 6.2 A schematic plot of ΔG versus r is shown and ΔG^* and r^* are indicated.

occur, thus heterogeneous nucleation can take over. From the slope of the line, we can calculate γ because the other physical parameters are known.

For heterogeneous nucleation, the basic concept is that an external surface can be wetted by the nucleus, and the effect of wetting is to reduce the volume of the critical nucleus so that the critical energy needed to form the critical nucleus is much reduced, in turn, the rate of heterogeneous nucleation is much enhanced. In Figure 6.3, a cap of the nucleus of β-phase is shown to wet an external surface of S. The wetting angle is θ, so according to Young's equation, we have

$$\gamma_{\alpha S} = \gamma_{\beta S} + \gamma_{\alpha\beta} \cos\theta$$

The volume and area of the cap are, respectively,

$$V = \frac{\pi r^3}{3}(2 - 3\cos\theta + \cos^3\theta)$$

$$A = 2\pi r^2(1 - \cos\theta)$$

The free energy of formation of the heterogeneous cap nucleus is

$$\Delta G = -\frac{\pi r^3}{3}(2 - 3\cos\theta + \cos^3\theta)\Delta G_V + 2\pi r^2(1 - \cos\theta)\gamma_{\alpha\beta} + \pi r^2 \sin^2\theta(\gamma_{\beta S} - \gamma_{\alpha S})$$

By substituting Young's equation into the above equation, we reduce it to

$$\Delta G = \left\{-\frac{4\pi}{3}r^3\Delta G_V + 4\pi r^2\gamma_{\alpha\beta}\right\}f(\theta) \tag{6.5}$$

where $f(\theta) = \frac{1}{4}(2 - 3\cos\theta + \cos^3\theta)$. Repeating the procedure of calculating the critical nucleus in homogeneous nucleation, we obtain

$$\Delta G^*_{\text{hetero}} = \Delta G^*_{\text{homo}} f(\theta) \quad \text{and} \quad r^*_{\text{hetero}} = r^*_{\text{homo}} \tag{6.6}$$

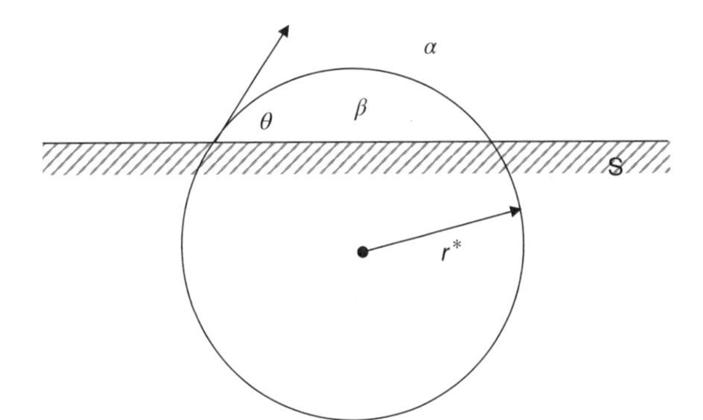

Figure 6.3 In heterogeneous nucleation, a cap of the nucleus of β-phase to wet an external surface of S.

The above finding can be presented graphically by drawing a part of a circle below the cap as shown in Figure 6.3. The radius of the circle (nucleus) is the same for homogeneous and heterogeneous nucleation. Yet in homogeneous nucleation, the critical volume is the full sphere, and in heterogeneous nucleation, the critical volume is the cap. In other words, if we divide the critical volume by atomic volume, a much smaller number of atoms is needed to form the critical nucleus in heterogeneous nucleation, and thus the frequency is much higher. Numerically, $\theta = \pi, f(\theta) = 1$; $\theta = \pi/2, f(\theta) = 1/2$; $\theta = \pi/3, f(\theta) = 3/16$; $\theta = 0, f(\theta) = 0$. The last is a complete wetting.

In solidification, we do not need to consider strain energy in nucleation. On the other hand, in solid state precipitation, we may have to consider strain energy in the formation of nucleus, especially if the nucleus has a coherent or semicoherent interface with the matrix. Nevertheless, comparing to surface energy or chemical bond energy, elastic strain energy is much smaller.

6.2.2 Kinetics of Nucleation

In the previous section, we considered the energy in forming a nucleus, especially the surface energy barrier in nucleation. If we divide the volume of the critical nucleus by atomic volume, we obtain the number of atoms that must come together to form the critical nucleus. We might tend to think that all these atoms, just by fluctuation or diffusion, come together instantaneously to form a nucleus; it seems to be an accidental event. Furthermore, we note that when we calculate the critical radius as shown in Figure 6.2, it is a maximum rather than a minimum in the curve. At the maximum, it is kinetically unstable and it can go forward or backward with equal chance. To become a stable nucleus, it has to gain at the least one or a few more atoms than those in the critical nucleus. These are the kinetic issues of nucleation.

A plausible kinetic picture of nucleation is that there exists a spectrum or a distribution of various sizes of nucleus in the metastable phase. We may call them embryos, and they are clusters of two atoms, three atoms, four atoms, up to the size of critical nucleus. The density of these embryos obeys the Boltzmann's distribution function given as

$$C_n = N_0 \exp\left(-\frac{\Delta G_n}{kT}\right) \tag{6.7}$$

where C_n is the density of embryos of n atoms per unit volume, N_0 is the total number of atoms per unit volume, and ΔG_n is the formation energy of an embryo of n atoms.

We consider the distribution of these embryos in a unit volume and unit time. Those that grow beyond the critical size control the rate of nucleation, I_n^*, which is the number of nuclei that have grown beyond the critical size per unit volume per unit time. Often, in a schematic diagram depicting the distribution of embryos, we assume that the distribution goes to zero just beyond the critical nucleus, which means that those larger than the critical nucleus are removed from the distribution.

Below, we consider an embryo containing n number of β atoms. The rate of change of this embryo depends on the rate of gaining a β atom minus the rate of losing a β atom. We express them as

$$n \xrightarrow{\beta_n} n+1 \quad n \xleftarrow[\alpha_{n+1}]{} n+1$$

where β_n means we take one atom from the solution and add it to embryo n to become embryo $n+1$; and α_{n+1} means we remove one atom from embryo $n+1$ and return the atom to the solution so the embryo $n+1$ becomes embryo n. We note that the process does not mean the exchange of one atom between a specific embryo n and embryo $n+1$. In other words, β_n and α_{n+1} are frequencies respectively of gaining and losing atoms among all embryos of n and $n+1$, and they are microscopic reversibility parameters; yet these reversible frequencies depends on the size of the embryo. In unit volume, there are many "n" and many "$n+1$" embryos. Let C_n be the number of embryos of n atoms/cm^3, and C_{n+1} be the number of embryos of $(n+1)$ atoms/cm^3. Therefore, we have

$$C_n \xrightarrow{\beta_n} C_{n+1} \quad C_n \xleftarrow[\alpha_{n+1}]{} C_{n+1}$$

The net change in n embryos is

$$I_n = \beta_n C_n - \alpha_{n+1} C_{n+1}$$

We note that the unit of I_n is the number of n embryos/cm^3-s, and it means the flux of embryos (not atoms) going through the size space of embryos. Actually we should have considered also the flux between embryo "$n-1$" and "n," but conceptually what we presented in the above is sufficient for our analysis. Now assume detailed balancing, at equilibrium we have $I_n = 0$. Hence,

$$\beta_n C_n^0 = \alpha_{n+1} C_{n+1}^0$$

where C_n^0, C_{n+1}^0 are equilibrium concentrations of embryos.

$$I_n = \beta_n \left[C_n - C_{n+1} \frac{C_n^0}{C_{n+1}^0} \right] = -\beta_n C_n^0 \left[\frac{C_{n+1}}{C_{n+1}^0} - \frac{C_n}{C_n^0} \right] \tag{6.8}$$

In differential form, we can express the following equation,

$$\frac{\dfrac{C_{n+1}}{C_{n+1}^0} - \dfrac{C_n}{C_n^0}}{\Delta n} = \frac{\partial \left(\dfrac{C_n}{C_n^0} \right)}{\partial n}$$

where we can take $\Delta n = 1$ because we are adding or losing one atom at a time. Hence,

$$I_n = -\beta_n C_n^0 \frac{\partial \left(\dfrac{C_n}{C_n^0} \right)}{\partial n} \tag{6.9}$$

Below we introduce the continuity equation in size space. First, we note that the concentration of embryos, C_n, will change with time if the time rate change of I_n to I_{n+1} is not equal to the time rate change of I_{n-1} to I_n. In other words, the time rate change of C_n is equal to how many I_{n-1} goes to I_n minus how many I_n goes to I_{n+1}, and this is the basic concept in a continuous change of a continuity equation. We recall that the unit of I is equal to the unit of βC, where C has the unit of number of embryos per unit volume per unit time and β frequency has the unit of 1/s. If we consider an interval of Δn in size space, as shown in Figure 6.4, specifically the middle interval of /n/ in … n−2/n−1/n/n+1/n+2 … , the net change of flux of embryos going in and out of the interval is the divergence in the interval. Therefore, we can use the continuity equation to express the change as

$$\frac{\partial C_n}{\partial t} = -\frac{\Delta I_n}{\Delta n} \tag{6.10}$$

We note that the unit of the above equation is number of embryo/cm^3-s, because Δn has no unit. Substituting I_n of Eq (6.9) into the above equation, we obtain the diffusion equation of embryos in the size space.

$$\frac{\partial C_n}{\partial t} = -\frac{\partial}{\partial n} \left[-\beta_n C_n^0 \frac{\partial}{\partial n} \left(\frac{C_n}{C_n^0} \right) \right] \tag{6.11}$$

The simplest solution of the above equation is the steady state solution below for the steady state nucleation rate. By taking $\frac{\partial C_n}{\partial t} = 0$, we have

$$-\beta_n C_n^0 \frac{\partial}{\partial n} \left(\frac{C_n}{C_n^0} \right) = \text{const} = I_{n^*}^{\text{S}} \tag{6.12}$$

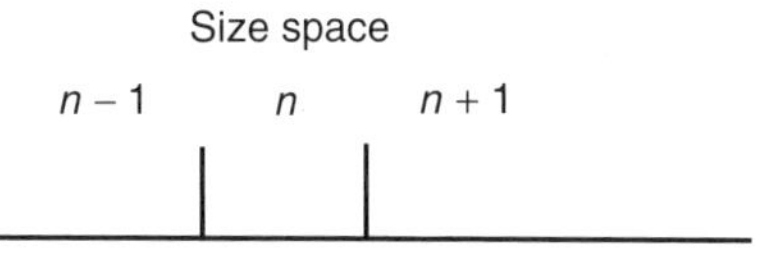

Figure 6.4 Change of embryos in middle interval in … $n-1/n/n+1$ … and the net change of flux of embryos going in and out of the interval.

Rearranging, we have

$$\int \frac{dn}{\beta_n C_n^0} = \frac{-1}{I_{n^*}^{\mathrm{S}}} \int d\left(\frac{C_n}{C_n^0}\right)$$

To consider the range of integration in the above equation, we note that the smallest n is unity, so at $n = 1$, $C_1 = C_1{}^0$, then at $n = \infty$, $C_\infty = 0$. In other words, $C_n/C_n{}^0 \to 1$ as $n \to 1$, and $C_n/C_n{}^0 \to 0$ as $n \to \infty$. So, we have

$$\int_1^\infty \frac{dn}{\beta_n C_n^0} = \frac{-1}{I_{n^*}^{\mathrm{S}}} \int_1^0 d\left(\frac{C_n}{C_n^0}\right) = \frac{-1}{I_{n^*}^{\mathrm{S}}}[0-1] = \frac{1}{I_{n^*}^{\mathrm{S}}} \tag{6.13}$$

$$I_{n^*}^{\mathrm{S}} = \left[\int_1^\infty \frac{dn}{\beta_n C_n^0}\right]^{-1} = \frac{1}{\displaystyle\int_1^\infty \frac{dn}{\beta_n C_n^0}} \tag{6.14}$$

In the above equation of steady state nucleation, we recall that the meaning of $C_n{}^0$ is the equilibrium concentration of embryo having n atoms. If we take N_0 to be the total number of atoms per unit volume, we have

$$C_n^0 = N_0 \exp\left(-\frac{\Delta G_n}{kT}\right) \tag{6.15}$$

where ΔG_n is the formation energy of an embryo of n atoms. In the expression of

$$\frac{1}{C_n^0} \propto \exp\left(\frac{\Delta G_n}{kT}\right)$$

it has the largest value when ΔG_n is maximum or when it is of the critical size. As $C_n{}^0$ is at equilibrium, there will be no nucleation event at the critical nucleation. We have to consider a nucleus having one or a few more atoms than the critical nucleus and in turn we need to know the formation energy of ΔG_{n+1} or ΔG_{n+2}. We can do so by taking a Taylor expansion at n that is slightly larger than n^*, or is slightly off the equilibrium value of the critical size. Also we can do so because we know the critical nucleus and ΔG_n^*. It is worth mentioning that this a beautiful case of the application of Taylor expansion.

We have the Taylor expansion shown below.

$$\Delta G_n = \Delta G_n^* + (n-n^*)\frac{\partial \Delta G_n}{\partial n}\Big|_{n=n^*} + \frac{1}{2}(n-n^*)^2 \frac{\partial^2 \Delta G_n}{\partial n^2}\Big|_{n=n^*} + \cdots$$

The second term in the expansion is zero because ΔG_n^* is a maximum.

$$\Delta G_n = \Delta G_n^* + \frac{1}{2}(n-n^*)^2 \frac{\partial^2 \Delta G_n}{\partial n^2}\Big|_{n=n^*}$$

Below, let us calculate the following term before we return to the above expansion.

$$\int_1^\infty \frac{dn}{\beta_n C_n^0}$$

First, we can assume β_n to be a constant near $n = n^*$ so that we can ignore it for a while, but we will remember to put it back later. Second, we take

$$\int_1^\infty \exp\left(\frac{\Delta G_n}{kT}\right) dn = \exp\left(\frac{\Delta G_n^*}{kT}\right) \cdot \int_1^\infty \exp\left(\frac{\Delta G_n - \Delta G_n^*}{kT}\right) dn$$

Therefore, we have

$$\int_1^\infty \frac{dn}{C_n^0} = \frac{1}{N_0}\int_1^\infty \exp\left(\frac{\Delta G_n}{kT}\right) dn$$
$$= \frac{\exp(\Delta G_n^*/kT)}{N_0}\int_1^\infty \exp\frac{(n-n^*)^2}{2kT}\left(\frac{\partial^2 \Delta G_n}{\partial n^2}\right)_{n^*} dn$$

The last integration is

$$\int_1^\infty \exp\frac{(n-n^*)^2}{2kT}\left(\frac{\partial^2 \Delta G_n}{\partial n^2}\right)_{n^*} dn = \int_0^\infty \exp[-a^2(n-n^*)^2]d(n-n^*) = \frac{\sqrt{\pi}}{2a}$$

where

$$a^2 = \frac{-1}{2kT}\left(\frac{\partial^2 \Delta G_n}{\partial n^2}\right)_{n^*}$$

and

$$\frac{\sqrt{\pi}}{2a} = \frac{1}{\sqrt{\frac{-1}{2\pi kT}\left(\frac{\partial^2 \Delta G_n}{\partial n^2}\right)_{n^*}}}$$

Now we can put every thing back, and we obtain the steady state nucleation rate as

$$I_{n^*}^{S} = \beta_{n^*} C_{n^*}^0 Z = \beta_{n^*}\left[N_0 \exp -\left(\frac{\Delta G_n^*}{kT}\right)\right]\left[\frac{-1}{2\pi kT}\left(\frac{\partial^2 \Delta G_n}{\partial n^2}\right)_{n^*}\right]^{1/2} \tag{6.16}$$

where β_{n^*} is the microscopic reversible jump frequency of adding an atom to the critical embryo of n^* atoms. $C_{n^*}^0$ is the equilibrium concentration of embryos of critical size and it is given by the first square bracket in the last equation, and Z is the Zeldovich factor of steady state nucleation, which is greater than 0. And

$$Z = \left[\frac{-1}{2\pi kT}\left(\frac{\partial^2 \Delta G_n}{\partial n^2}\right)_{n^*}\right]^{1/2} \tag{6.17}$$

Its physical meaning is that the nucleation process should be a smooth, gradual, and steady state transition rather than an abrupt transition. The steady state behavior is described by the Zeldovich factor. In a later section on epitaxial growth of silicide in nanowire of Si, we calculate Z for the homogeneous nucleation.

6.3 HETEROGENEOUS NUCLEATION IN GRAIN BOUNDARIES OF BULK MATERIALS

In phase transformations, atomic diffusion and the interface between the transformed and the untransformed region are the two most important kinetic parameters. It is essential not only to know what kind of atomic mechanism of diffusion is dominating the transformation process, but also to know what type of interface is involved in the transformation. This principle applies to spinodal decomposition as well as to nucleation and growth processes. For example, in cellular precipitation in a bulk material, nucleation occurs in grain boundaries and is dominated by grain boundary diffusion, not by lattice diffusion. The cellular growth is also controlled by atomic diffusion along a moving grain boundary. An actual example of cellular precipitation is the precipitation of lamellar β-Sn from a Pb-rich $Pb_{95}Sn_5$ alloy at room temperature. Thus, a basic question in cellular precipitation is that if we have a stationary grain boundary, how can the grain boundary initiate its motion for the precipitation? Indeed, after we anneal an alloy of Pb-rich PbSn sample and cool it down to room temperature to induce precipitation, all grain boundaries are stationary in the as-cooled sample. How can a stationary grain boundary become a moving grain boundary, especially if we consider a symmetrical tilt-type stationary grain boundary? The answer is in heterogeneous nucleation, and it is the central topic of this section.

Cellular or discontinuous precipitation occurs by nucleation and growth of cells from grain boundaries. The cells are typically hemispherical and consist of rather evenly spaced lamellae of precipitate (β phase) interspersed in the solute-depleted solid solution matrix (α phase). The cell boundary is the interface between the transformed and the untransformed regions. The precipitation proceeds as the cell boundary moves through the untransformed matrix leaving behind the transformed lamellar structure.

An example of cellular precipitation is the precipitation of Sn lamellae from a Pb-rich PbSn alloy near room temperature, for example, in Pb – 5 at.% Sn alloy. The binary Pb–Sn system has a eutectic phase diagram, with a eutectic temperature of 183 °C. Below the eutectic temperature, two primary phases of Pb(Sn) and Sn(Pb) coexist. Around room temperature, the Sn(Pb) phase has negligible solubility of Pb so we can regard it as more or less a pure Sn phase. More importantly, around room temperature, the lattice diffusion of Sn in the Pb(Sn) phase is negligible, nevertheless the grain boundary diffusion of Sn in the Pb(Sn) alloy is sufficiently fast for precipitation. However, the negligible lattice diffusion prevents Sn atoms in the matrix of Pb(Sn) alloy to diffuse to a grain boundary; in other words, the Sn atoms can be regarded as frozen in the lattice. While they may have sufficiently fast grain boundary diffusion, they cannot diffuse to a grain boundary, rather a grain boundary has to move to reach them. When a moving grain boundary meets a Sn atom, the latter

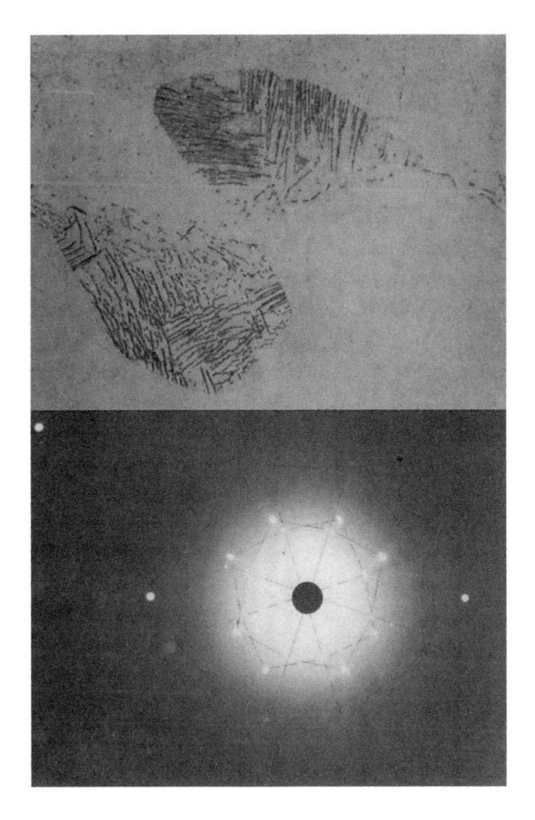

Figure 6.5 Optical micrograph showing the morphology of two cells in the 37-degree tilt-type grain boundary. *Source*: Reproduced with permission from [4b].

becomes activated and can diffuse along the former. Thus, the process of precipitation of β lamellae requires a moving grain boundary to sweep through the $\alpha(\beta)$ matrix.

Experimentally, [100] tilt-type bicrystals of Pb – 5.5 atomic % Sn alloy with the tilt angle of 37°, 23°, 19°, and 7° were grown by the horizontal Bridgeman method. The bicrystals had a cross-section of 1/2 in. × 5/8 in. and length of 6 in. Samples of 3/4 in. long were cut normal to the [100] tilt axis, and kept at room temperature for cellular precipitation. The cross-section was polished and etched for optical microscopic observation of precipitation from the tilt-type grain boundaries. The morphology of two cells, one grew to the top and the other grew to the bottom, from a 37-degree tilt-type grain boundary is shown in Figure 6.5. The X-ray Laue diffraction pattern was used to confirm the tilt angle of the bicrystal with [100] tilt axis. The habit plane (epitaxial interface) of the Sn lamellae is the {111} plane of Pb, and in each cell there are two sets of Sn lamellae normal to each other when they are viewed along the [100] direction.

This experimental study was performed in 1965. At that time, the epitaxial growth of Si on Si or epitaxial growth on compound semiconductors was not popular yet, hence habit plane rather than epitaxial growth was used to express the epitaxial orientation relationship between Sn and Pb. The (001) plane of body-centered tetragonal β-Sn lamella grows on the (111) plane of face-centered cubic Pb.

Figure 6.6 is a set of successive stages of cellular precipitation starting from a stationary bicrystal grain boundary and advancing to a moving grain boundary. The

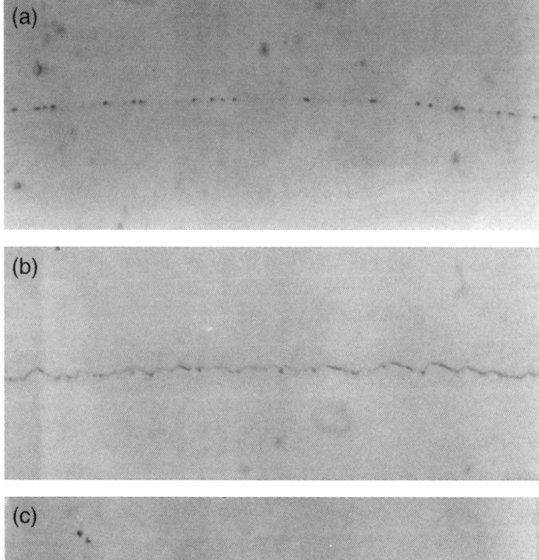

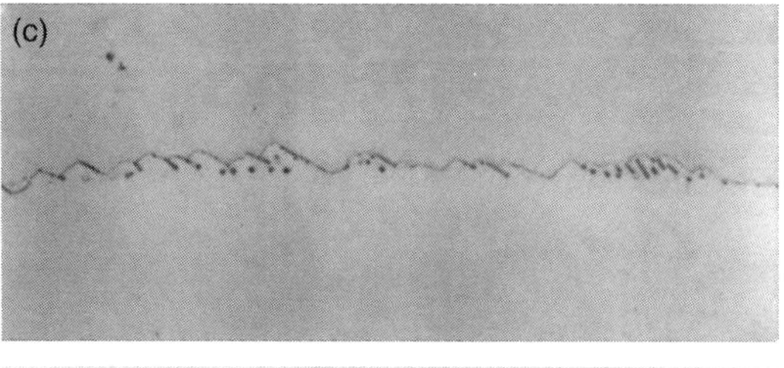

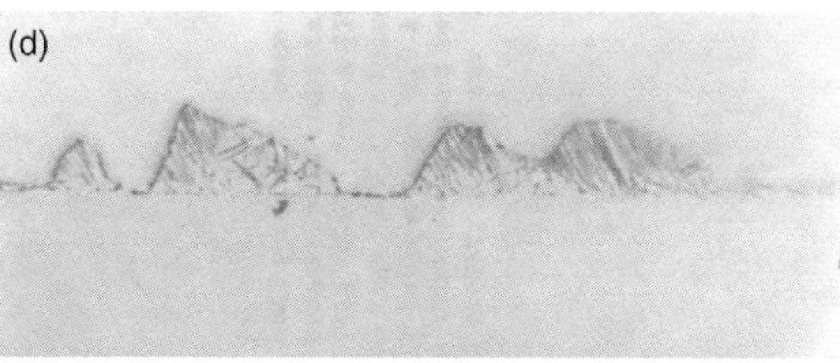

Figure 6.6 (a) Appearance of tin particles along the 29-degree bicrystal grain boundary after 30 min of quenching from 275 °C to room temperature. ×260. (b) Some particles growing to macroscopic size. ×260. (c) Separated lamellae lie along the 23-degree tilt-type grain boundary. (d) Four well-developed cells. These cells grow with the grain boundary moving into the upper grain. But the cell matrix has the same orientation with the lower grain. ×200.

question is if we assume that the tilt-type grain boundary is symmetrical and static in the beginning, how can it move either to the top or to the bottom?

Figure 6.6a shows the appearance of nano particles of Sn along a 29-degree tilt-type bicrystal grain boundary after 30 min of quenching from 275 °C to room temperature, magnification at ×260. In Figure 6.6b, some particles grew to macroscopic size, ×260. In Figure 6.6c, separated lamellae lie along the 29-degree tilt-type grain boundary. The lamellae are parallel to a (111) set of habit plane of the lower grain, ×300. In Figure 6.6d, it shows four well-developed cells. These cells grow with the grain boundary moving into the upper grain. But the matrix within the cell has the same crystallographic orientation with the lower grain, ×200.

6.3.1 Morphology of Grain Boundary Precipitates

To consider grain boundary heterogeneous nucleation, we depict in Figure 6.7 three types of two-dimensional nucleus in a static and symmetrical tilt-type grain boundary.

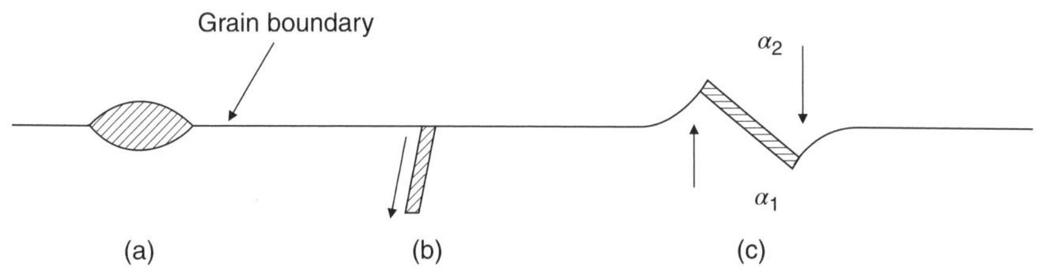

Figure 6.7 Three types of two-dimensional nucleus in a static and symmetrical tilt-type grain boundary. (a) The most common type of nucleus is the lens type, which is symmetrical and the two triple points at both ends are at equilibrium so the three surface tensions representing the surface energies are balanced at the triple point. (b) A platelet-type nucleus that grows into the lower grain, but with one end attached to the grain boundary. After the nucleation, the grain boundary remains stationary. (c) It is a platelet-type nucleus, but its nucleation has caused a deformation of the grain boundary, that is, by plucking or twisting the grain boundary locally. Importantly, the grain boundary has been moved locally by the plucking.

The most common type of nucleus is the lens type depicted in Figure 6.7a, which has a symmetrical shape, and the two triple points at both ends are at equilibrium so the three surface tensions representing the surface energies are balanced at the triple point. The key implication is that after the nucleation, the grain boundary remains static. Without lattice diffusion, there will be no Sn atoms in the matrix to be able to diffuse to the grain boundary to take part in the grain boundary nucleation and growth. We assume that there is a supersaturation of Sn in the grain boundary, which is sufficient for grain boundary nucleation. In Figure 6.7b, it depicts a platelet-type nucleus that has the morphology that it grows into the lower grain, but with one end attached to the grain boundary. Again, after the nucleation, the grain boundary remains stationary. In Figure 6.7c, it depicts a special type of nucleus. It is a platelet, but its nucleation has caused a deformation of the grain boundary, that is, by plucking or twisting the grain boundary locally. Importantly, the grain boundary has moved locally by the plucking.

6.3.2 Introducing an Epitaxial Interface to Heterogeneous Nucleation

Let us introduce the concept of habit plane or epitaxial interface in heterogeneous nucleation in order to explain the deformation as depicted in Figure 6.7c. In Figure 6.8a, we assume that the lens nucleus has a flat epitaxial interface with one of the {111} planes in the lower grain at the point A. The epitaxy changes the equilibrium condition at the left-hand side triple point. If the right-hand side triple point is at equilibrium, the left-hand side triple point is not. In Figure 6.8b, the three surface tensions at equilibrium at the right-hand side triple is depicted by the three solid arrows to form a closed triangle of abc, where $\sigma_{\alpha\alpha}$ (ab) is the surface tension of the grain boundary, and σ_1 (bc) and σ_2 (ac) are the surface tensions of the interface between the nucleus and grain 1 (lower) and grain 2 (upper), respectively. In a symmetrical tilt-type grain boundary, we can assume σ_1 and σ_2 are the same. Then if we add the three surface tensions of the left-hand side triple point to Figure 6.8b,

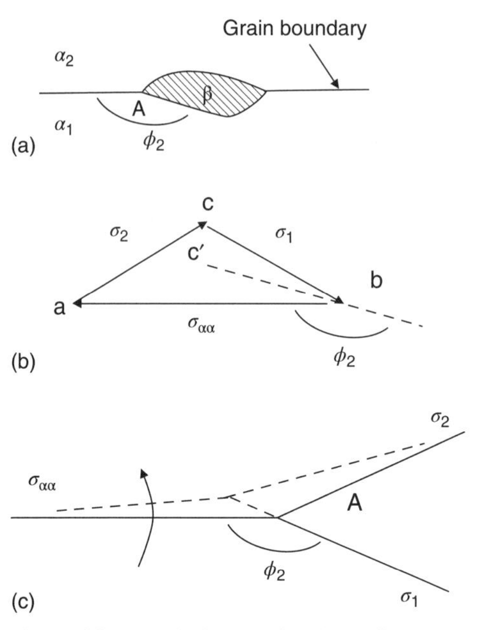

Figure 6.8 (a) The lens nucleus has a flat epitaxial interface with one of the {111} planes in the lower grain at the point A. The epitaxy changes the equilibrium at the left-hand side triple point. If the right-hand side triple point is at equilibrium, the left-hand side triple point is not. (b) The three surface tensions at equilibrium at the right-hand side triple are depicted by the three solid arrows to form a closed triangle of abc. If we add the three surface tensions of the left-hand side triple point to (b), we find that abc′ will not form a closed triangle. (c) In order to reach equilibrium to lower the free energy, the configuration of the left-hand side triple point will change from the broken curves to the solid curves, so a torque exerts a force to rotate the grain boundary. It starts the motion of the static grain boundary. The result of the torque is to produce a deformation of the local boundary. The deformation can be regarded as a pluck of the grain boundary locally, and it has moved the grain boundary. *Source*: Reproduced with permission from [4b].

where bc′ represents the surface tension of the epitaxial interface, we find that abc′ will not form a closed triangle. Consequently, in order to reach equilibrium to lower the free energy, the configuration of the left-hand side triple point will change as depicted in Figure 6.8c, from the broken curves to the solid curves, so a torque exerts a force to rotate the grain boundary. It starts the motion of the static grain boundary or the deformation.

By preserving the epitaxial interface, we follow the derivation of grain boundary torque and obtain the equation below from Figure 6.8c.

$$\sigma_1 - \sigma_3 \cos\varphi_3 - \sigma_{\alpha\alpha}\cos\varphi_2 + \frac{\partial\sigma_3}{\partial\varphi}\cos\varphi_3 + \frac{\partial\sigma_{\alpha\alpha}}{\partial\varphi}\cos\varphi_2 = 0 \tag{6.18}$$

The torque $(\partial\sigma_{\alpha\alpha}/\partial\varphi)$ exerts a force to initiate the motion of the grain boundary. As torque acts as a pair of forces, we should apply a similar argument to the other end of the grain boundary.

Physically, it is because of the anisotropy of the solid surfaces that a habit plane (epitaxial interface) of low energy exists between the α grain and the β nucleus. When a heterogeneous nucleus of β phase forms along an σ/σ grain boundary and favors a habit plane, in order to have the habit plane, the nucleus will exert a torque on the local boundary, and vice versa. The result of the torque is to produce a deformation of the local boundary, as illustrated in Figure 6.7c. The deformation can be regarded as a plucking of the grain boundary locally, and as a consequence, it has moved the grain boundary.

In the following, we consider the free energy of formation of such a nucleus including the deformation. We consider the formation of a disc with radius r and thickness t and it makes an inclination angle of φ to the original plane of the grain boundary, as depicted in Figure 6.7c.

$$\Delta G = -\pi r^2 t \Delta G_{\mathrm{V}} + \pi r^2 (\sigma_1 + \sigma_2 - \sigma_{\alpha\alpha}) + \pi r^2 \left(1 - \frac{1 - \sin\varphi}{\cos\varphi}\right) \sigma_{\alpha\alpha} + 2\pi r t \sigma_3$$
$$= -\pi r^2 t \Delta G_{\mathrm{V}} + \pi r^2 \left(\sigma_1 + \sigma_2 - \frac{1 - \sin\varphi}{\cos\varphi} \sigma_{\alpha\alpha}\right) + 2\pi r t \sigma_3 \tag{6.19}$$

where σ_1, σ_2, σ_3 are the interfacial free energy per unit area of the disc face that lies on the habit lane of grain 1, the opposite face, and the disc edge, respectively. Φ is the angle of inclination of the disc to the original plane of the boundary, and $\sigma_{\alpha\alpha}$ is the grain boundary energy per unit area. ΔG_{V} is the bulk free energy of the precipitate per unit volume. The dimensions of the critical nucleus are

$$r^* = +\frac{2\sigma_3}{\Delta G_{\mathrm{V}}} \tag{6.20}$$

$$t^* = +\frac{2f(\varphi)}{\Delta G_{\mathrm{V}}} \tag{6.21}$$

And the free energy of formation of the critical nucleus is

$$\Delta G^* = \frac{4\pi \sigma_3^2 f(\varphi)}{(\Delta G_{\mathrm{V}})^2} \tag{6.22}$$

where

$$f(\varphi) = \sigma_1 + \sigma_2 - \frac{1 - \sin\varphi}{\cos\varphi} \sigma_{\alpha\alpha}$$

In the above modeling, the formation of the nucleus is favored by the smallest value of σ_1 relative to σ_2, but it is hindered by the necessity to deform (or to increase the area) the grain boundary through the angle φ in order to match the disc face plane with the {111} plane of the α_1 grain. The nucleus will prefer to form with one of its faces lying in the grain boundary provided that the following condition is fulfilled;

$$\sigma_1 + \sigma_2 - \frac{1 - \sin\varphi}{\cos\varphi} \sigma_{\alpha\alpha} < 2\sigma_1$$

This condition is favored by a relatively small deformation angle and relatively large value of the grain boundary energy. When the grain boundary migrates to a position in parallel to the habit plane or a (111) plane, the nucleation of a new disc type nucleus will be in favor, which is discussed later.

In the above analysis, a very simple model of the deformation of grain boundary is given and no strain energy is taken into account. Comparing to the grain boundary energy per atom, the strain energy per atom is much smaller. Besides, while the platelet precipitate has one epitaxial face, the other face is nonepitaxial, so it will reduce the strain energy greatly. Furthermore, the platelet is very thin.

6.3.3 Replacive Mechanism of a Grain Boundary

After the nucleation, we need a mechanism to enable the nucleus to become embedded into the matrix with one of its tips still attached to the grain boundary so that it can grow with the moving grain boundary. The mechanism is the replacive motion of the grain boundary, which is discussed below. Figure 6.9a depicts a platelet along the deformed grain boundary after nucleation and having a certain amount of growth. We assume that its lower face is lying on a habit plane of the lower grain α_1 and forms an epitaxial interface having a low energy of σ_1. But the opposite face of the platelet does not lie on a habit plane of the upper grain α_2. Let σ_2 represent the interfacial energy between α_2 and the β platelet, then $\sigma_2 > \sigma_1$. The existence of this difference between the interfacial energies of the opposite faces of the platelet would induce a tendency to replace the high energy interface of σ_2 by the lower energy interface of σ_1. The replacement is accomplished first by the breakaway of the boundary at point B and followed by the grain boundary motion illustrated in Figure 6.9b, which carries the boundary to the other end of point A of the platelet. This motion of grain boundary from B to A embeds the β platelet into grain α_1 because this is the mechanism whereby a grain boundary platelet is transformed to a lamellar platelet in the matrix.

During the course of the replacive motion, a section of the grain boundary is pulled into a position such that it is almost parallel with the habit plane of the first platelet. Thus a second disc in the epitaxial orientation can nucleate on this section of the boundary, see Figure 6.9c, without appreciable grain boundary deformation.

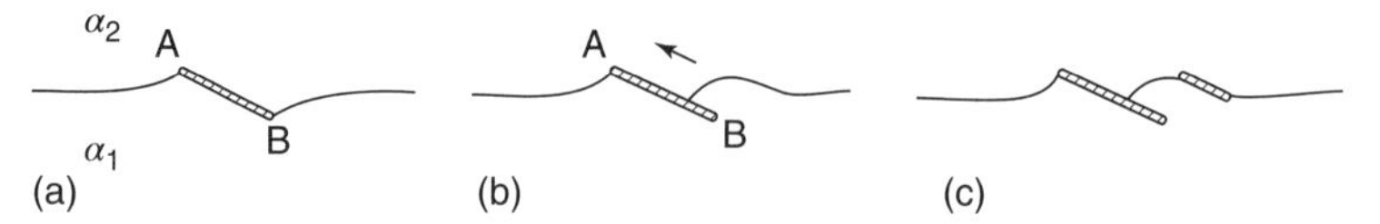

Figure 6.9 (a) A platelet along the deformed grain boundary after nucleation and having some considerable growth. We assume that if lower face is lying on a habit plane of the lower grain α_1 and forms an epitaxial interface having a low energy of σ_1. But the opposite face of the platelet does not lie on a habit plane of the upper grain α_2. Let σ_2 represent the interfacial energy between α_2 and the β platelet; then $\sigma_2 > \sigma_1$. (b) The replacement is accomplished first by the breakaway of the boundary at point B and followed by the grain boundary motion, which carries the boundary to the other end of point A of the platelet. (c) A second disc in the epitaxial orientation can nucleate on the section of the boundary without appreciable grain boundary deformation. *Source*: Reproduced with permission from [4b].

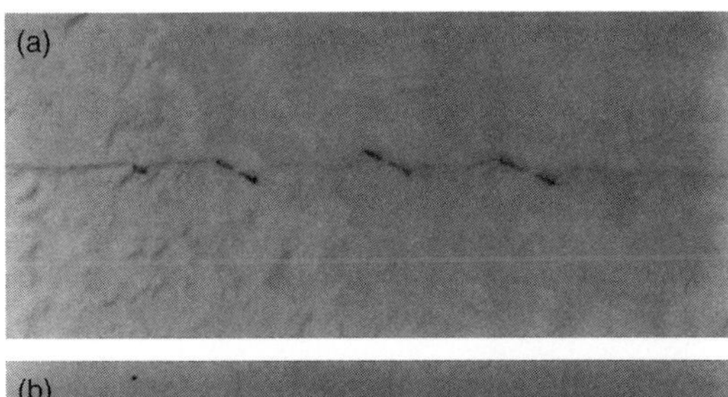

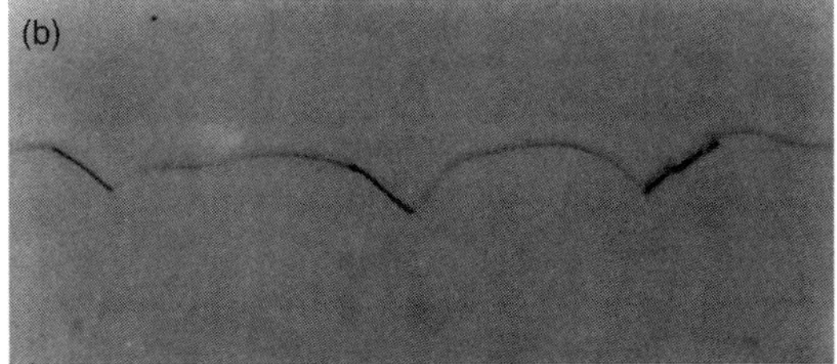

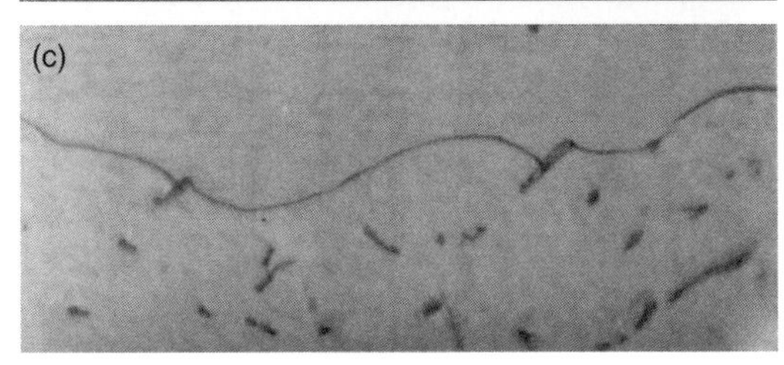

Figure 6.10 The optical microgrphs in (a) and (b) show the deformation of grain boundary due to nucleation and growth of platelet on the habit plane of the lower grain. (c) Replacive motion of grain boundary. ×750.

In other words, the term $\{1 - [(1 - \sin\varphi)/\cos\varphi]\}$ in the previous equation is almost zero. In this way, the growth of one lamella stimulates the nucleation and growth of another lamella on one side of it. This sidewise process can be repeated until a colony of lamellae or a cell is formed, because the boundary tends to be, after each replacive motion, parallel with the habit plane of the first lamella.

What has been described in the above is a repeating event of heterogeneous nucleation. It implies that the Sn lamellae are independent platelets, so they are not interconnected as in a branch. Indeed, when a piece of the PbSn alloy, after cellular precipitation, is dissolved in acid to remove the Pb, a large number of Sn platelets can be collected and they can be examined in transmission electron microscopy (TEM) because they are very thin.

In Figure 6.10a, b, optical micrographs of the deformation of grain boundary due to nucleation and growth of platelet on the habit plane of the lower grain are shown. Figure 6.10c shows the replacive motion of grain boundary, magnification at ×750. In Figure 6.11a–c, optical micrographs of the successive growth of a set of lamellae to form a cell in a [100] oriented 29-degree tilt-type grain boundary are shown. The elapsed time is 2.2 h between (a) and (b) and 2.5 h between (b) and (c), ×270.

The breakaway at point B is the important step to start the replacive motion. In the two-dimensional diagram as depicted in Figure 6.9, point B seems to be a configuration of local equilibrium, so the breakaway may require some extra driving

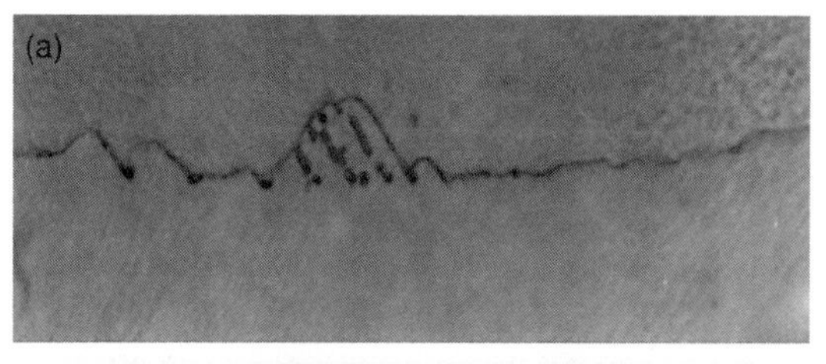

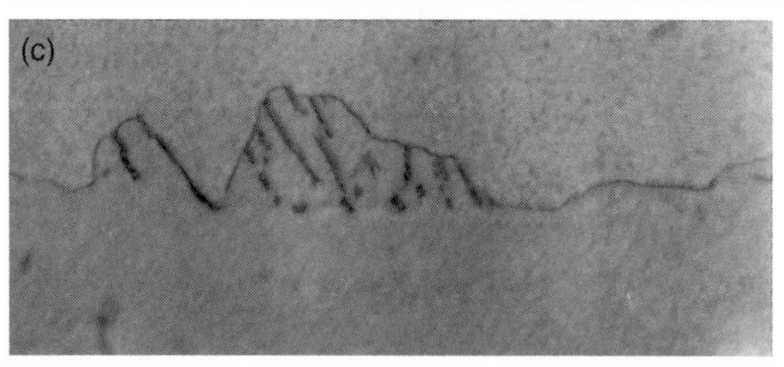

Figure 6.11 (a–c) The successive growth of a set of lamellae to form a cell in a 29-degree [100] tilt-type grain boundary. The elapsed time is 2.2 h between (a) and (b) and 2.5 h between (b) and (c). ×270. Reproduced with kind permission from [4b].

force to upset the metastable state. Yet in real case, the platelet is a disc, so the local deformation of the grain boundary at point B is the largest. Because the replacive motion reduces the deformation, it tends to start at point B.

6.4 NO HOMOGENEOUS NUCLEATION IN EPITAXIAL GROWTH OF Si THIN FILM ON Si WAFER

In Section 6.2, we have discussed that the rate of homogeneous nucleation is much less than that of heterogeneous nucleation. Events of nucleation in real world tend to be heterogeneous. In this section, we consider the homogeneous nucleation of a surface disc on a flat surface of Si, and we show that energetically it is indeed difficult for the homogeneous nucleation event to happen. In microelectronic device applications, epitaxial growth of Si thin film on Si wafer by vapor deposition is required, for example, the growth of n+-Si thin film on n-type Si wafer, and it relies on step-mediated growth on a 7°-cut Si surface so that the steps are present for epitaxial growth without the need of nucleation.

We consider first the vapor pressure above a small cluster of atoms on a surface as any surface kinetic process must be linked to the vapor pressure on the surface. Figure 6.12a, b depict respectively the nucleation of a circular disc on a flat surface and the schematic cross-sectional view of a single atomic layer of the disc on the surface. We note that there is no interface between the surface disc and the substrate

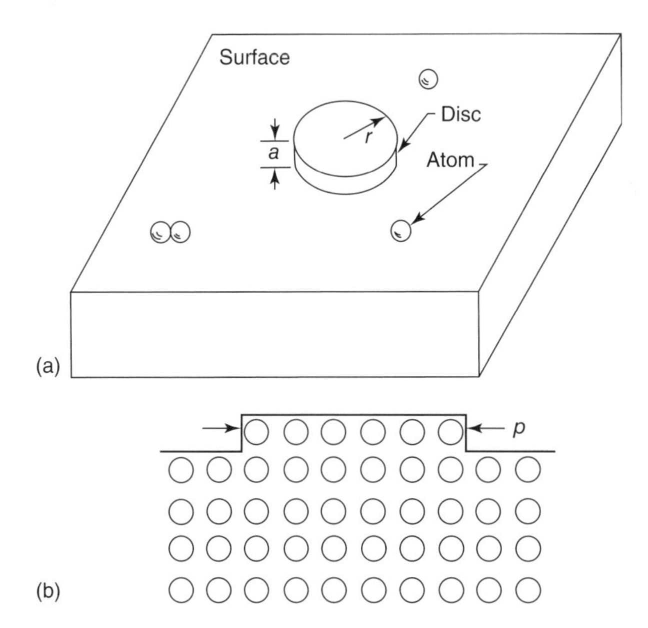

Figure 6.12 (a, b) The nucleation of a circular disc on a flat surface and the schematic cross-sectional view of a single atomic layer of disc on the surface. *Source*: Reproduced with kind permission from [7].

as shown in Figure 6.12b. In homogeneous nucleation in a bulk material, the nucleus is a sphere, but the homogeneous nucleus in two-dimension is a single atomic layer of a circular disc, and the heterogeneous nucleus in two-dimension will be a slice of the circular disc. The energy barrier of the nucleation is due to the formation of the circumference of the disc. Before we consider the energies involved in the nucleation, we must first consider the change in equilibrium condition of the vapor on the surface when a disc exists.

The equilibrium pressure on a planar surface is p_0. At equilibrium, the flux condensing on or desorbing from the surface are equal. The condensing flux can be given as

$$J_c = Cv = nv_a$$

where $C = n(= p_0/kT)$ is the density of atom in the vapor, or the number of atoms per unit volume of the vapor, and v_a is root mean square velocity of the atoms in the vapor. For desorbing flux, we express

$$J_d = J_0 = N_0 \nu_{des} = N_0 \nu_S \exp\left(-\frac{\Delta G_{des}}{kT}\right)$$

where $J_d = J_0$ is the equilibrium desorption flux, N_0 is the number of adatoms per unit area of the planar surface, ν_{des} is desorption frequency, and ν_S is surface vibration frequency, and ΔG_{des} is the activation energy of desorption. At equilibrium, we have $J_c = J_d$, or $N_0 \nu_{des} = nv_a$.

To nucleate a disc on the planar surface, we need supersaturation, $J_c \gg J_d$. The question is how much supersaturation is needed. First we consider the energy change in the nucleation of a disc. The surface energy of the circumference is given as

$$E_{\mathrm{d}} = 2\pi ra\gamma \tag{6.23}$$

where "a" is atomic layer thickness and γ is surface energy per unit area of the circumference. The surface energy of the circumference is positive, and thus it tends to shrink the disc and in turn exerts the Laplace pressure on the disc, which is

$$p_{\mathrm{S}} = \frac{1}{A}\frac{dE_{\mathrm{d}}}{dr} = \frac{2\pi a\gamma}{2\pi ar} = \frac{\gamma}{r} \tag{6.24}$$

where A is the area of the circumference of the disc. Under Laplace pressure, the energy of each atom in the disc is increased by the amount of

$$p_{\mathrm{S}}\Omega = \frac{\gamma}{r}\Omega \tag{6.25}$$

This is the chemical potential energy increase due to the formation of the circumference surface of the disc. Because of the increase in energy, the atoms in the disc can sublime more easily,

$$J_{\mathrm{d}}^{*} = N_0 \nu_{\mathrm{S}} \exp\left(-\frac{\Delta G_{\mathrm{des}}}{kT} + \frac{p_{\mathrm{S}}\Omega}{kT}\right) \tag{6.26}$$

$$\frac{J_{\mathrm{d}}^{*}}{J_0} = \exp\left(\frac{p_{\mathrm{S}}\Omega}{kT}\right) = \exp\left(\frac{\gamma\Omega}{rkT}\right) = \frac{n\nu}{n_0\nu} = \frac{p_{\mathrm{S}}}{p_0} \tag{6.27}$$

The last equation is Gibbs–Thompson equation for the vapor pressure above a cylindrical disc of radius r. It means that if we keep the disc under the equilibrium pressure of p_0, because $p_{\mathrm{S}} > p_0$, the disc tends to evaporate and shrink. So the disc will reduce its radius and the pressure increases, and it will continue to shrink until it disappears. However, if we want to keep the disc stable or to grow the disc, we must increase the vapor pressure or the supersaturation or the condensation rate. What is the supersaturation needed so that it becomes stable? Or what is the supersaturation needed in order to nucleate a new disc of the critical size on the surface?

The energy change in nucleating the disc is given as

$$\Delta E_{\mathrm{disc}} = 2\pi ra\gamma - \pi r^2 a\Delta E_{\mathrm{S}} \tag{6.28}$$

where ΔE_{S} is the latent heat of condensation per unit volume and γ is the surface energy per unit area of the circumference of the disc. We define r_{crit} such that at $r = r_{\mathrm{crit}}$,

$$\frac{d\Delta E_{\mathrm{disc}}}{dr} = 0$$

Thus $2\pi a\gamma - 2\pi ra\Delta E_{\mathrm{S}} = 0$, and

$$r_{\mathrm{crit}} = \frac{\gamma}{\Delta E_{\mathrm{S}}} \tag{6.29}$$

The net energy change in nucleating the critical disc is

$$\Delta E_{\text{crit}} = \pi r_{\text{crit}} a \gamma \tag{6.30}$$

The critical disc at r_{crit} is metastable because ΔE_{crit} is a maximum in the plot of ΔE versus r. Any slight deviation from r_{crit} will lead to a decrease of energy. Now, considering the vapor pressure on the critical disc, we have

$$\frac{p_{\text{crit}}}{p_0} = \exp\left(\frac{\gamma\Omega}{r_{\text{crit}} kT}\right)$$

Thus,

$$r_{\text{crit}} = \frac{\gamma\Omega}{kT \ln\left(\frac{p_{\text{crit}}}{p_0}\right)}$$

Then,

$$\Delta E_{\text{crit}} = \pi r_{\text{crit}} a \gamma = \frac{\pi\gamma^2 a\Omega}{kT \ln\left(\frac{p_{\text{crit}}}{p_0}\right)} = \frac{\pi\gamma^2 a^4}{kT \ln\left(\frac{p_{\text{crit}}}{p_0}\right)} \tag{6.31}$$

In the last step of the above equation, we have taken $\Omega = a^3$. Knowing the critical energy or activation energy of formation of the critical disc, we can calculate the probability of nucleation of the disc or the density of nucleus, that is, the number of nuclei per unit area, as given below.

$$N_{\text{crit}} = J_c \tau_0 \exp\left(-\frac{\Delta E_{\text{crit}}}{kT}\right) \tag{6.32}$$

In the above equation, we recall that J_c is the condensation flux, having a unit of the number of atoms per unit area per unit time, and τ_0 is the residence time of adatoms on the surface. So we have

$$N_{\text{crit}} = J_c \tau_0 \exp\left[-\frac{\pi\left(\gamma a^2\right)^2}{(kT)^2 \ln\left(\frac{p_{\text{crit}}}{p_0}\right)}\right] \tag{6.33}$$

On the basis of the last equation, we can estimate the nucleation rate and we show that it is extremely low. For the condensation flux, we can take $J_c = 10^{15}$ atoms/cm^2 s, which means we deposit one monolayer per second, or we deposit 100 nm thick film in 5 min, which is a typical rate of deposition.

For the residence time, we recall that

$$\tau_0 = \frac{1}{\nu_S} \exp\left(\frac{\Delta G_{\text{des}}}{kT}\right)$$

We take $\Delta G_{\mathrm{des}} = 1.1$ eV, $T = 1223$ K, which means that $kT = 0.1$ eV, then

$$\tau_0 = \frac{1}{10^{13}} \exp\left(\frac{1.1}{0.1}\right) = 10^{-13} 10^{11/2.3} = 10^{-8}\mathrm{s}$$

For the activation energy, we take the surface energy per atom of Si, $\gamma a^2 = 0.6$ eV/atom, and $kT = 0.1$ eV at 1223 K, we have

$$\Delta E_{\mathrm{crit}} = \frac{\pi(\gamma a^2)^2}{kT \ln\left(\dfrac{p_{\mathrm{crit}}}{p_0}\right)} = \frac{3.14 \times 3.6}{\ln\left(\dfrac{p_{\mathrm{crit}}}{p_0}\right)} \tag{6.34}$$

If we give

$$\frac{p_{\mathrm{crit}}}{p_0} = 10, \quad \ln 10 = 2.3, \quad \Delta E_{\mathrm{crit}} = 4.9\,\mathrm{eV}$$

$$\frac{p_{\mathrm{crit}}}{p_0} = 100, \quad \ln 100 = 4.6, \quad \Delta E_{\mathrm{crit}} = 2.5\,\mathrm{eV}$$

$$\frac{p_{\mathrm{crit}}}{p_0} = 1000, \quad \ln 1000 = 6.9, \quad \Delta E_{\mathrm{crit}} = 1.6\,\mathrm{eV}$$

If we input the above values into N_{crit}, for the case of $p_{\mathrm{crit}}/p_0 = 1000$, we find that

$$N_{\mathrm{crit}} = J_c \tau_0 \exp\left(-\frac{\Delta E_{\mathrm{crit}}}{kT}\right) = 10^{15} \times 10^{-8} \times \exp\left(-\frac{1.6}{0.1}\right) = 10^7 \times 10^{-7} = 1$$

It means that we have one nucleus on a unit area of 1 cm^2 even under the supersaturation of 1000, which is indeed a very low rate of nucleation. Clearly, homogeneous nucleation is a rare event.

Historically, in bulk single crystal growth, it was found that the measured crystal growth rate cannot be explained when homogeneous nucleation is assumed, especially when the growth of each atomic layer requires an independent homogeneous nucleation event or when it requires repeating events of homogeneous nucleation. Thus, the spiral growth model around a screw dislocation was proposed by F. C. Frank. In the model no nucleation is needed. Also, as we have discussed in this section on epitaxial growth of Si on Si, we assume that there is no screw dislocation in the Si; instead we buy Si wafers having a 7-degree cut surface for stepwise epitaxial growth. However, we present below that in epitaxial growth of silicide in nanowires of Si, repeating homogeneous nucleation is found.

6.5 REPEATING HOMOGENEOUS NUCLEATION OF SILICIDE IN NANOWIRES OF Si

In the kinetic theory of nucleation, we have obtained the Zeldovich factor of steady state nucleation, yet it is for homogeneous nucleation only. In real materials, heterogeneous nucleation is encountered very often because of the much smaller activation energy of nucleation. Thus there is a gap between our theoretical understanding of homogeneous nucleation and our experimental findings of heterogeneous nucleation.

To promote homogeneous nucleation, we have to remove all heterogeneities that may enhance heterogeneous nucleation. Nanoscale materials have the advantage for us to do so. We discuss here the repeating events of homogeneous nucleation in epitaxial growth of $CoSi_2$ and NiSi silicides in the axial direction of nanowires of Si. The growth of every single atomic layer of the silicides in nanowires of Si requires nucleation. It is homogeneous nucleation because heterogeneous nucleation has been suppressed not only because of nanosize, but also because of the surface oxide of the Si nanowire. The oxide has prevented heterogeneous nucleation from occurring because the energy of the oxide/Si interface is lower than that of the oxide/silicide interface. Thus, the oxide does not provide a low energy surface for heterogeneous nucleation.

Owing to the repeating events, the distribution of incubation time of homogeneous nucleation can be measured. The incubation time is the time between two events of nucleation. Knowing the incubation time, we can calculate the steady state rate of homogeneous nucleation, that is, the number of stable critical nucleus per unit area per unit time, when the activation energy of nucleation is measured. As it is homogeneous nucleation, we can apply Zeldovich factor to calculate the number of molecules in forming the critical nucleus in the steady state process. In this section, we correlate the theory and experiment of homogeneous nucleation.

One-dimensional nanostructures, such as nanowires of Si, have been attractive for nanotechnology because their morphology, size, and electronic properties make them attractive to serve as the basic components in electronic and optoelectronics devices, especially biosensors. Well-defined nanoscale building blocks such as ohmic contacts and gates on Si nanowires must be developed in order to be assembled into functional device structures. To process these building blocks, it requires a systematic study of the chemical reactions in nanoscale to control the formation of them.

6.5.1 Point Contact Reactions in Nanowires

When two nanowires cross and touch each other, a point contact forms. The point contact reactions between a Si nanowire and a metal nanowire of Ni or Co have been studied by *in situ* high-resolution transmission electron microscopy (HRTEM) [6–8]. The samples were prepared by putting droplets of solutions containing Si and metal nanowires of Ni (or Co) on a Cu grid covered with a 50 nm thick amorphous SiO_2 film, which is transparent to the electron beam in the microscope and does not interfere much with the images of Si and metal nanowires in HRTEM. The samples were dried at room temperature. Both the Si nanowires and metal nanowires of Ni and Co have diameters ranging from 20 to 70 nm and lengths of a few microns. *In situ* annealing for point contact reactions and high-resolution lattice imaging were performed in a JEOL 2000V ultrahigh vacuum TEM. The annealing temperature of Co and Si samples was at 800 °C and that of Ni and Si samples was from 450 to 700 °C. The vacuum in the sample stage was about 3×10^{-10} Torr.

Both NiSi and $CoSi_2$ were found to grow epitaxially and axially in the [111] oriented Si nanowires. The axial growth of silicide occurs atomic layer by atomic

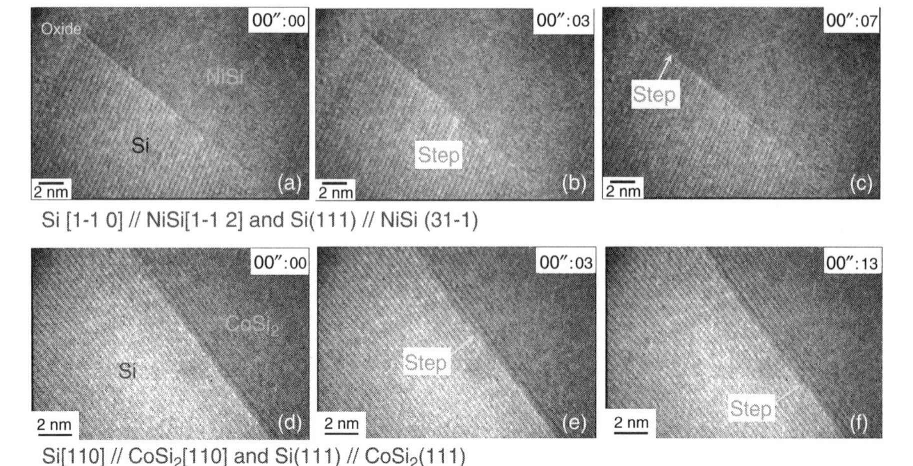

Figure 6.13 The upper row shows a set of three HRTEM images of the motion of one NiSi atomic layer across the NiSi/Si interface and the stepwise growth direction. The lower row shows three HRTEM images of the step motion on the interface between Si and $CoSi_2$. *Source*: Reproduced with kind permission from [7].

layer, with the moving of steps or kinks across the epitaxial interface. The epitaxial interface is a moving interface. The growth rate of the silicide, at 450–700 °C for NiSi and at 800 °C for $CoSi_2$, was measured from *in situ* HRTEM video. In the upper row of Figure 6.13, the three HRTEM images show the step motion of one NiSi atomic layer across the NiSi/Si interface, and the stepwise growth direction is in the radial direction of the wire. Similarly, in the lower row of Figure 6.13, the three HRTEM images show of the step motion on the epitaxial interface between Si and $CoSi_2$.

Surprisingly, it was found that there is a long period of stagnation of motion between the growths of two successive atomic layers. This is true for both $CoSi_2$ and NiSi. When the stagnation period and the stepwise growth period were plotted from the video recording, it shows the stair-type curves as shown in Figure 6.14a, b. The stepwise growth rate of each of the silicide atomic layer is about the same, which is approximately 0.17 s per layer for $CoSi_2$ and approximately 0.06 s per layer for NiSi, and we note that it is just the width of the vertical line in the stair-type curves. In between the vertical lines, the horizontal part of the steps in Figure 6.14a, b, is the stagnation period, which we define as the incubation time of nucleation of a new layer. The average value of the incubation time of $CoSi_2$ is about 6 s and that of NiSi is about 3 s. This is the most important experimental measurement of the homogeneous nucleation event in the study. Every step represents an event of homogeneous nucleation.

In the stair-type growth curves, the nucleation event and the growth event can be studied separately. The lateral or radial growth rate (parallel to the epitaxial interface) of $CoSi_2$ has been measured from the HRTEM images to be about 135 nm/s, which is about 3700 times faster than that of the axial growth rate (normal to the epitaxial interface) of 0.0365 nm/s in the axial direction of the nanowire. This is because the

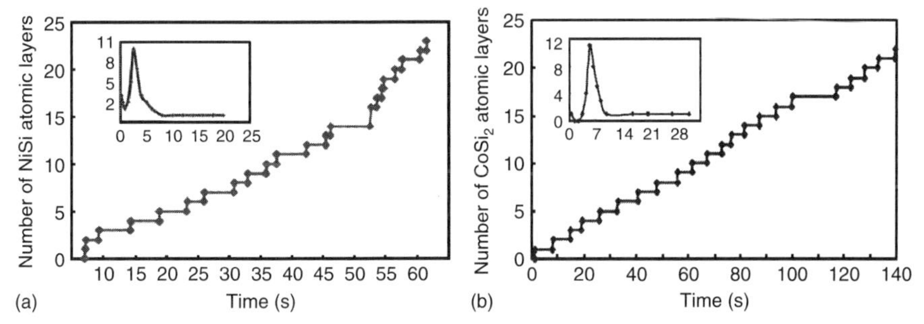

Figure 6.14 (a) and (b) Stair-type curves for NiSi and $CoSi_2$, respectively. *Source*: Reproduced with kind permission from [7].

measured average axial growth rate has to include the incubation time of nucleation of every layer. Without the incubation time, the axial growth rate of each $CoSi_2$ atomic layer, on the basis of the measured radial growth rate of 135 nm/s, would have been 1.82 nm/s. This value was obtained by using the conventional equation of growth in molecular beam epitaxy, $V = Nvh$, where V is axial growth rate, v is radial growth rate, N is the number of steps per unit length, and h is the height of the step. However, this conventional stepwise growth equation fails in calculating correctly the epitaxial growth rate in nanowires when there is a long period of stagnation between layers.

The HRTEM videos show that the overall axial growth rate of the silicide layers is linear. Actually the linear curve can be decomposed into many stair-steps with the step height equaling to an atomic layer thickness and the step width equaling to the incubation time of nucleating a new step. After a step is nucleated, it propagates very rapidly across the Si/silicide epitaxial interface. So the overall reaction rate is limited by the incubation time of nucleation. We ask where the nucleation of the new step is.

The fast radial growth of $CoSi_2$ and NiSi starts from the center rather than from the edge of the nanowire. This is owing to the fact that the step always moves toward the edge, rather than away from the edge. Five cases of video recording were obtained to substantiate the observation that NiSi and $CoSi_2$ atomic layers grow toward the edge of the oxide surface of Si nanowires, and some were observed to have initiated from the center of the Si nanowire.

Owing to the native surface oxide of the Si nanowire, we assume that the energy of the oxide/silicide interface is higher than that of the oxide/Si interface. This is a reasonable assumption because it was found that when a step approaches the edge of the Si nanowire, it slows down before it transforms the edge from the oxide/Si interface to the oxide/silicide interface due to a higher energy barrier. This is shown in Figure 6.15a, b, where we can see a curvature of the untransformed Si near the oxide/Si edge because several of the silicide layers have greatly slowed down their growth in approaching the edge within an atomic distance. The drawings in in Figure 6.15c, d are sketchs of the observed curvature of the silicide layers. Furthermore, when the electron beam was kept at the edge region and waited for heterogeneous nucleation to take place, but no nucleation events was observed.

The heterogeneous nucleation of a step at the edge is depicted in Figure 6.16a. It must replace the low energy oxide/Si interface by the high energy oxide/silicide

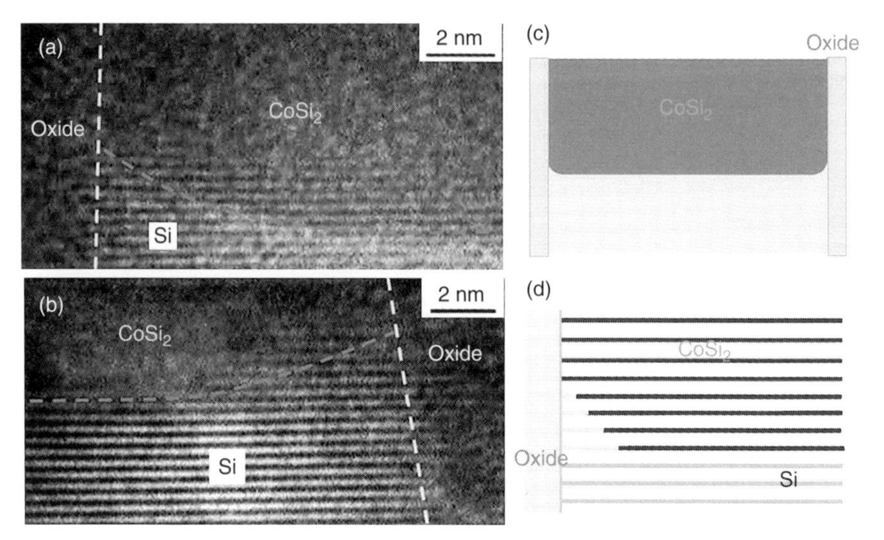

Figure 6.15 Curvature of the untransformed Si near the oxide/Si edge is shown in (a) and (b). (c) and (d) are sketches of the observed curvature of the silicide layers. *Source*: Reproduced with kind permission from [7].

interface. There is no microreversibility, so the heterogeneous nucleation is suppressed. In Figure 6.16b, a schematic diagram of a heterogeneous nucleus is assumed with a wetting angle larger than 90°. At the triple point, we consider

$$\gamma_{\text{silicide/oxide}} \geq \gamma_{\text{Si/oxide}} + \gamma_{\text{Si/silicide}} \cos(180 - \theta)$$

where γ represents the surface energy per unit area of the interfaces. We note that the epitaxial interface between Si and silicide is a low energy interface. When the inequality is satisfied and $\theta = 180°$, heterogeneous nucleation will not occur and homogeneous nucleation of a circular disc in the center of the nanowire becomes possible, and the cross-section is depicted in Figure 6.16c. Figure 6.16d redraws Figure 6.12a on an silicide/Si epitaxial interface in the nanowire.

6.5.2 Homogeneous Nucleation of Epitaxial Silicide in Nanowires of Si

In homogeneous nucleation of a circular disc, the energy change is

$$\Delta G_{\text{disc}} = 2\pi r a \gamma - \pi r^2 a \Delta G_s \tag{6.35}$$

where ΔG_s is the gain in the free energy of formation of the silicide per unit volume, and γ is the interfacial energy per unit area of the circumference of the disc, r is the radius of the disc, and a is the atomic height. At the critical nucleus of size r_{crit}, we have $r_{\text{crit}} = -\gamma/\Delta G_s$. The net energy change in nucleating the critical disc is $\Delta G^* = \pi r_{\text{crit}} a \gamma$.

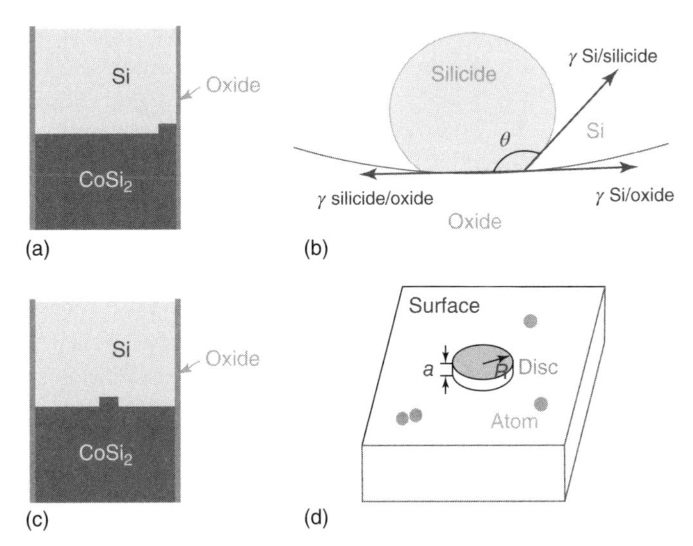

Figure 6.16 (a) Heterogeneous nucleation of a step at the edge. (b) A schematic diagram of a heterogeneous nucleus assumed with a wetting angle larger than 90°. (c) The cross-section of homogeneous nucleation of a circular disc in the center of the nanowire. (d) A redraw of Figure 6.12a for the circular disk on the epitaxial silicide/Si interface.

Knowing the critical energy or activation energy of formation of the critical disc, we can calculate the probability of nucleation of the critical nucleus, that is, the number of critical nuclei per unit area per unit time. Experimentally what we have measured in the upper row Figure 6.13 is one stable critical nucleus on the cross-section of the Si/silicide nanowire in the period of one incubation time. Thus we have the nucleation rate of

$$I_{\text{stable-crit}} = \frac{1}{\pi R^2 \tau_{\text{i}}} \tag{6.36}$$

where R is the radius of the Si nanowire, and τ_{i} is the incubation time. Taking the diameter of Si nanowire to be 30 nm, and the incubation time to be 3 s, we have

$$I_{\text{stable-crit}} = 4.7 \times 10^{10} \text{ stable nuclei/cm}^2\text{-s for the case of NiSi}$$

After one stable critical nucleus forms in each period of incubation, the growth of an atomic layer of silicide follows. It is a steady state, indicating that during one period of incubation, it must have dissolved from the point contact the same amount of Ni atoms to supply the growth of one atomic layer of the silicide. For simplicity, we assume that there are 10^{15} atoms per atomic layer per centimeter square, and the flux of Ni needed to grow an atomic layer is

$$J_{\text{Ni}} = \frac{10^{15}}{2 \times 3} = 1.67 \times 10^{14} \text{atoms/cm}^2\text{s}$$

where the factor of 2 in the denominator is because the concentration of Ni in NiSi is half and the factor 3 is from the incubation time. While we can regard this to be a flux of Ni atoms being deposited onto the silicide/Si interface, we note that not all these Ni atoms will involve directly in the nucleation process. This is because, similar to thin film deposition on a substrate, we should consider the adatoms on the interface, and we assume that only the adatoms are taking part in the nucleation process. However, the adatoms have a residence time, τ_{des}, on the interface because of desorption.

$$\tau_{\text{des}} = \frac{1}{\nu_{\text{s}}} \exp \frac{\Delta G_{\text{des}}}{kT} \tag{6.37}$$

where υ_{s} is the vibrational frequency of an adatom, ΔG_{des} is the activation energy of desorption of an adatom, and kT is thermal energy. Thus $J_{\text{Ni}}\tau_{\text{des}}$ is the effective number of adatoms per unit area involved in the nucleation process.

Then, the equilibrium concentration of critical nucleus can be given as

$$C_{\text{crit}} = J_{\text{Ni}}\tau_{\text{des}} \exp\left(-\frac{\Delta G^*}{kT}\right) \tag{6.38}$$

where C_{crit} has the unit of the number of nuclei per unit area.

On the basis of assuming a thermally activated process of fluctuation of subcritical nucleus, the steady state homogeneous nucleation rate has been given in Section 6.4 as

$$I_{n^*}^{\text{S}} = \beta_n^* C_{\text{crit}} Z = \beta_n^* C_o e^{-\frac{\Delta G_n^*}{kT}} \left[-\frac{1}{2\pi kT} \left(\frac{\partial^2 \Delta G_n}{\partial n^2} \right)_{n^*} \right]^{1/2} \tag{6.39}$$

where $\beta_n{}^*$ is the reversible frequency of atomic jump toward a critical nucleus and to convert it into a stable nucleus, and $C_{\text{crit}} = C_0 \exp(-\Delta G_n^*/kT)$ is the equilibrium concentration of critical size nucleus, which we note is the same as Eq (6.38). The Zeldovich factor,

$$Z = \left[-\frac{1}{2\pi kT} \left(\frac{\partial^2 \Delta G_n}{\partial n^2} \right)_{n^*} \right]^{1/2} \tag{6.40}$$

has been included in the nucleation rate equation as a kinetic factor because the nucleus that has overcome the nucleation barrier may not definitely become a stable nucleus until one more atom has joined it; otherwise most of them may shrink back to subcritical size. If all the critical nuclei crossing the nucleation barrier succeed in becoming stable nuclei, Zeldovich factor is 1. However, Zeldovich factor is less than 1 in all real cases. In other words, Zeldovich factor is a kinetic factor stands for the percentage of critical size nuclei that become stable.

The second order derivative in the Zeldovich factor can be rewritten for the nano silicide nucleation in nanowire of Si on the basis of the shape of the nucleus. By assuming a circular disc shape of atomic height, the Zeldovich factor is changed to

$$Z = \left[\frac{1}{4\pi kT} \cdot \frac{\Delta G^*}{(n^*)^2} \right]^{1/2} \tag{6.41}$$

TABLE 6.1 The Required Molecules to Form a Stable Silicide Nucleus

	$CoSi_2$ at 800 °C			NiSi at 700 °C		
Z	1	0.1	0.05	1	0.1	0.05
n^*	1.6	16	31	1.1	11	22

n^* is the number of molecules required to form a stable nucleus.

where ΔG^* and n^* are respectively the activation energy in forming the critical nucleus and the number of molecules in it.

Knowing the activation energy of NiSi to be 1.25 eV/atom, we can calculate n^* at a given value of Z. Table 6.1 shows the n^* of NiSi at 700 °C and $CoSi_2$ at 800 °C with different Zeldovich factors. For $CoSi_2$, we took the activation energy from thin film study at 800 °C. Typical experimental value of Z factor is about 0.05. Table 6.1 lists the value of n^* to be about 10 for both silicides. The n^* value of $CoSi_2$ is higher than that of NiSi, and it may be one of the reasons that the temperature of reaction of $CoSi_2$ is higher than that of NiSi.

Because we know the experimentally measured steady state nucleation rate as given by Eq (6.36), we can check it by using Eqs (6.39) and (6.41). We have

$$I_{\text{stable-crit}} = \beta_n^* J_{\text{Ni}} \tau_{\text{des}} \exp\left(-\frac{\Delta G^*}{kT}\right) Z = \beta_n^* J_{\text{Ni}} \frac{1}{\nu_s} \exp\left(-\frac{\Delta G^* - \Delta G_{\text{des}}}{kT}\right) Z$$

$$= \nu_0 J_{\text{Ni}} \frac{1}{\nu_s} \exp\left(-\frac{\Delta G^* - \Delta G_{\text{des}} + \Delta G_\beta}{kT}\right) Z$$

where we assume that $\beta_n^* = \upsilon_0 \exp(-\Delta G_\beta / kT)$, and υ_0 is the Debye frequency of vibration and ΔG_β is the activation energy of adding an atom to the critical nucleus. It is worthwhile mentioning that the basic nature of the parameter of β_n^* is microreversibility. In order to maintain the equilibrium distribution of subcritical size embryos in nucleation, the frequency of adding and subtracting atoms among the embryos is high. Hence we can assume that $\Delta G^* \gg \Delta G_\beta$, so we can ignore ΔG_β.

To evaluate the products on the right-hand side of the above equation, we cancel υ_0 against υ_s owing to the fact that both are Debye frequency of atomic vibration. For ΔG_{des}, it is known from epitaxial growth of Si on Si, where $\Delta G_{\text{des}} = 1.1$ eV/atom. For the desorption of Ni, the activation energy should be lower and we assume that $\Delta G_{\text{des}} = 0.7$ eV/atom. Then we take the measured $\Delta G^* = 1.25$ eV/atom and $Z = 0.1$. As $J_{\text{Ni}} = 1.67 \times 10^{14}$ atoms/cm^2-s, the products on the right-hand side at $T = 700$ °C is 3×10^{10} nuclei/cm^2-s, which is in good agreement with the measured nucleation rate of 4.7×10^{10} nuclei/cm^2-s.

We caution that there is some uncertainty about ΔG_{des}. While we have no measured data, we note that even if we give it a high uncertainty by taking $\Delta G_{\text{des}} = 0.7 \pm 0.2$ eV/atom, it will only change the outcome by a factor about 10. As this is an attempt to correlate theory and experiment on homogeneous nucleation, it is expected to have a large uncertainty.

Nucleation requires supersaturation. We can calculate the supersaturation in the nucleation of NiSi. The solubility of Ni in Si at 700 °C is about $10^{15}-10^{16}$ Ni atoms/cm^3. As there are 2.5×10^{22} Si-atoms/cm^3, the equilibrium concentration of Ni in Si is about $10^{-7}-10^{-8}$. When we have dissolved half of a monolayer of Ni (which has a layer thickness of 0.3 nm) into a Si nanowire of 3 μm long before homogeneous nucleation occurs, the concentration of Ni is 0.5×10^{-4}, so the supersaturation is approximately 10^3, which seems very large. As we dissolve Ni into the nanowire of Si, the solubility can be increased because of the Gibbs–Thomson effect by a factor of $\exp(\gamma\Omega/rkT)$. To calculate this factor, we take $\Omega = a^3$ and $a = 0.3$ nm as atomic diameter, $r = 15$ nm, and $kT = 0.084$ eV at 973 K. When we let $\gamma a^2 = 1-2$ eV, we obtain the factor to be 1.25–1.58, respectively, which is small compared to the estimated supersaturation of 1000. We recall that in the last Section we have shown that in the homogeneous nucleation of Si on Si by vapor deposition, the supersaturation is indeed very high.

REFERENCES

1. Porter DA, Easterling KE. *Phase Transformations in Metals and Alloys*. 2nd ed. London: Chapman and Hall; 1992.
2. Christian JW. *The Theory of Transformations in Metals and Alloys*. 2nd ed. Oxford: Pergamon Press; 1975.
3. Aaranson HI, editor. *Phase Transformations*. Ohio: ASM Metals Park; 1970.
4. a. Tu KN, Turnbull D. Morphology of cellular precipitation of tin from lead-tin bicrystals. Acta Met 1967;15:369. b. Tu KN, Turnbull D. Morphology of cellular precipitation of tin from lead-tin bicrystals-II. Acta Met 1967;15:1317.
5. Cahn JW. The kinetics of cellular segregation reactions. Acta Met 1959;7:18–28.
6. Chou Y-C, Wu W-W, Cheng S-L, Yoo B-Y, Myung N, Chen LJ, Tu KN. In situ TEM observation of repeating events of nucleation in epitaxial growth of nano $CoSi_2$ in nanowires of Si. Nano Lett 2008;8:2194–2199.
7. Chou Y-C, Wu WW, Chen LJ, Tu KN. Homogeneous nucleation of epitaxial $CoSi_2$ and NiSi in Si nanowires. Nano Lett 2009;9:2337–2342.
8. Chou Y-C, Wu W-W, Lee C-Y, Chen LJ, Tu KN. Heterogeneous and homogeneous nucleation of epitaxial $NiSi_2$ in [110] Si nanowires. J Phys Chem C 2011;115:397–401.

PROBLEMS

6.1. In man-made rain, how can we enhance the nucleation of rain drops?

6.2. When a bilayer thin film of Ni/Zr is annealed at 400 C, an amorphous alloy of NiZr is formed between the Ni and Zr. Since it is a slow heating not a rapid quenching process, can the amorphous phase formation occur by nucleation and growth of the amorphous phase? How can an amorphous nucleus be formed?

6.3. Assume homogeneous nucleation of a spherical β nucleus occurs in the matrix of α. Let $\gamma_{\alpha\beta}$ be the interfacial energy per unit area and $\Delta H_{\alpha\beta}$ be the heat of transformation per unit volume of the nucleus. Calculate the radius and the activation energy of nucleation of the critical nucleus. What is the volume of the critical nucleus and what is the number of atoms in the critical nucleus?

6.4. Compare homogeneous nucleation of a sphere and a cube. Evaluate the heights of nucleation barriers at the same driving force per atom of Δg, same interface energy of γ, and atomic volume of Ω.

6.5. Consider a nucleus of intermetallic compound, i = AB, to form at the initial interface between A and B. Assume that it has a shape of a disc with radius r and height h, and further assume that half of which is in A and the other half is in B. Take the thermodynamic driving force per atom to be Δg, and surface energy to be $\gamma_{\alpha\beta}, \gamma_{ai}, \gamma_{\beta i}$ for the corresponding interface. Find the nucleation barrier by assuming that the nucleus has the possibility to obtain the optimal shape at each fixed volume.

6.6. In nucleation theory, the Zeldovich factor is given as

$$Z = \left[-\frac{1}{2\pi kT} \left(\frac{\partial^2 \Delta G_n}{\partial n^2} \right)_{n*} \right]^{1/2}$$

Let us consider the nucleation in the crystallization of a pure phase, the formation energy of a nucleus of n atoms is given as

$$\Delta G_n = -n\Delta g_{ls} + bn^{2/3}\gamma_{ls}$$

where Δg_{ls} is the free energy change per atom from liquid to solid, and γ_{ls} is surface energy per surface atom of the nucleus, and b is a geometrical factor. Calculate the Zeldovich factor as

$$Z = \left[\frac{1}{3\pi kT} \frac{\Delta G*}{(n^*)^2} \right]^{1/2}$$

Note that ΔG^* and n^* mean critical nucleus, and the second derivative in the Zeldovich factor is evaluated at $n = n^*$.

6.7. Consider the two-dimensional nucleation of an epitaxial NiSi layer on a nanowire of Si. It is given that Δg is the free energy gain per atom in the change from the formation of the NiSi. γ is surface energy per surface atom of the nucleus.

(a) Find the critical nucleus size or the critical number of atoms n^* needed to from the two-dimensional critical nucleus, and find the critical energy ΔG^* needed to formed the critical nucleus.

(b) Find the Zeldovitch factor in terms of ΔG^* and n^* . We recall that in general the Zeldovitch factor is given as.

$$Z = \left[-\frac{1}{2\pi kT} \left(\frac{\partial^2 \Delta G_n}{\partial n^2} \right)_{n=n^*} \right]^{1/2}$$

CHAPTER 7

CONTACT REACTIONS ON Si; PLANE, LINE, AND POINT CONTACT REACTIONS

7.1 INTRODUCTION

In the last chapter, we have discussed the difference in nucleation behaviors among bulk, thin film, and nanoscale materials. In this chapter, we discuss the difference among them in growth behaviors, specifically in the growth of intermetallic compounds in contact reactions. For example, it is well known that in bulk materials and thin films, we have diffusion-controlled growth and interfacial-reaction-controlled

---

Kinetics in Nanoscale Materials, First Edition. King-Ning Tu and Andriy M. Gusak.

growth. In nanomaterials, we have instead a supply-controlled growth in point contact reactions between nanowires.

Interfacial reaction or contact reaction has been a very important subject in wafer-based semiconductor technology [1–6]. In very-large-scale-integration of circuits on a piece of Si chip, from the point of view of integration, the building blocks to form a field-effect transistor are the p–n junction, the gate and side-wall oxides, the metallic silicide source/drain and gate contacts. The processing of these building blocks in two-dimensional integration is based on planar reactions. The diffusion of dopant or the ion implantation of dopant to fabricate p–n junction is carried out through an area, same as the oxidation and silicidation of Si. However, when we consider making devices on Si nanowires or Si nanodots, the contact reaction is not over an area, rather it occurs across a contact line or a contact point. In this chapter, we discuss the effect of interfacial geometry on the kinetics of interfacial reactions.

In Chapter 3, we discussed interdiffusion to form continuous solid solutions (or alloys) in bulk diffusion couples such as copper and nickel. Interdiffusion leads to compositional change but no structural change. For example, Cu, Ni, and CuNi alloys have a face-centered cubic structure. Structure continuity is a prerequisite of compositional continuity. This applies to spinodal decomposition too. But it is not in intermetallic compound formation.

In this chapter, we discuss interdiffusion to form intermetallic compounds in contact reactions. These compounds tend to have a very narrow composition range and their crystal structure different from that of their constituents. The difficulty in dealing with intermetallic compound formation is that more than one of them may be formed. If there are at the least two unknowns for each of the compounds (i.e., the intrinsic diffusion coefficient of the two constituents, if we ignore the interfacial-reaction coefficient), there are $2N$ unknowns for N compounds. The problem is intractable because of the large number of unknowns. Besides, because the compound interface are discontinuities in structure and composition, it makes theoretical analysis difficult.

To simplify the problem, we consider a special case where the binary system can form only one intermetallic compound. This assumption is important and extremely useful because in contact reactions in thin films on Si wafers for very large scale integration (VLSI) device manufacturing, they tend to form only one intermetallic compound, in contrast to bulk couples where a multiple of compounds are formed.

We consider first an example of a bulk couple of gold–aluminum interdiffused at 460 °C for 100 min; all five intermetallic compounds as given in the equilibrium binary Au–Al phase diagram were found, as depicted in Figure 7.1. They formed in the correct order in composition change as expected from the Au–Al phase diagram; $Au/Au_4Al/Au_5Al_2/Au_2Al/AuAl/AuAl_2/Al$. However, if this couple is annealed at a different temperature such as 200 °C, $AuAl_2$ and AuAl are absent.

Instead, if we interdiffuse a bimetallic thin film couple of Au/Al at 200 °C, there is only one intermetallic compound, Au_2Al, formed. The other phases are missing and do not form simultaneously as in the bulk couple. However, the other phases will form sequentially one by one on further annealing, as depicted in Figure 7.2. In a thin film couple that has a thicker Al, $AuAl_2$ will form after Au_2Al. But in a couple that has a thicker Au, Au_5Al_2 will form and be followed by Au_4Al. Comparing the

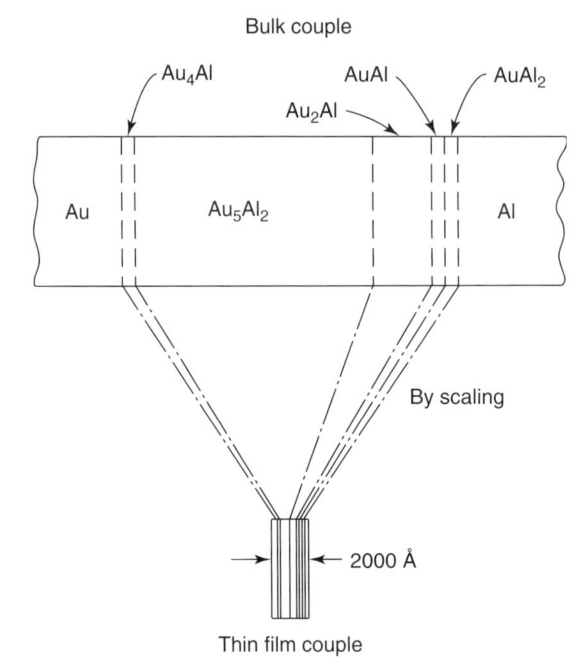

Figure 7.1 A gold–aluminum bulk couple interdiffused at 460 °C for 100 min; all five intermetallic compounds as given in the binary Au–Al phase diagram.

reaction products between the bulk couple and the thin film couple, we note that we cannot scale down the dimension from bulk to thin film and expected to find the same reaction products, as inferred in Figure 7.1.

Another example for comparison is the binary system of Ni/Si [7]. Figure 7.3 shows the optical cross-sectional micrograph of a bulk couple of Ni and Si annealed at 850 °C for 8 h. It shows that Ni_5Si_2, Ni_2Si, Ni_3Si_2, and NiSi were formed. But, if we deposit a Ni thin film on a Si wafer and anneal the couple at 250 °C for 1 h, we find that only Ni_2Si forms. Figures 7.4 and 7.5 are, respectively, Rutherford backscattering spectrum and glancing incidence X-ray diffraction spectrum taken to identify the formation of Ni_2Si in the thin film sample. On further annealing to around 400 °C, Ni_2Si will transform to NiSi, and if the sample is annealed about 750 °C, $NiSi_2$ will replace NiSi. This sequence of formation is depicted in the right-hand side of the phases shown in Figure 7.6. On the other hand, if we deposit a Si thin film on a thick Ni foil, the first phase formation near 250 °C is still Ni_2Si, but on annealing to 400 °C, Ni_5Si_2 will form, followed by Ni_3Si formation at temperatures above 450 °C, as depicted in the left-hand side of the phases shown in Figure 7.6.

On nanoscale reactions between Ni and Si, in the last chapter, we have discussed the point contact reaction between Ni and Si nanowires. On annealing, the nanowires in temperature range from 550 to 750 °C, and only NiSi is formed. Although it is a single phase formation, the phase formed is different from that in the thin film case.

The major difference in the growth behavior among bulk, thin films, and nanowires is the missing phase phenomenon. Why only one of them can grow in thin film case as well as in some nanowire cases? What is the mechanism behind

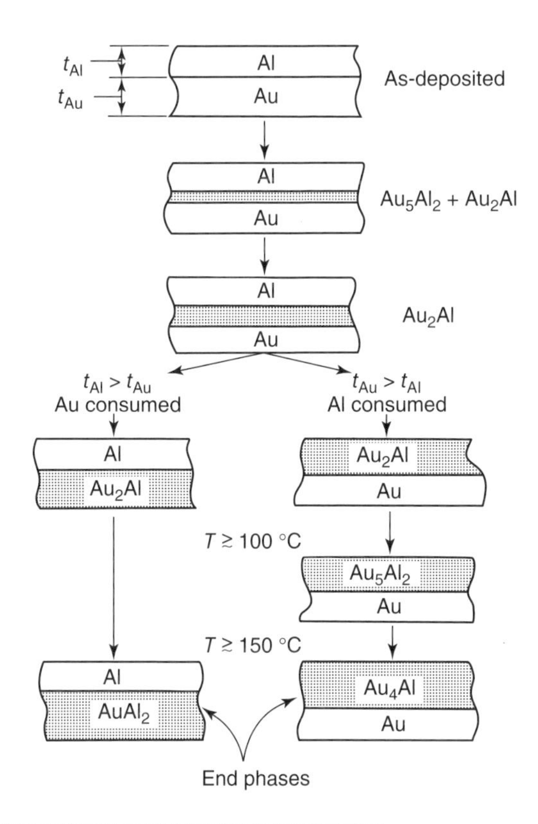

Figure 7.2 In a thin film couple that has a thicker Al, $AuAl_2$ will form after Au_2Al. But in a couple that has a thicker Au, Au_5Al_2 will form and be followed by Au_4Al.

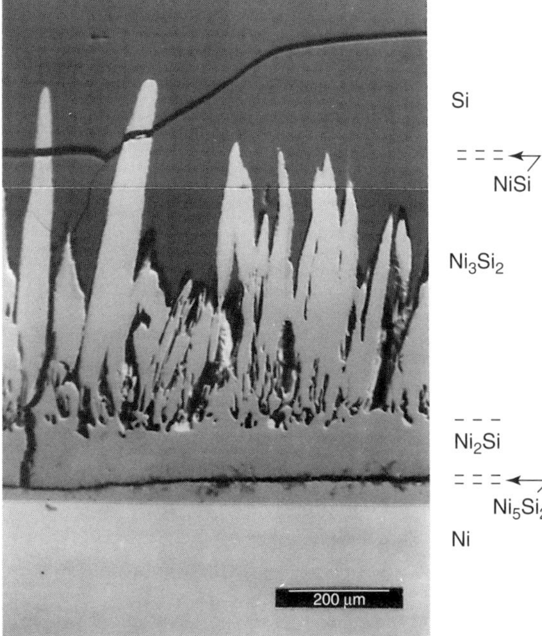

Figure 7.3 Optical cross-sectional micrograph of a bulk couple of Ni and Si annealed at 850 °C for 8 h, showing the formation of Ni_5Si_2, Ni_2Si, Ni_3Si_2, and NiSi.

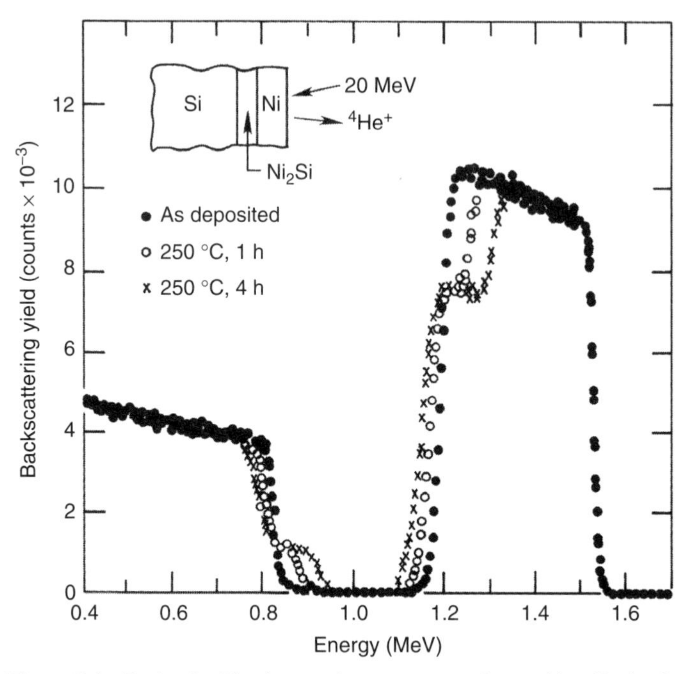

Figure 7.4 Rutherford backscattering spectrum taken to identify the formation of Ni_2Si in the thin film sample.

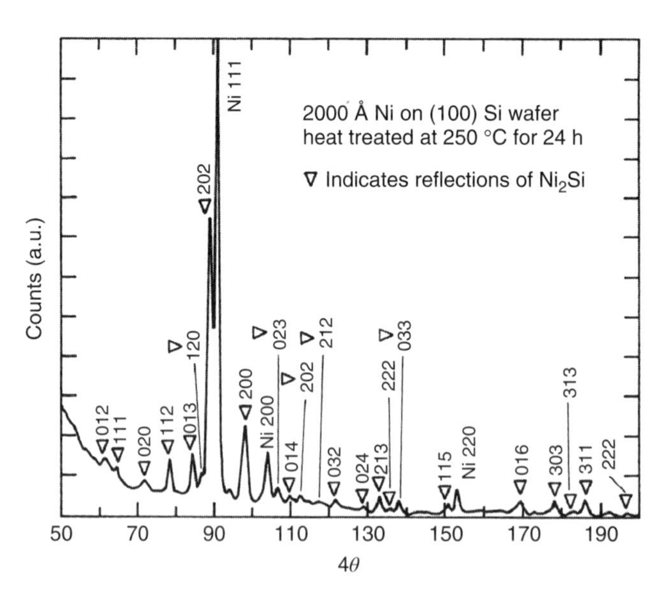

Figure 7.5 Glancing incidence X-ray diffraction spectrum taken to identify the formation of Ni_2Si in the thin film sample.

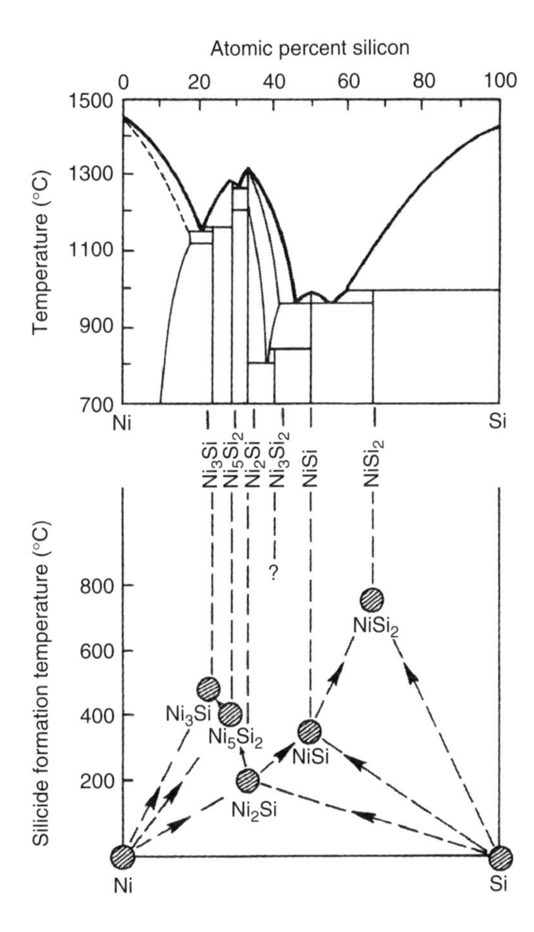

Figure 7.6 Schematic diagram depicting the sequence of phase formation in the interfacial reaction between Ni and Si. The first phase formation is Ni_2Si. The subsequent phase will depend on whether there is excess Ni or excess Si.

the selection of a specific phase to form among all the equilibrium phases? A kinetic mechanism of growth competition is proposed to explain the missing phase phenomenon.

7.2 BULK CASES

7.2.1 Kidson's Analysis of Diffusion-Controlled Planar Growth

Figure 7.7 depicts the growth of a layered intermetallic compound phase between two pure elements, for example, the growth of Ni_2Si between Si and Ni. We represent Si, Ni_2Si, and Ni by α, β, and γ, respectively. The thickness of Ni_2Si is ξ_β and the position of its interface with Si and Ni is defined by $\xi_{\alpha\beta}$ and $\xi_{\beta\gamma}$, respectively. Across the interfaces, there is an abrupt change in concentration. For example, the composition across the $\xi_{\alpha\beta}$ interface changes from $C_{\alpha\beta}$ to $C_{\beta\alpha}$. In a diffusion-controlled growth of ξ_β, the concentrations at its interfaces are assumed to have the equilibrium value, which is independent of time and position.

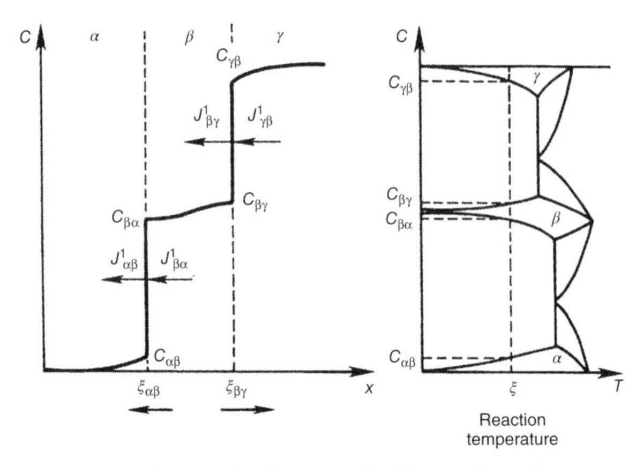

Figure 7.7 Schematic diagram of a binary phase diagram on the right for a single intermetallic compound, and a sketch of the concentration profile during the compound growth, showing the abrupt concentration change across the interfaces.

To consider a diffusion-controlled growth of a layered phase of ξ_β in Figure 7.7, we use Fick's first law in one-dimension to describe the flux within the layer,

$$J = -D\frac{dC}{dx} \tag{7.1}$$

where J is atomic flux, having the unit of number of atoms/cm^2-s, D is atomic diffusivity, cm^2/s, C is concentration, number of atoms/cm^3, and x is length, cm. At the interface, $\xi_{\alpha\beta}$, by conservation of matter, we have

$$(\Delta C)v = \Delta J \tag{7.2}$$

$$(C_{\alpha\beta} - C_{\beta\alpha})\frac{d\xi_{\alpha\beta}}{dt} = J_{\alpha\beta} - J_{\beta\alpha} = -D\frac{\partial C}{\partial x}\Big|_{\alpha\beta} + D\frac{\partial C}{\partial x}\Big|_{\beta\alpha} \tag{7.3}$$

where $v = d\xi_{\alpha\beta}/dt$ is the velocity of the interface $\xi_{\alpha\beta}$. Rearranging, we obtain the expression of velocity of the $x_{\alpha\beta}$ interface as

$$\frac{d\xi_{\alpha\beta}}{dt} = \frac{1}{C_{\alpha\beta} - C_{\beta\alpha}}\left[\left(-D\frac{\partial C}{\partial x}\right)_{\alpha\beta} - \left(-D\frac{\partial C}{\partial x}\right)_{\beta\alpha}\right] \tag{7.4}$$

To overcome the unknown concentration gradients in the square bracket in the above equation, we can make a transformation by combining the two variables of x and t into one, that is, Boltzmann's transformation,

$$C(x,t) = C(\eta)$$

where

$$\eta = \frac{x}{\sqrt{t}}$$

and so

$$\frac{\partial C}{\partial x} = \frac{1}{\sqrt{t}} \frac{dC}{d\eta} \tag{7.5}$$

Because the concentrations at the interface, that is, $C_{\alpha\beta}$ and $C_{\beta\alpha}$, can be assumed to remain constant with respect to time and position, because we can take them as the equilibrium values under the assumption of a diffusion-controlled growth [8], we have

$$\frac{dC(\eta)}{d\eta} = f(\eta) \tag{7.6}$$

where $f(\eta)$ is constant if η is constant, independent of time and position, at the interfaces for a diffusion-controlled process. Therefore, the equation of velocity can be rewritten as

$$\frac{d\xi_{\alpha\beta}}{dt} = \frac{1}{C_{\alpha\beta} - C_{\beta\alpha}} \left[-\left(D\frac{\partial C}{\partial \eta} \right)_{\alpha\beta} + D\left(\frac{\partial C}{\partial \eta} \right)_{\beta\alpha} \right] \frac{1}{\sqrt{t}} \tag{7.7}$$

The quantity within the square bracket is independent of time, after we have taken the factor of time out of the square bracket. Integration of the above equation gives

$$\xi_{\alpha\beta} = A_{a\beta}\sqrt{t} \tag{7.8}$$

where

$$A_{\alpha\beta} = 2\left[\frac{(DK)_{\beta\alpha} - (DK)_{\alpha\beta}}{C_{\alpha\beta} - C_{\beta\alpha}} \right]$$

$$K_{ij} = \left(\frac{dC}{d\eta} \right)_{ij}$$

Following the similar approach, we obtain at the other interface of $\xi_{\beta\gamma}$,

$$\xi_{\beta\gamma} = A_{\beta\gamma}\sqrt{t} \tag{7.9}$$

By combining the two interfaces, we have the width of the β phase to be

$$W_{\beta} = \xi_{\beta\gamma} - \xi_{\alpha\beta} = (A_{\beta\gamma} - A_{\alpha\beta})\sqrt{t} = B\sqrt{t} \tag{7.10}$$

that shows that the β phase has a parabolic growth rate or diffusion-controlled growth. We note that the above is a very simple derivation of a diffusion-controlled growth of a layered phase, or a relationship of $w^2 \propto t$ for a layered growth with abrupt change of composition at its interfaces by using only Fick's first law.

A fundamental nature of a diffusion-controlled layer growth is that the layer will not disappear or it cannot be consumed in the competition of growth in a multilayered structure because its velocity of growth is inversely proportional to its thickness. As the thickness "w" approaches 0,

$$\lim \left.\frac{dw}{dt}\right|_{w\to 0} = \frac{B}{w} \to \infty \tag{7.11}$$

The growth rate will approach infinity, or the chemical potential gradient to drive the growth will approach infinity. Thus, it cannot be consumed.

Therefore, in a multilayered structure, for example, Si/NiSi/Ni_2Si/Ni, when both NiSi and Ni_2Si coexist and have diffusion-controlled growth, they will coexist and grow together. For this reason, in a sequential growth of Ni_2Si followed by NiSi, we cannot assume that both of them can nucleate and grow together by diffusion-controlled process; otherwise they will coexist and we cannot have the growth of a single intermetallic compound phase. To overcome this difficulty, we introduce interfacial-reaction-controlled growth, which is discussed in Section 7.3.

7.2.2 Steady State Approximation in Layered Growth of Multiple Phases

Kidson's analysis of the layer growth of multiple phases by considering only interface movement cannot give a self-consistent solution of kinetics of layered growth. Generally speaking, the flux balance equation, Eq (7.3), at each moving interface should be treated as the boundary condition of the Fick's second law of diffusion inside each phase layer. In mathematics, this is called *Stephan problem*. When a system has several layers with moving interfaces and with concentration dependent diffusion coefficient within each layer, only numerical analysis becomes possible. Nevertheless, in most practical cases, when all the intermetallic compound phases have almost stoichiometric composition (having a very narrow concentration range), the rigorous solution is not needed at all. Then, the rate of concentration change within each phase is close to zero. And as $\partial C/\partial t = -\partial J_x/\partial x$, the interdiffusion flux is almost the same in each point of the layer. In other words, when $\partial C/\partial t = 0$, we have $J_x = \text{const}$. We have for the "i" phase,

$$J^{(i)} = -\widetilde{D}^{(i)} \frac{C_{\mathrm{R}}^{(\mathrm{i})} - C_{\mathrm{L}}^{(\mathrm{i})}}{x_{\mathrm{R}}^{(\mathrm{i})} - x_{\mathrm{L}}^{(\mathrm{i})}} \cong -\widetilde{D}^{(\mathrm{i})} \frac{\Delta C^{(\mathrm{i})}}{\Delta x^{(\mathrm{i})}}$$

If the phase growth is diffusion-controlled, the concentrations at the two interfaces, $C_{\mathrm{R}}^{(\mathrm{i})}, C_{\mathrm{L}}^{(\mathrm{i})}$, are constants and correspond to the concentrations in equilibrium phase diagram.

To further illustrate this simple picture of the growth of "i" phase between A and B with negligible solubility, as shown in Figure 7.8, we have, respectively, at the left and the right boundaries of the "i" phase,

$$(C_{\mathrm{L}}^{(\mathrm{i})} - 0)\frac{dx_{\mathrm{Ai}}}{dt} = -\widetilde{D}^{(\mathrm{i})} \frac{\Delta C^{(\mathrm{i})}}{\Delta x^{(\mathrm{i})}} - 0$$

$$(C^{\mathrm{B}} - C_{\mathrm{R}}^{(\mathrm{i})})\frac{dx_{\mathrm{iB}}}{dt} = 0 - \left(-\widetilde{D}^{(\mathrm{i})} \frac{\Delta C^{(\mathrm{i})}}{\Delta x^{(\mathrm{i})}}\right)$$

Rearranging them and changing the "i" phase to "β" phase, we have

$$\frac{d(x_{iB} - x_{Ai})}{dt} = \left(\frac{1}{C^B - C_R^{(i)}} + \frac{1}{C_L^{(i)}} \right) \widetilde{D}^{(i)} \frac{\Delta C^{(i)}}{\Delta x^{(i)}}$$

Thus,

$$(\Delta x_\beta)^2 = 2\frac{(C_L^{(i)} + C^B - C_R^{(i)})\Delta C^{(i)}}{(C^B - C_R^{(i)})C_L^{(i)}} \widetilde{D}^{(i)} t = 2\frac{C^B \Delta C^{(\beta)}}{C^{(\beta)}(C^B - C^{(\beta)})} \widetilde{D}^{\beta} t \tag{7.12}$$

In the above equation, we drop $(C_L^{(i)} - C_R^{(i)})$ because it is small as compared to C^B. Also, we take $C_L^{(i)} \approx C_R^{(i)} \approx C^{(\beta)}$.

As an example of the application of Eq (7.12), we consider the growth of Ni_2Si between a Ni thin film and a Si wafer. We take Si Ni_2Si and Ni to be A, i, and B phase as depicted in Figure 7.8. If we assume that the atomic volume of Ni is Ω, we have $C^B = 1/\Omega$, and $C^{(\beta)} = (2/3)(1/\Omega)$, if the average volume per atom in the intermetallic phase is the same as atomic volume of pure Ni. However, we cannot determine $\Delta C^{(\beta)} \cong C_R^{(i)} - C_L^{(i)}$, because it is extremely small. This is the difficulty when we assume that the intermetallic compound phase has a stoichiometric composition (having a very narrow concentration range). To overcome it, we have to introduce Wagner diffusivity, which is presented in Section 7.2.5.

7.2.3 Marker Analysis

If it is known that the thickness or the position of the interface of a compound layer is proportional to $(\text{time})^{1/2}$, the growth of the layer is diffusion-controlled. Thus it is

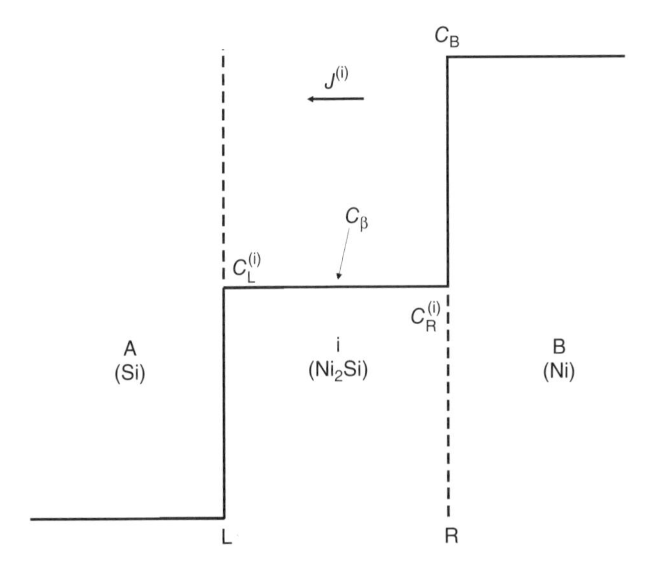

Figure 7.8 Formation of a single intermetallic compound of i-phase between A and B.

possible to measure the interdiffusion coefficient by taking

$$(\Delta x_\beta)^2 = 2\frac{C^{\mathrm{B}}\Delta C^{(\beta)}}{C^{(\beta)}(C^{\mathrm{B}} - C^{(\beta)})}\widetilde{D}^{\beta}t \tag{7.13}$$

As there are two unknowns of the intrinsic diffusivities of A and B in the compound phase, we need one more equation to solve them. Besides Eq (7.13), we need the marker motion equation. On experimental marker analysis, because only in thin film (not in bulk couples) contact reaction that a single intermetallic compound layer forms. The experimental results discussed below were from marker analysis in Ni_2Si formation between a Ni thin film deposited on a (100) Si wafer. Figure 7.9 shows Rutherford backscattering spectra before and after annealing at 250 °C of a Ni thin film on a Si wafer having Xe marker implanted near the Ni/Si interface before annealing. After annealing to form Ni_2Si, the Xe marker was found to have moved outward, indicating that Ni is the dominant diffusing species. Figure 7.10 is a plot of the marker position as a function of annealing time. The plot shows that Ni is the dominant diffusing species; the measured marker position is in good agreement with the calculated marker position assuming Ni to be the diffusing species.

If the marker is placed at the original interface between A and B, each B atom passing the marker will form a molecule of $A_\beta B$ on the other side. If we ignore partial

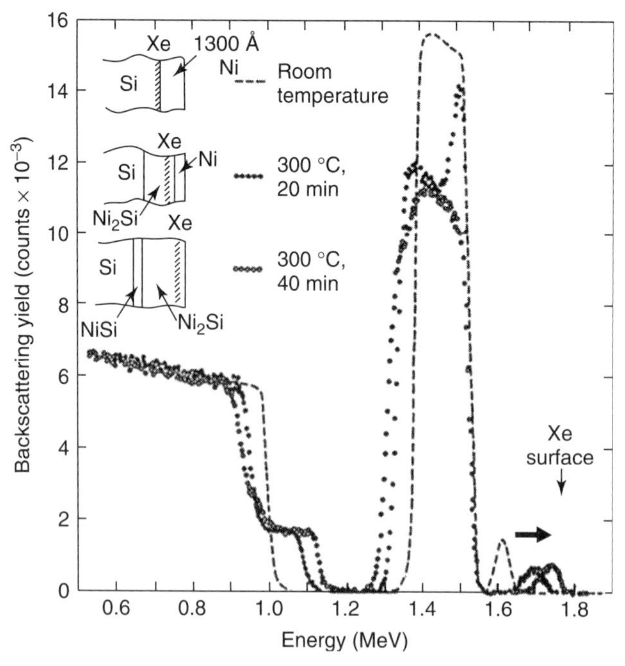

Figure 7.9 Rutherford backscattering spectra before and after annealing at 250 °C of a Ni thin film on a Si wafer having Xe marker implanted near the Ni/Si interface before annealing.

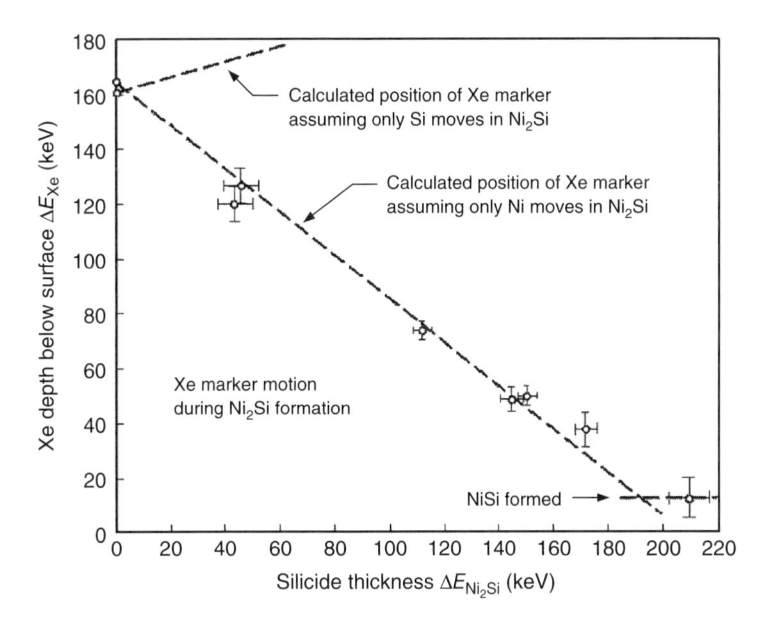

Figure 7.10 A plot of the marker position as a function of annealing time.

molar volume change, we have

$$\frac{J_B}{\beta J_A} = \frac{x_m - x_1}{x_2 - x_m} \tag{7.14}$$

And from the flux equations, we have

$$J_A = -D_\beta^A \frac{\partial C_A}{\partial x} \quad \text{and} \quad J_B = -D_\beta^B \frac{\partial C_B}{\partial x}$$

Then we have a pair of equations relating the intrinsic diffusion coefficient,

$$\frac{D_\beta^B}{D_\beta^A} = \frac{\beta(x_m - x_1)}{x_2 - x_m} \tag{7.15}$$

$$(x_2 - x_1)^2 = 2\frac{C^B \Delta C^{(\beta)}}{C^{(\beta)}(C^B - C^{(\beta)})}\widetilde{D}^\beta t \tag{7.16}$$

where

$$\widetilde{D}^\beta = X_B D_A^\beta + X_A D_B^\beta = \frac{1}{1+\beta} D_A^\beta + \frac{\beta}{1+\beta} D_B^\beta$$

By measuring the marker displacement x_m and knowing the composition of $A_\beta B$, we can determine the ratio of the diffusion coefficients in Eq (7.15). From Eq (7.16), we may take ln and we can determine the activation energy of $\widetilde{D}^\beta$ when the thicknesses of the intermetallic have been measured at several temperatures. Still, even after knowing the thickness of $(x_2 - x_1)$, the time, and the equilibrium concentrations at the boundaries, we cannot proceed as we cannot determine $\Delta C^{(\beta)}$ until we introduce Wagner diffusivity, which is discussed in Section 7.2.5.

In Figure 3.7, we depict a single intermetallic compound of $A_\beta B$ (Ni_2Si forms between A (Si) and B (Ni)). The Matano plane is defined such that the two shaded area are equal. In principle, the interdiffusion coefficient, at the point C' in the compound, can be calculated from the slope at C' and the shaded area between $C=0$ and $C=C'$. Again, the measurement of the slope is difficult.

The Matano plane occurs, as depicted in Figure 3.7, which we have discussed in Section 3.2.2 in Chapter 3, at the location that is 2/3 from the Ni_2Si/Ni interface and 1/3 from the Si/ Ni_2Si interface, and it is the original interface between Ni and Si before the reaction or interdiffusion.

7.2.4 Interdiffusion Coefficient in Intermetallic Compound

In Figure 7.11, we depict the free energy diagram of A(α-phase) and B(β-phase) and an intermetallic compound (i-phase). We begin the analysis by considering the mixing of A and B atoms. First, we assume it is a mechanical mixing without chemical interaction. We have the Gibbs free energy of mixture as

$$G = \mu_A N_A + \mu_B N_B$$

where μ_A and μ_B are chemical potential of A and B atoms in pure metals, respectively, so they are constants. However, in the analysis below for an intermetallic compound, they vary with the concentration or fraction of concentration. N_A and N_B are the number of A and B atoms, respectively. We have $N_A + N_B = N$, which is the total number

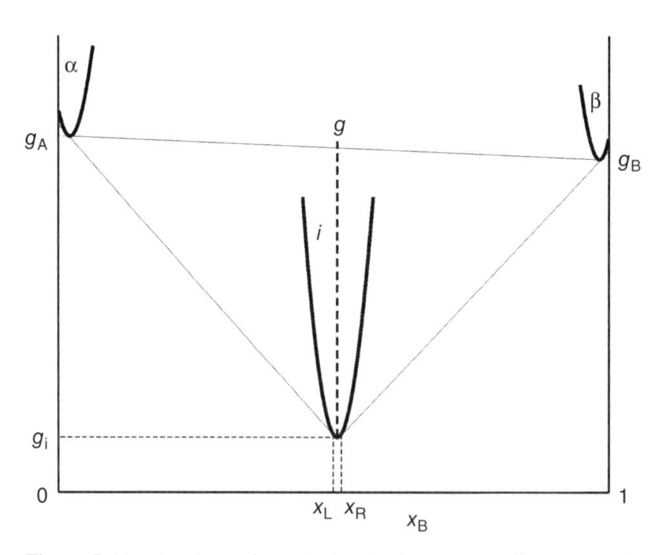

Figure 7.11 A schematic to depict the free energy diagram of A(α-phase) and B(β-phase) and an intermetallic compound (i-phase).

of atoms in the mixing. Then, $X_A = N_A/N$ and $X_B = N_B/N$, which are the fraction of A and B in the mixing, respectively. We take

$$G = gN \quad \text{or} \quad g = \frac{G}{N}$$

We have

$$g = \mu_A X_A + \mu_B X_B \tag{7.17}$$

where g is the Gibbs free energy per atoms. In Eq (7.17), μ_A is a function of X_A, and μ_B is a function of X_B. In Figure 7.11, we draw a common tangent line to the α-phase and β-phase, and the tangent meets the concentration line of X_i at the point g, which is shown in Figure 7.11. As

$$dG = -SdT + Vdp + \sum_{i=1}^{2} \mu_i dN_i$$

dividing the above equation by N, we have

$$dg = -sdT + \Omega dp + \mu_A dX_A + \mu_B dX_B$$

As $X_B = 1 - X_A$, we have $dX_A = -dX_B$. At constant temperature and constant pressure, we have

$$dg|_{T,p} = (\mu_A - \mu_B)dX_A$$

or

$$\frac{\partial g}{\partial X_A} = \mu_A - \mu_B \tag{7.18}$$

Similarly, we have

$$\frac{\partial g}{\partial X_B} = \mu_B - \mu_A$$

We multiple Eq (7.18) by X_B and add it to Eq (7.17). Because $X_A + X_B = 1$, we obtain

$$\mu_A = g + X_B \frac{\partial g}{\partial X_A} \tag{7.19}$$

Substituting Eq (7.19) into Eq (7.18), we have

$$\mu_B = g - X_A \frac{\partial g}{\partial X_A} \tag{7.20}$$

Eqs (7.19) and (7.20) indicate that if we know the function of g, we can calculate the chemical potential of A and B. Furthermore, we take the differential of μ_A with respect to X_A,

$$\frac{\partial \mu_A}{\partial X_A} = \frac{\partial g}{\partial X_A} + \frac{\partial X_B}{\partial X_A}\frac{\partial g}{\partial X_A} + X_B \frac{\partial^2 g}{\partial X_A^2} = X_B \frac{\partial^2 g}{\partial X_A^2} = X_B g''$$

In the above equation, we note that $\partial X_B/\partial X_A = -1$.

We shall define a thermodynamic factor "φ" as below, and the reason of having such a thermodynamic factor for interdiffusion coefficient is explained later.

$$\varphi = \frac{X_A}{kT}\frac{\partial \mu_A}{\partial X_A} = \frac{X_A X_B}{kT} g'' \tag{7.21}$$

In Figure 7.11, we draw the common tangent of the α-phase and the i-phase, and the tangential point on the latter is X_L. Similarly, the common tangent of the β-phase and the i-phase has the tangential point on the latter at X_R. And we define $\Delta X = X_R - X_L$. Then we express

$$\frac{\left.\frac{\partial g}{\partial X}\right|_{X_R} - \left.\frac{\partial g}{\partial X}\right|_{X_L}}{X_R - X_L} \approx \frac{\partial^2 g}{\partial X^2} = g''$$

From the above equation, we have

$$\begin{aligned} g''\Delta X &= \left.\frac{\partial g}{\partial X}\right|_{X_R} - \left.\frac{\partial g}{\partial X}\right|_{X_L} = \frac{g_B - g_i}{1 - X_i} - \frac{g_i - g_A}{X_i - 0} = \frac{[g_B X_i + g_A(1 - X_i)] - g_i}{(1 - X_i)X_i} \\ &= \frac{g - g_i}{(1 - X_i)X_i} = \frac{\Delta g_i}{X_A X_B} \\ \Delta g_i &= X_A X_B g'' \Delta X \end{aligned} \tag{7.22}$$

In the above equation, X_i is located roughly between X_R and X_L. The square bracket term is equal to "g" by the lever rule, and we can take $g - g_i = \Delta g_i$ to be the intermetallic compound formation energy per atom. The purpose of the above derivation is to show that

$$\widetilde{D} = (X_A D_B^* + X_B D_A^*)\varphi = (X_A D_B^* + X_B D_A^*)\frac{X_A X_B}{kT} g'' = (X_A D_B^* + X_B D_A^*)\frac{\Delta g_i}{kT\Delta X}$$

$$\widetilde{D}\Delta X = (X_A D_B^* + X_B D_A^*)\frac{\Delta g_i}{kT}$$

We note that although ΔX is unmeasurable, the product of $\widetilde{D}\Delta X$, in terms of Δg_i, becomes measurable. Because of the chemical effect, Darken's interdiffusion coefficient is modified by the thermodynamic factor φ. Following the derivation of Eq (3.17) in Chapter 3, we derive the above interdiffusion coefficient. First, we have in the moving coordination,

$$j_B = C_B \frac{D_B^*}{kT}\left(-\frac{\partial \mu_B}{\partial x}\right) = -C_B \frac{D_B^*}{kT}\frac{\partial \mu_B}{\partial C_B}\frac{\partial C_B}{\partial x} = -D_B^*\left(\frac{X_B}{kT}\frac{\partial \mu_B}{\partial X_B}\right)\frac{\partial C_B}{\partial x} = -D_B^*\varphi\frac{\partial C_B}{\partial x}$$

Similarly, we have

$$j_A = -D_A^*\varphi\frac{\partial C_A}{\partial x}$$

Then, in the laboratory frame or fixed frame, we have

$$
\begin{aligned}
J_B &= j_B + C_B v = -D_B^* \varphi \frac{\partial C_B}{\partial x} - C_B \frac{1}{C}(j_A + j_B) \\
&= -\frac{C_A + C_B}{C} D_B^* \varphi \frac{\partial C_B}{\partial x} - X_B \left(-D_A^* \varphi \frac{\partial C_A}{\partial x} - D_B^* \varphi \frac{\partial C_B}{\partial x} \right) \\
&= -(X_A D_B^* + X_B D_A^*) \varphi \frac{\partial C_B}{\partial x} = -\widetilde{D} \frac{\partial C_B}{\partial x}
\end{aligned}
$$

Thus, we have

$$
\widetilde{D} = (X_A D_B^* + X_B D_A^*) \varphi \tag{7.23}
$$

To understand the physical meaning of φ, we need to calculate "g" by using a regular solution. We take the free energy per atom "g" of the regular solution to be

$$
g = \frac{Z}{2}(X_A^2 \varepsilon_{AA} + X_B^2 \varepsilon_{BB} + 2X_A X_B \varepsilon_{AB}) + kT(X_A \ln X_A + X_B \ln X_B)
$$

where Z is the number of nearest neighbors in the solid solution, and ε_{AA}, ε_{BB}, and ε_{AB} are the AA, BB, and AB bond energies, respectively. We need to differentiate g, and we recall that

$$
\begin{aligned}
\frac{\partial X_A}{\partial X_B} &= -1 \\
\frac{\partial^2}{\partial X_B^2}(X_B^2) &= \frac{\partial}{\partial X_B}(2X_B) = 2 \\
\frac{\partial^2}{\partial X_B^2}(X_A^2) &= \frac{\partial}{\partial X_B}(-2X_A) = +2 \\
\frac{\partial^2}{\partial X_B^2}(X_A X_B) &= \frac{\partial}{\partial X_B}[(-1)X_B + X_A] = (-1) + (-1) = -2 \\
\frac{\partial^2}{\partial X_B^2}(X_B \ln X_B) &= \frac{\partial}{\partial X_B}\left(\ln X_B + X_B \frac{1}{X_B} \right) = \frac{1}{X_B} + 0 \\
\frac{\partial^2}{\partial X_B^2}(X_A \ln X_A) &= \frac{\partial}{\partial X_B}\left[(-1)\ln X_A + X_A \left(\frac{-1}{X_A} \right) \right] = (-1)\frac{-1}{X_A} + 0 = \frac{1}{X_A} \\
g'' = \frac{\partial^2 g}{\partial X_B^2} &= \frac{Z}{2}[2\varepsilon_{AA} + 2\varepsilon_{BB} + 2(-2)\varepsilon_{AB}] + kT\left(\frac{1}{X_A} + \frac{1}{X_B} \right) \\
&= 2Z\left[\frac{1}{2}(\varepsilon_{AA} + \varepsilon_{BB}) - \varepsilon_{AB} \right] + kT\frac{X_A + X_B}{X_A X_B} = -2ZE^{\text{mix}} + \frac{kT}{X_A X_B}
\end{aligned}
$$

where $E^{\text{mix}} = -\varepsilon_{\text{AB}} - (1/2)(\varepsilon_{\text{AA}} + \varepsilon_{\text{BB}})$. Therefore, we have

$$\varphi = \frac{X_A X_B}{kT} g'' = \frac{X_A X_B}{kT}\left(-2ZE^{\text{mix}} + \frac{kT}{X_A X_B}\right) = 1 + \frac{2Z(-E^{\text{mix}})X_A X_B}{kT} \tag{7.24}$$

We see that for an ideal solid solution, when $E^{\text{mix}} = 0$, $\varphi = 1$. When there is chemical interaction to form a regular solution or an intermetallic compound, and E^{mix} is negative, $\varphi > 1$.

7.2.5 Wagner Diffusivity

In the kinetic analysis of the layered growth of a compound, the driving force is chemical potential gradient, rather than concentration gradient. Yet in the flux equation or growth equation of the compound layer, we have used ΔC or ΔX as in Eq (7.18), yet ΔC or $\Delta X\ (= \Omega \Delta C)$ is extremely small for a stoichiometric compound. How can we calculate the flux and the growth rate? Conceptually, this is a difficult problem for layered compound growth. However, in the analysis, the $\widetilde{D}$ and ΔC always appear together as a product, and the product is called *Wagner diffusivity* in the literature. We consider the product below.

$$\begin{aligned} \widetilde{D}\Delta X &= (X_A D_B^* + X_B D_A^*)\varphi \Delta X = (X_A D_B^* + X_B D_A^*)\frac{X_A X_B}{kT} g'' \Delta X \\ &= (X_A D_B^* + X_B D_A^*)\left(\frac{\Delta g_i}{kT}\right) = \Omega(C^A D_B^* + C^B D_A^*)\left(\frac{\Delta g_i}{kT}\right) \end{aligned} \tag{7.25}$$

It becomes a measurable product because all the terms in the last part of the above equation can be measured independently. We recall that Δg_i is the formation energy per atom of the intermetallic compound.

By substituting Eq (7.25) into Eq (7.12), we have

$$(x_2 - x_1)^2 = 2\frac{C^B}{C^{(\beta)}(C^B - C^{(\beta)})\Omega}(X_A D_B^* + X_B D_A^*)\left(\frac{\Delta g_i}{kT}\right)^t \tag{7.26}$$

In Eq (7.15), we can replace the ratio of intrinsic diffusivities of D_β^B / D_β^A by the ratio of D_B^* / D_A^* in the β-phase, and we have

$$\frac{D_B^*}{\beta D_A^*} = \frac{(x_m - x_1)}{x_2 - x_m} \tag{7.27}$$

By combining Eq (7.27) with Eq (7.26), we can determine the intrinsic diffusivities in the β compound. For Ni_2Si, the formation enthalpy has been measured to be −10.5 kcal/gram-atom [5, 9].

Below, we determine the intrinsic diffusivity of Ni and Si in Ni_2Si. As depicted in Figure 7.8, we take Si as phase A, Ni as phase B, and Ni_2Si as phase i. Thus

$X_A = 1/3$ and $X_B = 2/3$. Also, we have $C^B = 1/\Omega$, and $C^{(\beta)} = (2/3)(1/\Omega)$, and $\beta = 1/2$. As shown in Figure 7.9, we measure the thickness of $\Delta x_\beta = x_2 - x_1 = 200\,\text{nm}$ after $t = 40$ min at $T = 300\,°\text{C}$. Substituting these values in Eq (7.26), we have

$$(\Delta x_\beta)^2 = 2\frac{1/\Omega}{(2/3)(1/\Omega)[(1/\Omega) - (2/3)(1/\Omega)]\Omega}\left[\frac{1}{3}D^*_{\text{Ni}} + \frac{2}{3}D^*_{\text{Si}}\right]\frac{\Delta g_i}{kT}t$$

$$(\Delta x_\beta)^2 = [3D^*_{\text{Ni}} + 6D^*_{\text{Si}}]\frac{\Delta g_i}{kT}t \tag{7.28}$$

About Δg_i for Ni_2Si, we note that the value of -10.5 kcal/g-atom is the heat of formation of 1 mol of the compound divided by the number of atoms in the chemical formula [5]. We have $\Delta g_i = (10.5 \times 3)/(23 \times 3) = 0.46$ eV/atom, and $kT = 0.05$ eV/atom at 300 °C.

About the marker motion measurement, the results as depicted in Figure 3.10 show that the flux of Ni is dominant. By considering the uncertainty, we can take the ratio in Eq (7.27) to be about 20/1. Thus we have $D^*_{\text{Si}} = 0.1D^*_{\text{Ni}}$, and after it is substituted into Eq (7.28), we obtain

$$D^*_{\text{Ni}} = \frac{(2 \times 10^{-5})^2\text{cm}^2}{2400 \times 3.6 \times 9.2} = 0.5 \times 10^{-14}\text{cm}^2/\text{s} \quad \text{and} \quad D^*_{\text{Si}} = 0.5 \times 10^{-15}\text{cm}^2/\text{s}$$

The chemical effect is to slow down the tracer diffusivities in the silicide by one order of magnitude.

7.3 THIN FILM CASES

7.3.1 Diffusion-Controlled and Interfacial-Reaction-Controlled Growth

Contact reaction between a Si wafer and metallic films to form silicide intermetallic compound phases has been studied widely because of the application of silicides as source/drain and gate contacts in VLSI technology [1–6]. Single phase formation of a specific silicide on Si to serve as ohmic contacts and gates in field-effect transistor devices has been a very important technological issue. There are millions or even billions of silicide contacts and gates on a Si chip of the size of our finger nail. These contacts and gates must have the same physical properties. In other words, we cannot allow contacts to have different silicide phases or a mixture of silicide phases. Therefore, the device application demands single phase formation, which in principle is against thermodynamics. Thus, a kinetic rather than a thermodynamic reason has to be given for single phase formation in thin film contact reactions and in turn the reason for the selection of the specific single phase among the choice of multiple phases from the phase diagram. The kinetics of single phase growth has been analyzed, assuming a layered model of competition of growth of coexisting phases by combining diffusion-controlled growth and interfacial-reaction-controlled growth [10].

7.3.2 Kinetics of Interfacial-Reaction-Controlled Growth

In order to explain the "missing" phase phenomenon and in turn the single phase growth in thin film reactions, we can treat the layered growth of a compound phase in a more general manner by considering both diffusion-controlled and interfacial-reaction-controlled kinetic processes together. If we define J as atomic flux and C as concentration, if the ratio of

$$\frac{J}{-\frac{\partial C}{\partial x}} = D = \text{constant}$$

We are in a diffusion-controlled regime. On the other hand, if

$$\frac{J}{C} = v = \text{constant}$$

We are in an interfacial-reaction-controlled regime. Here we recall that, in Chapter 2, we showed that the last two equations can be linked by

$$J = Cv = CMF = C\frac{D}{kT}\left(-\frac{\partial \mu}{\partial x}\right)$$

where $M = D/kT$ is the atomic mobility and F is the driving force. For comparison, in a diffusion-controlled growth, the interfacial velocity has a $(\text{time})^{1/2}$ dependence, as shown by Kidson's analysis, but in an interfacial-reaction-controlled growth, the interfacial velocity is a constant, independent of time. We can take $J/C = K$, where K is defined as a reaction-controlled interfacial constant. It has the dimension of velocity and is a measure of the mobility of an interface. In Figure 7.12, a schematic diagram of energy barrier and jump frequency near an interface between A and $A_\beta B$ (or Ni and Ni_2Si) for (a) At equilibrium and (b) During reaction, are shown.

Figure 7.13 depicts the growth of a layered intermetallic compound phase between two pure elements, for example, the growth of Ni_2Si between Ni and Si. We represent Ni, Ni_2Si and Si by $A_\alpha B$, $A_\beta B$, and $A_\gamma B$, respectively. The thickness of Ni_2Si is x_β and the position of its interface with Ni and Si is defined by $x_{\alpha\beta}$ and $x_{\beta\gamma}$, respectively. Across the interfaces, there is an abrupt change in concentration as shown in Figure 7.13a, where the concentration change of Ni across the interfaces is shown.

In a diffusion-controlled growth of x_β, the concentrations at its interfaces are assumed to have equilibrium values, represented by the broken curve in the x_β layer in Figure 7.13a. In an interfacial-reaction-controlled growth, the concentrations at its interface are assumed to be nonequilibrium, represented by the solid curve in Figure 7.13a.

To formulate interfacial-reaction-controlled growth, we assume that the concentrations at the interfaces are not the equilibrium values as shown by the solid curve in x_β in Figure 7.13a. Why? Physically, we consider for the moment that $A_\beta B$ is a liquid solution and is dissolving A atoms from $A_\alpha B$, which is a pure phase of A. If the dissolution rate is extremely high and is only limited by how fast the A atoms can

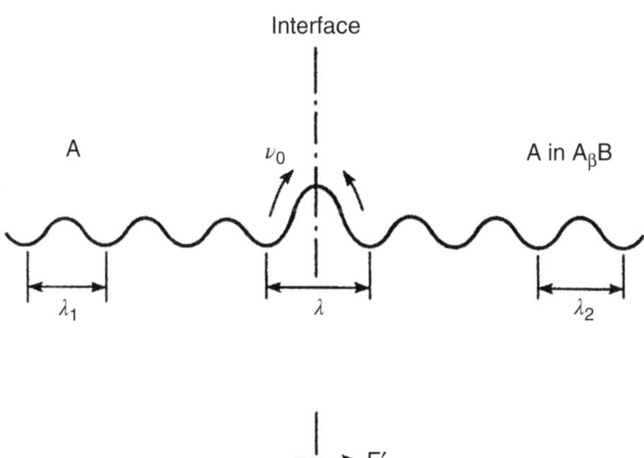

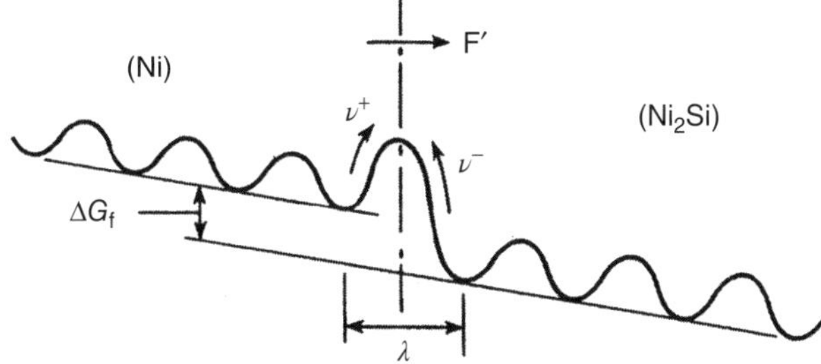

Figure 7.12 Schematic diagram of energy barrier and jump frequency near the interface between A and $A_\beta B$.

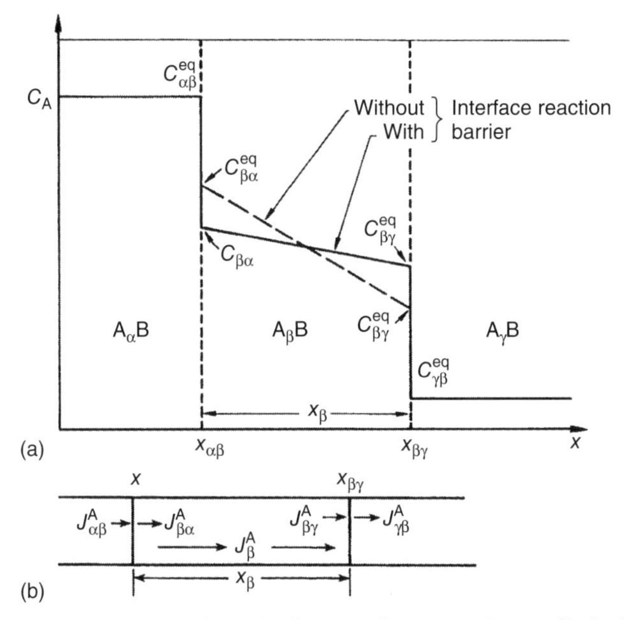

Figure 7.13 (a) Schematic diagram of concentration profile in the growth of a layered intermetallic compound phase between two pure elements, for example, the growth of Ni_2Si between Ni and Si. Both diffusion-controlled and interfacial-reaction-controlled growths are considered. (b) Fluxes in the growth process.

diffuse away, then the liquid will be able to maintain the equilibrium concentration of A near the interface even though A atoms are being drained away to the other end through the $A_\beta B$ phase. On the other hand, if the process of breaking A atoms from the $A_\alpha B$ surface is slow or the dissolution of A is slow, the liquid will not be able to maintain the equilibrium concentration at the interface because whenever an A atom is dissolved, it can diffuse away quickly. The interfacial-reaction-controlled process is the slower one, so the concentration of A near the interface will be undersaturated; it leads to $C_{\alpha\beta} < C_{\alpha\beta}^{eq}$, as indicated at the $x_{\alpha\beta}$ interface in Figure 7.13a. We assume that there is sluggishness in dissolving A atoms from the surface of $A_\alpha B$, so that the concentration $C_{\beta\alpha}$ is less than the equilibrium value.

At the other end of the $A_\beta B$ phase, A atoms are incorporated into the $A_\gamma B$ surface for the growth of the latter. If the incorporation can take place as soon as the atoms arrive at the interface, the equilibrium concentration can be maintained. However, at the $x_{\beta\gamma}$ interface, if there is sluggishness in accepting the incoming A atoms, there is a build-up of A atoms. It becomes supersaturated. The process is therefore interfacial-reaction-controlled, and the concentration of $C_{\beta\gamma}$ is greater than the equilibrium value, and we have $C_{\beta\gamma} > C_{\beta\gamma}^{eq}$.

The rate of the growth of the $A_\beta B$ phase does not depend on diffusion across itself, but rather it depends on the interfacial-reaction processes at the two interfaces. Now we consider the interface $x_{\alpha\beta}$, which is moving with a velocity $v = dx_{\alpha\beta}/dt$,

$$(C_{\alpha\beta}^{eq} - C_{\beta\alpha})\frac{dx_{\alpha\beta}}{dt} = J_{\alpha\beta}^{A} - J_{\beta\alpha}^{A} = \left(-\widetilde{D}_{\alpha}\frac{dC_{\alpha}^{A}}{dx}\right) - \left(-\widetilde{D}_{\beta}\frac{dC_{\beta}^{A}}{dx}\right)$$

$$= \widetilde{D}_{\beta}\frac{dC_{\beta}^{A}}{dx} = -J_{\beta}^{A} \tag{7.29}$$

where J is the atomic flux having the unit of the number of atoms/cm^2-s; D is the atomic diffusivity, cm^2/s; C is the concentration, number of atoms/cm^3; x is length, cm; and v is the velocity of the moving interface, cm/s. In Figure 7.13b, the fluxes are depicted.

The term $J_{\alpha\beta}^{A}$ goes to zero because we have assumed that $A_\alpha B$ is pure A, so the concentration of A is flat and its gradient is zero. The last equality in the last equation is the definition of a flux equation and it shows that $J_{\beta\alpha}^{A} = J_{\beta}^{A}$. In the compound of $A_\beta B$, we can assume a linear concentration gradient that

$$\frac{dC_{\beta}^{A}}{dx} = \frac{C_{\beta\alpha} - C_{\beta\gamma}}{x_{\beta}}$$

Therefore, we have

$$(C_{\alpha\beta}^{A} - C_{\beta\alpha})\frac{dx_{\alpha\beta}}{dt} = \widetilde{D}_{\beta}\frac{C_{\beta\alpha} - C_{\beta\gamma}}{x_{\beta}} = -J_{\beta}^{A} \tag{7.30}$$

If we consider the flux from the viewpoint of a reaction-controlled process, we have

$$J_\beta^A = (C_{\beta\alpha}^{eq} - C_{\beta\alpha})K_{\beta\alpha} \tag{7.31}$$

where $K_{\beta\alpha}$ is defined as the interfacial-reaction coefficient of the $x_{\alpha\beta}$ interface. It has the unit of velocity, cm/s, and it infers the rate of removal of A atoms from the $A_\alpha B$ surface. If there is no interface sluggishness, $C_{\beta\alpha}$ will approach $C_{\beta\alpha}^{eq}$. Because of sluggishness, however, the actual concentration at the interface is lower than the equilibrium value, so $K_{\beta\alpha}$ is a measure of the actual flux J_β^A leaving the interface with respect to the concentration change at the interface. The physical meaning of interfacial-reaction coefficient K (velocity) is that of interfacial velocity, which is a product of interfacial driving force and interfacial mobility.

Similarly, at the $x_{\beta\gamma}$ interface, we have

$$J_\beta^A = (C_{\beta\gamma} - C_{\beta\gamma}^{eq})K_{\beta\gamma} \tag{7.32}$$

From Eq (7.31) and Eq (7.32), we have, respectively,

$$\frac{J_\beta^A}{K_{\beta\alpha}} = C_{\beta\alpha}^{eq} - C_{\beta\alpha}$$

$$\frac{J_\beta^A}{K_{\beta\gamma}} = C_{\beta\gamma} - C_{\beta\gamma}^{eq}$$

By adding the last two equations, we have

$$J_\beta^A\left(\frac{1}{K_{\beta\alpha}} + \frac{1}{K_{\beta\gamma}}\right) = (C_{\beta\gamma} - C_{\beta\alpha}) + (C_{\beta\alpha}^{eq} - C_{\beta\gamma}^{eq})$$

Let

$$\frac{1}{K_\beta^{eff}} = \frac{1}{K_{\beta\alpha}} + \frac{1}{K_{\beta\gamma}}$$

As we have from Eq (7.30)

$$C_{\beta\gamma} - C_{\beta\alpha} = -\frac{J_\beta^A x_\beta}{\widetilde{D}_\beta}$$

If we define

$$\Delta C_\beta^{eq} = C_{\beta\alpha}^{eq} - C_{\beta\gamma}^{eq}$$

we obtain

$$J_\beta^A = \frac{\Delta C_\beta^{eq} K_\beta^{eff}}{\left(1 + \dfrac{x_\beta K_\beta^{eff}}{\widetilde{D}_\beta}\right)} \tag{7.33}$$

Now to calculate the thickening rate of $A_\beta B$, we take

$$\frac{dx_\beta}{dt} = \frac{d}{dt}(x_{\beta\gamma} - x_{\alpha\beta}) = \left(\frac{1}{C_{\beta\gamma} - C_{\gamma\beta}^{eq}} - \frac{1}{C_{\alpha\beta}^{eq} - C_{\beta\alpha}}\right) J_\beta^A = G_\beta J_\beta^A \tag{7.34}$$

By substituting Eq (7.33) into the last equation, we have

$$\frac{dx_\beta}{dt} = \frac{G_\beta \Delta C_\beta K_\beta^{eff}}{1 + x_\beta \dfrac{K_\beta^{eff}}{\widetilde{D}_\beta}} = \frac{G_\beta \Delta C_\beta K_\beta^{eff}}{1 + \dfrac{x_\beta}{x_\beta^*}} \tag{7.35}$$

where, we recall, $\frac{1}{K_\beta^{eff}}$ is the effective interfacial-reaction coefficient of the β-phase.

If we define a "changeover" thickness of

$$x_\beta^* = \frac{\widetilde{D}_\beta}{K_\beta^{eff}} \tag{7.36}$$

$$\frac{dx_\beta}{dt} = \frac{G_\beta \Delta C_\beta^{eq} K_\beta^{eff}}{1 + \dfrac{x_\beta}{x_\beta^*}}$$

For a large changeover thickness, or $x_\beta / x_\beta^* \ll 1$, that is, $\widetilde{D}_\beta \gg K_\beta^{eff}$, under the condition that the interdiffuion coefficient is much large than the effective interfacial-reaction coefficient, we obtain

$$\frac{dx_\beta}{dt} = G_\beta \Delta C_\beta K_\beta^{eff} \quad \text{or} \quad x_\beta \propto t \tag{7.37}$$

The process is interfacial-reaction-controlled, and the growth rate is constant. For a small changeover thickness, $x_\beta / x_\beta^* \gg 1$, that is, $\widetilde{D}_\beta \ll K_\beta^{eff}$, the process is diffusion-controlled. Then

$$\frac{dx_\beta}{dt} = G_\beta \Delta C_\beta^{eq} \frac{\widetilde{D}_\beta}{x_\beta} \quad \text{or} \quad (x_\beta)^2 \propto t \tag{7.38}$$

The above demonstrates the well-known relationship that in an interfacial-reaction-controlled growth, the layer thickness is linearly proportional to time, but in a diffusion-controlled growth, the layer thickness is proportional to the square root of time. Furthermore, a reaction-controlled growth will always change over to a diffusion-controlled growth when the layer thickness has grown sufficiently large, as shown in Figure 7.14.

There are two significant consequences in a linear growth. First, the linear growth cannot keep on going forever; when the layer grows to a certain thickness, diffusion across the thicker layer will be rate-limiting and the growth will change to diffusion-controlled or its time dependence will change from linear to parabolic, as

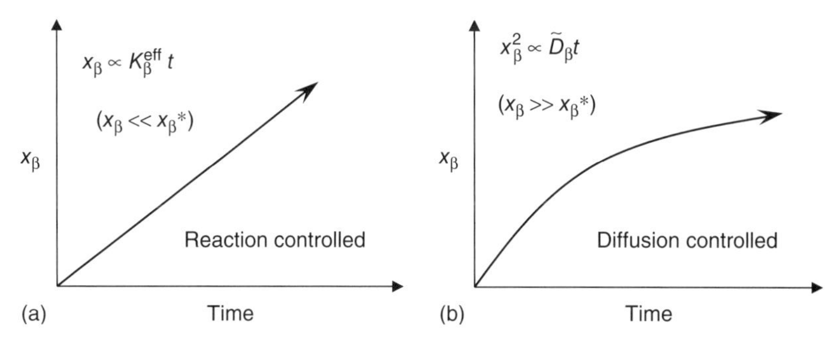

Figure 7.14 (a) A reaction-controlled growth will always change over to (b) A diffusion-controlled growth when the layer thickness has grown sufficient large.

shown in Figure 7.14. Second, and more importantly, a layer of linear growth cannot compete with a neighboring layer having a diffusion-controlled growth, especially when the latter is very thin having a very high growth rate. The latter will consume the former and become a single layer growth, which is discussed below.

7.3.3 Kinetics of Competitive Growth of Two-Layered Phases

We recall that an interfacial-reaction-controlled growth has a constant growth rate, independent of layer thickness. Thus a phase that is very thin and has a slow interfacial-reaction-controlled growth can be easily consumed by a very thin neighboring layer that has a faster interfacial-reaction-controlled growth or a diffusion-controlled growth.

In considering the competition of growth between two coexisting phases, we may have three combinations [10]. They are (i) Both are diffusion-controlled, (ii) Both are interfacial-reaction-controlled, and (iii) One is diffusion-controlled and the other is interfacial-reaction-controlled. Here we analyze the kinetics of the third case.

In Figure 7.15 we consider the competing growth of two-layered phases of $A_\beta B$ and $A_\gamma B$ between A and B. We assume that the growth of $A_\beta B$ is interfacial-reaction-controlled (i.e., $x_\beta \ll x_\beta^*$) and has a velocity v_1. The growth of $A_\gamma B$ is diffusion-controlled (i.e., $x_\gamma \gg x_\gamma^*$) and has velocity v_2. When their thickness is small, the magnitude of v_2 can be quite large because of the inverse dependence on layer thickness, so we can assume $v_2 \gg v_1$ and the rapid growth of $A_\gamma B$ can consume all of $A_\beta B$. Now we have obtained a single phase growth. Quantitatively, we have

$$J_\beta^A \cong \Delta C_\beta^{eq} K_\beta^{eff}$$

$$J_\gamma^A = \frac{\Delta C_\gamma^{eq} \widetilde{D}_\gamma}{x_\gamma}$$

Then the flux ratio is

$$\frac{J_\beta^A}{J_\gamma^A} = \frac{\Delta C_\beta^{eq} K_\beta^{eff}}{\Delta C_\gamma^{eq} \widetilde{D}_\gamma} x_\gamma \tag{7.39}$$

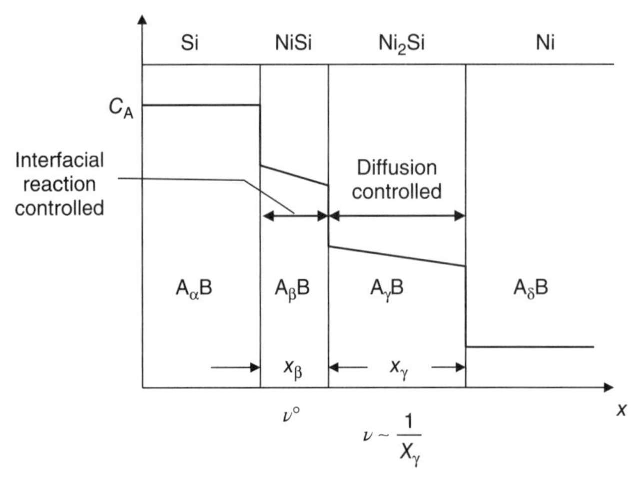

Figure 7.15 We consider the competing growth of two layered phases of $A_\beta B$ and $A_\gamma B$ between A and B. We assume that the growth of $A_\beta B$ is interfacial-reaction-controlled (i.e. $x_\beta \gg x_\beta^*$) and has a velocity v_0. The growth of $A_\gamma B$ is diffusion-controlled (i.e. $x_\gamma \gg x_\gamma^*$) and has velocity v which is proportional to $1/x_\gamma$.

When the above ratio is small, which means that $J_\beta^A \ll J_\gamma^A$, the flux in the diffusion-controlled growth of $A_\gamma B$ is very large (when it is very thin), so its growth rate is very fast and it can be much larger than the constant growth rate of $A_\beta B$. If the thickness of the latter is small, it can be taken over or consumed by the rapid growth of $A_\gamma B$.

We can rewrite the last equation as

$$x_\gamma^{\text{crit}} = \frac{\Delta C_\gamma^{\text{eq}} \widetilde{D}_\gamma}{\Delta C_\beta^{\text{eq}} K_\beta^{\text{eff}}} \frac{J_\beta^A}{J_\gamma^A} \tag{7.40}$$

The physical meaning of x_γ^{crit} is that when the thickness of x_γ is below the critical thickness, it will be able to consume x_β and achieve the single phase formation. The critical thickness is estimated to be of the order of the magnitude of microns. As most thin film diffusion couples have thicknesses in hundreds of nanometer, we observe only a single phase growth when the above criterion is satisfied. The conclusion is that by combining diffusion-controlled growth and interfacial-reaction-controlled growth, we can explain the phenomena of single phase growth in thin film reactions.

7.3.4 First Phase in Silicide Formation

Single phase formation in the reaction between Si and many metallic films to form intermetallic compound phases of silicide has been studied widely. For example, in the reaction between Si and Pt, Pd, Ni, or Co, the first phase formed is M_2Si, where

M represents the metals. In the case of Ni film on Si wafer, the sequential formation is that after Ni_2Si, the phase of NiSi and $NiSi_2$ will follow to form one by one. The same happens in Co–Si. In the case of Pt and Pd, the reaction stops at PtSi and PdSi.

For transition metals such as Ti, the first phase formed has been determined to be TiSi. Yet, sometimes, amorphous TiSi has been the first phase to form.

For refractory metals such as Mo and W, the first phase formation has been $MoSi_2$ and WSi_2, respectively. As they are the most Si-rich phase, they are also the last phase.

For rare-earth metals, such as Dy, the first phase formed is the disilicide.

Table 7.1 lists the phase formation of the three kinds of metal thin film silicides. We note that for the noble and near-noble metals, the transition metals, and the refractory metals, the first phase formation temperature is around 200, 400, and 600 °C, respectively. Actually, Pd_2Si can be formed around 100 °C; it has the lowest temperature of formation on Si wafers. Table 7.2 lists silicide formation sequence of various metals on Si and their free energy of formation.

In reacting a metal film and a Si wafer, the most important kinetic step is how to break the covalent bonds in the single crystal of Si. The formation temperature indicates that the mechanism of silicide formation of various metals is not the same because of the difference in formation temperature. Typically, to break the Si covalent bonds, thermal energy near 550 °C is required as indicated by the crystallization temperature of amorphous Si. This is also the lowest silicide formation temperature of transition metals and refractory metals.

On the other hand, it is known that noble and near-noble metals diffuse interstitially in Si. Transition metals and refractory metals diffuse substitutionally in Si. The different diffusion behavior affects their silicide formation temperature. Marker analysis has shown that in Ni_2Si formation, Ni is the dominant diffusing species. When a Ni atom dissolves interstitially in Si near the silicide/silicon interface, owing to charge transfer, the saturated Si covalent bonds are converted into unsaturated metallic bonds, so the bonds in Si around an interstitial Ni are weakened, so that they can be broken at a low temperature for silicide formation down to 100 °C [11].

TABLE 7.1 Comparison of the Three Transition Metal Silicide Classes

Characteristics	Near noble metal (Ni, Pd, Pt, Co, ...)	Refractory metal (W, Mo, V,Ta, ...)	Rare earth metal (Eu, Gd, Dy, Er, ...)
First phase formed	M_2Si	MSi_2	MSi_2
Formation temperature (°C)	~200	~600	~350
Growth rate	$x^2 \propto t$	$x \propto t$	?
Activation energy of growth	1.1–1.5 eV	>2.5 eV	?
Dominant diffusion species	Metal	Si	Si
Barrier height to n-Si (eV)	0.66–0.93	0.52–0.68	~0.40
Resistivity (μΩ-cm)	20–100	13–1000	100–300

TABLE 7.2 Heat of Formation ΔH of Silicides

Silicide	ΔH (kcal/g-atom)	Silicide	ΔH (kcal/g-atom[1])	Silicide	ΔH (kcal/g-atom[1])
Mg_2Si	6.2	Ti_5Si_3	17.3	V_3Si	6.5
		TiSi	15.5	V_5Si_3	11.8
FeSi	8.8	$TiSi_2$	10.7	VSi_2	24.3
$FeSi_2$	6.2				
		Zr_2Si	16.7	Nb_5Si_3	10.9
Co_2Si	9.2	Zr_5Si_3	18.3	$NbSi_2$	10.7
CoSi	12	ZrSi	18.5, 17.7	Ta_5Si_3	9.5
$CoSi_2$	8.2	$ZrSi_2$	12.9, 11.9	$TaSi_2$	8.7, 9.3
Ni_2Si	11.2, 10.5	HfSi		Cr_3Si	7.5
NiSi	10.3	$HfSi_2$		Cr_5Si_3	8
				CrSi	7.5
Pd_2Si	6.9			$CrSi_2$	7.7
PdSi	6.9			Mo_3Si	5.6
				Mo_5Si_3	8.5
Pt_2Si	6.9			$MoSi_2$	8.7, 10.5
PtSi	7.9			W_5Si_3	5
RhSi	8.1			WSi_2	7.3

In single silicide phase growth on (001) Si wafer, one of the very unique findings is the epitaxial growth of silicide on Si. The typical examples are the growth of $NiSi_2$ and $CoSi_2$ on Si [12, 13]. These two silicides have cubic CaF_2 structure and lattice parameter about 0.54 nm, which is very close to that of Si. It is worth mentioning that Schottky barrier height of epitaxial silicide is lower than that of nonepitaxial silicde. During the epitaxial growth, the epitaxial interface is a moving interface. Epitaxial growth of silicide in the nanowire of Si is discussed in the next section.

This principle of breaking the Si covalent bonds should be applied to silicide formation in Si nanowires. For those Si nanowires that have a very small radius, Gibbs–Thomson effect on contact reactions should be considered. In nanowire cases of contact reactions, which are discussed below, we note that the oxide on the Si nanowire and the easy oxidation of metallic nanowires, for example, Ti nanowires, are issues to be carefully studied.

7.4 NANOWIRE CASES

There are three kinds of contact reactions on a Si nanowire. If a Si nanowire crosses and touches a metal wire, it forms a point contact. If a Si nanowire lies on a pair of metal pads, the contacts are a line. If a Si nanowire is coated with a metal film by electroplating or if we deposit a metal film over a Si nanowire, the contact is a surface. In the last case, if Si diffuses out during the reaction, we can have nanotube formation, as discussed in Chapter 1.

7.4.1 Point Contact Reactions

As we have discussed the contact reactions in bulk and thin films of Ni and Si, we have chosen now to analyze the point contact reaction between a Si nanowire and a Ni nanowire, in which the Si nanowire is transformed into a grain of single crystal NiSi. The reaction was performed *in situ* in a high-resolution transmission electron microscope (HRTEM) and an ultrahigh vacuum transmission electron microscopy (TEM) chamber at 550–700 °C. To drive the reaction, Ni atoms dissolve into the Si nanowire from the location of the point contact. The NiSi silicide and Si form an atomically flat epitaxial interface that moves as the reaction progresses. Migration of the interface is linear with time, as recorded by *in situ* TEM video. While a mismatch should exist at the interface, no misfit dislocations were found. The point contact reaction has enabled us to fabricate single crystal NiSi/Si/NiSi heterostructures for nanoscale devices as well as nanogap of Si in between the NiSi electrodes.

The potential of nanoscale transistor devices based on Si nanowires is of wide interest for sensor applications [14, 15]. To realize this potential, nanoscale device elements such as ohmic contacts and gates must be developed. The formation of these circuit elements requires a systematic study of chemical reactions in the nanoscale. One good reason for the study of Ni and Si is that the monosilicide NiSi is one of the three silicides, NiSi, $CoSi_2$ and C-54 $TiSi_2$, having the lowest resistivity for applications in shallow junction devices. Compared to the other two, the monosilicide of NiSi consumes less Si. In point contact reactions between nanoscale Ni and Si, we observed a reactive epitaxial growth of NiSi on Si in which the epitaxial interface moves. This growth mode differs from conventional molecular beam epitaxy by deposition in which the epitaxial interface is static.

Si nanowires were prepared on a Si wafer by the vapor–liquid–solid (VLS) method using nano Au dots as nucleation sites for single crystal Si nanowires with a [111] growth direction. Polycrystalline Ni nanowires were synthesized via the anodic aluminum oxidation (AAO) method and stored in isoproponal. The Ni nanowires and Si nanowires ranged in diameter from 10 to 40 nm with lengths of a few microns.

To prepare point contact samples using the Si and Ni nanowires stored in solutions, droplets of both solutions were placed on Si grids having a square opening covered with a window of glassy Si_3N_4 films. The thickness of the glassy film is about 20 nm so that it is transparent to the electron beam in the microscope and does not interfere with the images of Si and Ni nanowires. The samples were dried under light bulbs. Figure 7.16 is a typical TEM image of randomly oriented Si and Ni nanowires on the window. The Si nanowire is a single crystal with a thin surface oxide about 1–5 nm thick. The Ni nanowire is polycrystalline and also has an oxidized surface. There are bend contour images in the Ni nanowires.

In situ annealing for point contact reactions was performed in a JEOL 2010 TEM with a GATAN 628 sample holder that can be heated to 1000 °C at adjustable rates. High-resolution lattice images were taken via a JEOL 3000 F HRTEM, and the vacuum in the sample stage is better than 10^{-6} Torr. Most of the experiments were conducted in a JEOL 2000 V ultrahigh vacuum transmission electron microscope (UHV-TEM), where the vacuum in the sample stage is about 3×10^{-10} Torr [16, 17].

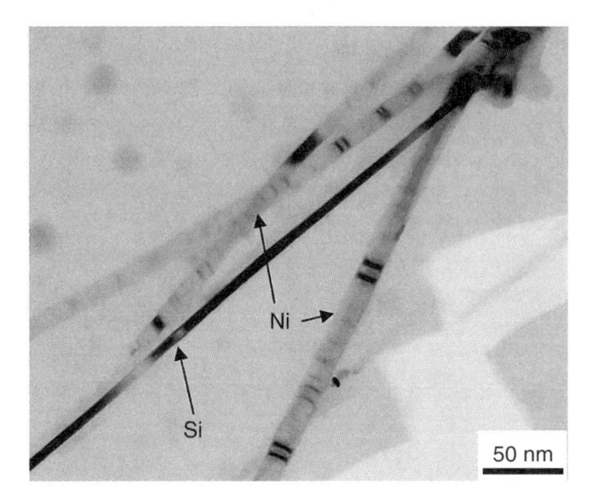

Figure 7.16 TEM image of randomly oriented Si and Ni nanowires on the window. The Si nanowire is a single crystal with a thin surface oxide about 1–5 nm thick. The Ni nanowire is polycrystalline and has an oxidized surface. *Source*: Reproduced with permission from [16].

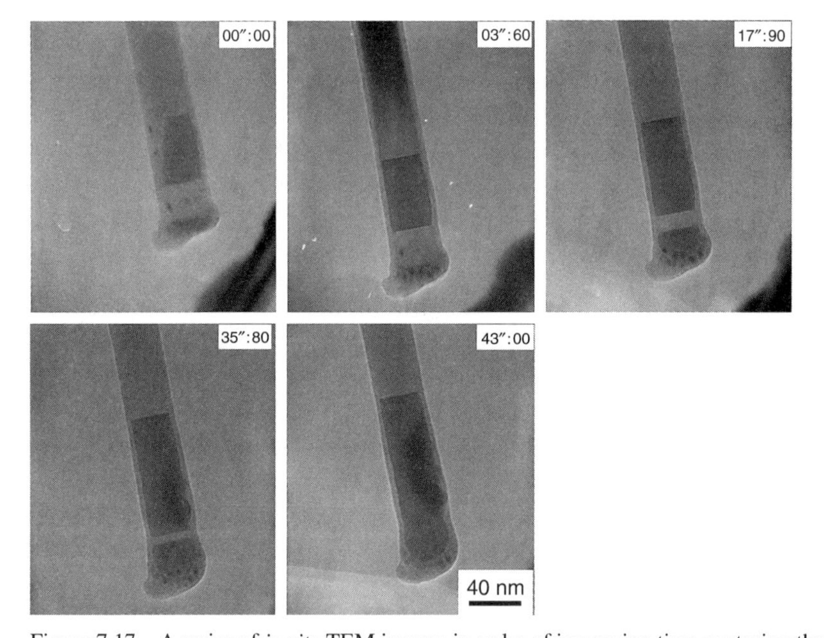

Figure 7.17 A series of *in situ* TEM images in order of increasing time capturing the growth of a bamboo-type grain of NiSi silicide within the straight Si nanowire at 700 °C. *Source*: Reproduced with permission from [16].

Figure 7.17 shows a series of *in situ* TEM images in order of the increasing time, capturing the growth of a bamboo-type grain of NiSi silicide within the straight Si nanowire at 700 °C. Using the HRTEM images and selected area diffraction patterns, the silicide is verified to be a single crystal NiSi.

Figure 7.18 is a schematic diagram depicting the growth of NiSi, in which the Ni atoms dissolve into the Si, diffuse and accumulate at the tip of the Si nanowire, thereby

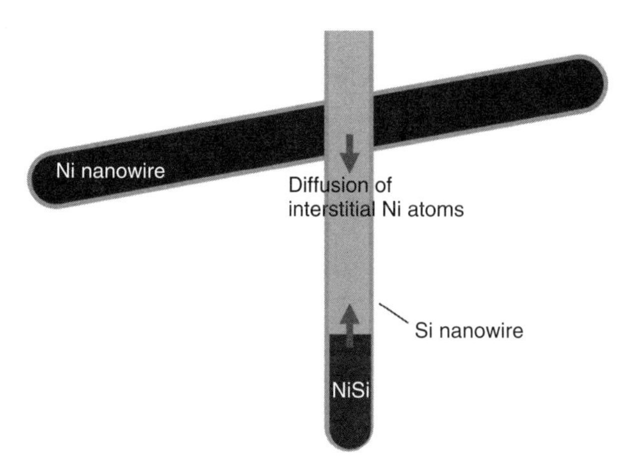

Figure 7.18 Schematic diagram depicting the growth of NiSi, in which the Ni atoms dissolve into the Si, diffuse and stop at the tip of the Si nanowire, thereby nucleating growth of NiSi.

nucleating the growth of NiSi. It is known that Ni atoms diffuse interstitially in Si. Therefore, the flux of interstitial Ni atoms supplied from the point contact diffuses toward the Si/silicide interface, and the interfacial reaction leads to the growth of a single crystal grain of silicide from the tip toward the point contact. The Si/silicide growth interface is flat.

To record the growth kinetics of NiSi in Si nanowires, dynamic observation using *in situ* TEM videos was conducted. *In situ* TEM, which can associate atomic-scale microstructure change with time, is capable of recording the lattice imaging of growth processes of the epitaxial interface. On the basis of the dynamic information, we found that the single crystal NiSi silicide grows and transforms the Si nanowire linearly with time. In addition, the interface between NiSi and Si appears atomically flat during the growth, having no misfit dislocations. Using the *in situ* video, we are able to measure the growth or transformation rate. Figure 7.19 shows the linear growth behavior of the NiSi nanowire in the straight Si nanowires in the temperature range of 500–650 °C. The activation energy of the growth has been determined to be 1.25 eV/atom. In addition, knowing the diameter of the wire and the growth rate, we can estimate the total number of Ni atoms in a given volume of the wire because we know the unit cell volume of NiSi and the number of Ni atoms per unit cell. Thus, we have determined that the time needed to incorporate one Ni atom on the growth interface is about 7×10^{-4} s.

We can estimate the time for Ni atoms to diffuse in a Si nanowire from $x^2 = Dt$ where x is the length of the Si nanowire between the point contact and the Si/silicide interface, D is interstitial diffusivity of Ni atoms in Si nanowires, and t is the time. If we take D to be 10^{-6} cm^2/s, and x to be 1 μm, we have $t = 10^{-2}$ s. Because the diffusion time is much longer than what was estimated for the growth of one atom on the interface, the reaction cannot be interface-controlled even though a linear rate of growth was measured. We conclude that the steady state reaction is limited by the rate of dissolution of Ni into the Si. It is a supply-controlled growth and is a unique feature of point contact reactions in nanowires.

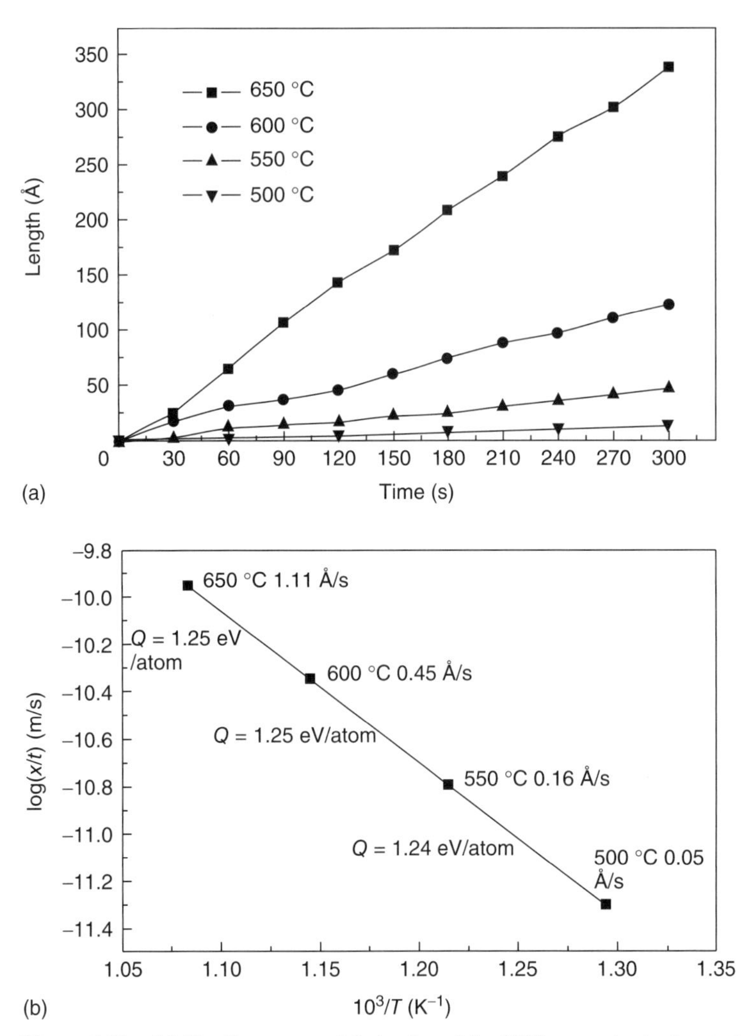

Figure 7.19 (a) The linear growth behavior of the NiSi nanowire in the straight Si nanowire with a 37.5 nm diameter. The average growth rate is calculated to be 1.1 Å/s. (b) The measured activation energy of the linear growth is Q = 1.25 eV/atom. *Source*: Reproduced with permission from [16].

Actually, the situation can be even more interesting and interrelated; each new atomic layer of silicide is formed after the successful nucleation of a two-dimensional island on the previous layer. In its turn, the nucleation rate is controlled by the supersaturation of interstitial Ni atoms that are supplied from the point contact. Moreover, the successful nucleation and fast lateral growth of each new layer leads to a substantial depletion of Ni solute atoms that should be replenished by supply-limited transfer of Ni atoms from the point contact.

Figure 7.20a–d shows a set of HRTEM images of the NiSi/Si interface taken as the interface advances from the NiSi into the Si. The interface is parallel to the (111) plane of Si as well as the (311) plane of NiSi. Thus, the growth direction of NiSi

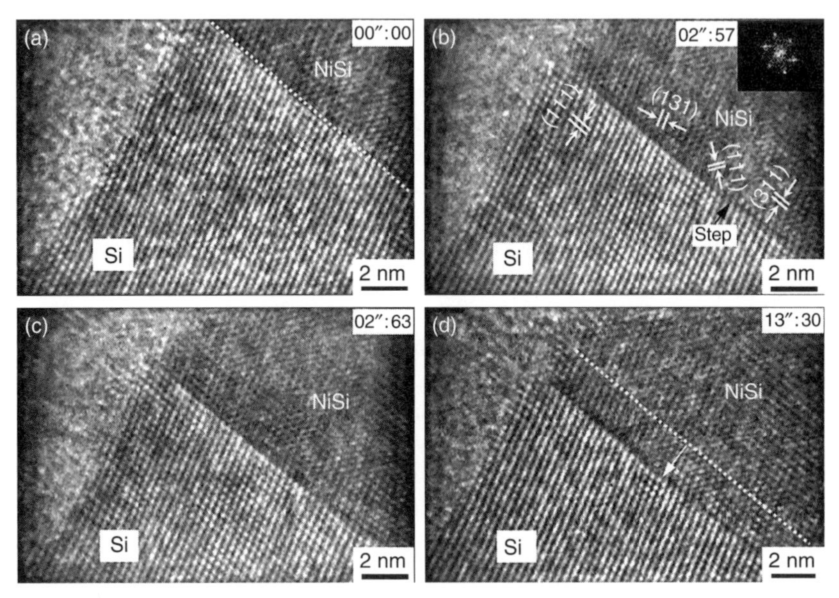

Figure 7.20 (A) to (D) show a set of high-resolution TEM images of the NiSi/Si interface taken as the interface advances from the NiSi into the Si. The interface is parallel to the (111) plane of Si as well as the (311) plane of NiSi.

is normal to the (311) plane. To determine the crystallographic orientation relationships between them, we note that Si has a diamond lattice with a lattice constant of $a = 5.431$ Å, and NiSi has an orthorhombic lattice with lattice constants of $a = 5.62$ Å, $b = 5.18$ Å, and $c = 3.34$ Å. We found that

$$[1-10] \text{ Si}//[1-12] \text{ NiSi} \quad \text{and} \quad (111) \text{ Si}//(31-1) \text{ NiSi}$$

From the lattice image across the epitaxial interface, we can find the misfit on the basis that the interplanar spacing of (111) Si and twice (131) NiSi to be 3.1355 Å and 2.9594 Å, respectively. The misfit is calculated as

$$f = \frac{d_{\text{Si}} - d_{\text{NiSi}}}{d_{\text{Si}}} = \frac{3.1355 - 2.9594}{3.1355} = \frac{0.1761}{3.1355} \cong 5.62\%$$

If misfit dislocations exist at the interface, the spacing between them should be about 18×0.31 nm = 5.58 nm, which we note is smaller than the approximately 10 nm diameter of the Si nanowire. However, we were not able to find any misfit dislocations at the NiSi/Si epitaxial interface.

It is surprising that there is no obvious molar volume change or strain in the transformation of Si into NiSi. In other words, the added volume of Ni atoms is absorbed without producing any obvious strain. It is possible that at 650–700°C, the surface oxide could yield, thereby relieving part of the strain.

The atomistic mechanism of the epitaxial growth involving a moving interface is of interest. Reactive epitaxial growth is unlike molecular beam epitaxy by deposition, in which atomic layers are built on the substrate surface and the growth requires no breaking of the substrate bonds. In the present case, the covalent Si–Si bonds must be broken and transformed into metallic Ni–Si bonds in order for NiSi to grow and have a moving interface. Although the thermal energy at 700 °C is sufficient to break the covalent Si–Si bonds, the interstitial Ni atoms are crucial in the Si bond-breaking process [11]. The atomistic growth process at the epitaxial NiSi/Si interface is unclear. For example, whether it obeys a model similar to stepwise growth on a free surface having kinks at a step, or whether it proceeds with a single step climb of a misfit dislocation across the interface, requires more study.

It is known that in the interfacial reactions between metal thin films and single crystal Si wafers, near-noble metals such as Ni and Pd can react with Si to form silicide at temperatures as low as 100 °C. Near-noble metals diffuse through the silicide (not through Si as we reported here), dissolve interstitially in Si, and assist the breaking of the Si–Si covalent bonds. Furthermore, native oxide on the Si surface is not an effective diffusion barrier and therefore the metal atoms can diffuse through the oxide and react with the Si substrate. For this reason, while we do not know the precise chemical reaction, we assume that the dissolution of Ni into the Si nanowire can take place at the location of point contact.

Because NiSi grows inside the oxidized Si nanowire, we must assume that the diffusion of Ni occurs within Si. It is possible that some Ni atoms may diffuse along the oxide surface; however, they have to diffuse through the oxide to react. We will need some very clear high-resolution images of the contact area to determine the reaction at the point contact location.

To conclude, we have observed point contact reactions between Si and Ni nanowires that transform a single crystal Si nanowire into a single crystal NiSi nanowire. *In situ* TEM videos show that single crystal of NiSi has a linear growth rate, which is a supply-controlled growth owing to the slow dissolution of Ni atoms into Si. Single crystal NiSi/Si/NiSi heterostructures as well as nanogap structure can be made with this method, which is discussed below.

7.4.2 Line Contact Reactions

According to point contact reaction discussed in the previous section, the surface oxide surrounding Si nanowires is an important factor in silicide formation at nanoscale. Platinum, being chemically stable in ambient or oxidizing environment, is an interesting contact material for nanoelectronics [18, 19]. In addition, platinum silicide (PtSi) can be used as ohmic contact to p-channel Si nanowire transistors. Therefore, it is of interest to study the controlled reactions between lithographically defined Pt pads and Si nanowires with and without oxide under *in situ* TEM for investigating the oxide effect on line contact formation of single crystal PtSi and PtSi/Si nanowire heterostructures.

Silicon nanowires were prepared on a p-type Si wafer by the VLS method using nano Au dots as nucleation sites. The resultant [111] single crystal Si nanowires had thin surface oxide (~1–5 nm thick), with lengths of a few microns and diameters ranging from 10 to 40 nm.

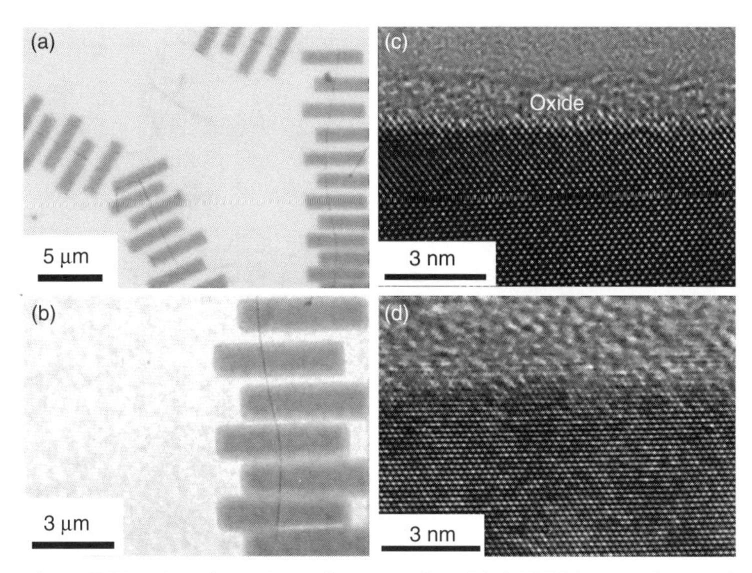

Figure 7.21 Overview of sample preparation. (a) A TEM image of a sample with multiple sets of a Si nanowire across Pt contact pads. (b) A TEM image showing a closer look at one of the Si NW-Pt pad sets. (c) A HRTEM image of a Si nanowire with 2 nm oxide and with smooth Si/oxide interfaces. (d) A HRTEM image of a Si nanowire without oxide and with rough Si/oxide interfaces.

The samples of Si nanowire covered with Pt contact pads were fabricated on Si/Si_3N_4 substrate using e-beam lithography and e-beam evaporation, as shown in Figure 7.21a, b. As each of the samples contains multiple Si nanowires and as there are many patterned Pt pads, as shown in Figure 7.21a, b, numerous contacts between Si and Pt can be found in one sample and many samples have been studied. Prior to Pt deposition, some of the Si nanowire samples were etched in buffered hydrofluoric acid for 5 s to remove native oxide at the contact region while some were not. The etched samples were HF-dipped again prior to being loaded in UHV-TEM to prevent native oxide at regions other than contact areas. Figure 7.21c, d are HRTEM images showing the samples with and without oxide, respectively; thereby, the effect of native oxide on kinetics of silicide formation can be investigated. For nanowires with HF dipping, no oxide was seen and the Si/oxide interface was rough; for nanowires without HF dipping, the surface oxide is of 1–2 nm and the Si/oxide interface was smooth. TEM examinations were conducted in a JEOL 2000 V UHV-TEM under a base pressure of 3×10^{-10} Torr, where sample can be heated to 1000 °C. On heating in UHV-TEM, Pt reacts with Si nanowires to form PtSi nanowires, identified by electron diffraction pattern and energy dispersion X-ray spectrometer (EDS) analysis.

In using electron beam lithography and e-beam evaporation to define and deposit the Pt pads on Si nanowires, depicted in Figure 7.22a, some of the samples were etched in buffered hydrofluoric acid for five seconds to remove native oxide prior to Pt deposition. After Pt deposition, the samples are loaded into TEM and annealed *in situ* at 500 °C to form PtSi, which is depicted in Figure 7.22b.

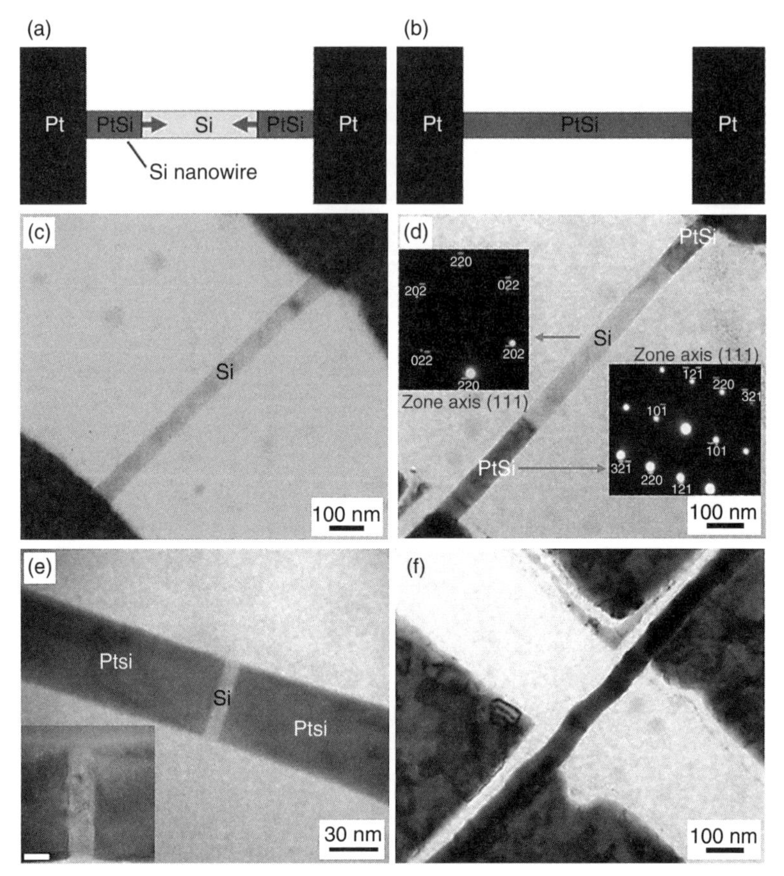

Figure 7.22 Formation of single crystal PtSi nanowire and PtSi/Si/PtSi nano-heterostructures with varying lengths of the Si region. (a) A schematic illustration depicting the growth of a PtSi/Si/PtSi nano-heterostructure. (b) A schematic illustration showing full transformation from a Si nanowire to a PtSi nanowire. (c) *In situ* TEM image of a Si nanowire before reaction. (d, e) *In situ* TEM images showing PtSi/Si/PtSi nano-heterostructures in which the Si regions are 575 and 8 nm in length, respectively. The darker region is PtSi, while the brighter region is Si. The insets in (d) are the corresponding selected area diffraction patterns of Si and PtSi, respectively. The inset in (e) is the magnification of the gap region and the scale bar is 5 nm. (f) *In situ* TEM image of a PtSi nanowire after reaction and full transformation from Si into PtSi. *Source*: Reproduced with permission from [19].

Figure 7.22c shows the TEM image of an etched Si nanowire with Pt pads on both sides before reaction. After annealing at 500 °C for about 5–10 min, higher contrast sections with sharp interface with the Si start to emerge from both ends of the nanowire near the Pt pads, suggesting the formation of PtSi, as shown in Figure 7.22d. This is attributed to the fact that many Pt atoms are able to dissolute into silicon through the long line contact between the Si nanowire and the Pt pad, so that supersaturation can be reached; thereby, nucleation and growth of PtSi occurs below the

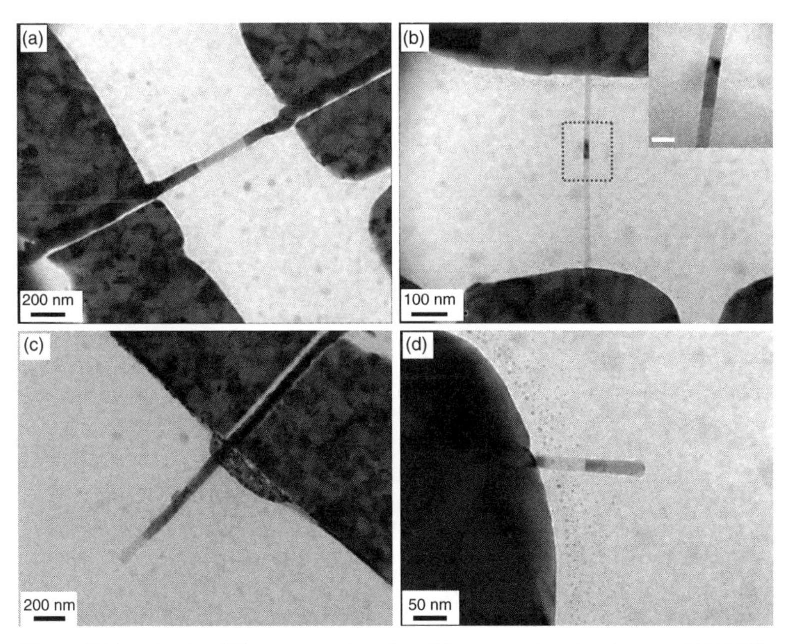

Figure 7.23 *In situ* TEM images showing different heterostructures of PtSi/Si based on different growth mechanisms due to oxide effect. (a) *In situ* TEM image of a PtSi/Si/PtSi heterostructure within a Si nanowire without surface oxide. The bright area in the middle is Si and the two dark end areas are PtSi. (b) *In situ* TEM image of a Si/PtSi/Si heterostructure within a Si nanowire with surface oxide. The bright area is Si and the middle dark area is PtSi. The inset in (b) is magnification of the heterostructure in which the scale bar is 20 nm. (c) *In situ* TEM image showing the formation of PtSi within an over-hang Si nanowire without oxide from a single Pt pad. The PtSi grows from the contact towards the end of the Si. (d) *In situ* TEM image showing the formation of PtSi within an over-hang Si nanowire with surface oxide. The PtSi grows from the end of the Si nanowire towards the Pt pad.

line contact. Utilizing this phenomenon, nanowire heterostructure of PtSi/Si/PtSi in which the length of the middle Si can be controlled precisely to sub-10 nanometer regime, as shown in Figure 7.22e, to form a nanogap of Si. On further annealing, all the Si is consumed completely and transformed into a PtSi nanowire, as shown in Figure 7.22f. Detachment of wires from pads is following strain resulting from volume expansion.

On the basis of electron diffraction pattern corresponding to the platinum silicide region, which is shown in the right part of Figure 7.22d, the silicide material is identified to be a single crystal PtSi with an orthorhombic crystal structure having lattice constants $a = 0.5567$ nm, $b = 0.3587$ nm, and $c = 0.5927$ nm. The epitaxial relationships between the Si and PtSi has been determined as Si(20−2)//PtSi(10−1) and Si[111]//PtSi[111], according to the diffraction patterns in Figure 7.23d.

The effect of surface oxide on PtSi formation is shown in Figure 7.23 [19]. In the TEM images of Figure 7.23a, c, Si nanowires were HF-dipped so that they were free of oxide. At 500 °C, PtSi formed within the Si nanowire from the contact

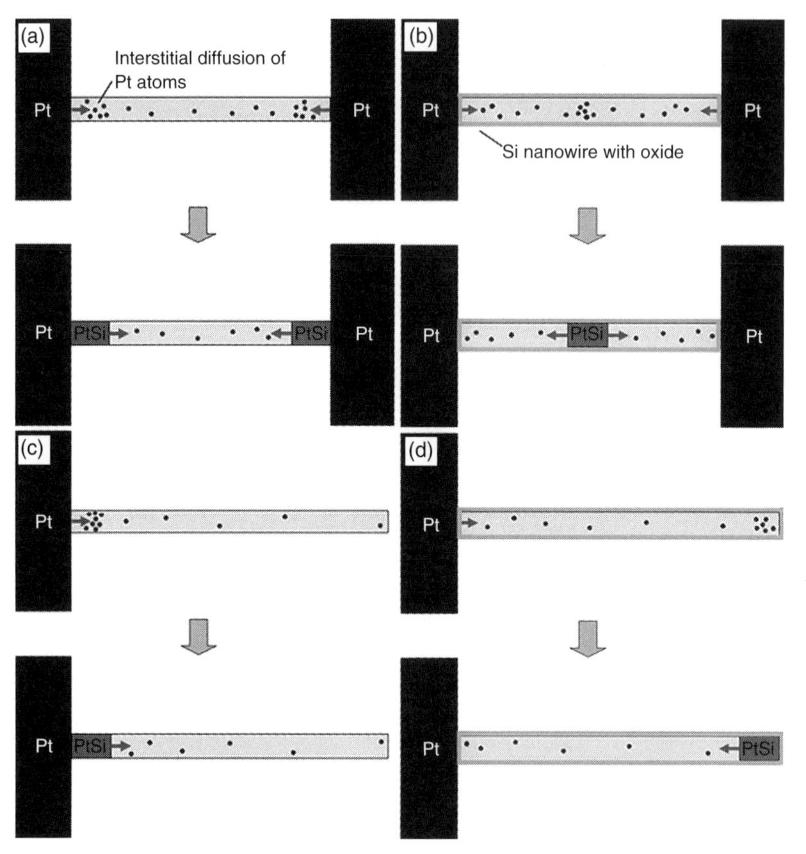

Figure 7.24 Schematic illustrations of oxide effect on kinetics of nano-silicide formation. (a) A schematic illustration corresponding to Fig. 7.23(a), showing growth of a heterostructure of PtSi/Si/PtSi within a Si nanowire without oxide. (b) A schematic illustration corresponding to Fig. 7.23(b), showing growth of a heterostructure of Si/PtSi/Si within a Si nanowire with surface oxide. (c) A schematic illustration corresponding to Fig. 7.23(c), showing the PtSi formation within an over-hang Si nanowire without oxide. The growth occurs from the Pt pad to an end of the Si nanowire. (d) A schematic illustration corresponding to Fig. 7.23(d), showing PtSi formation within an over-hang Si nanowire with surface oxide. The growth occurs from an end of the Si nanowire to the Pt pad.

area between the Pt pad and the Si nanowire. However, in Figure 7.24b, d, where Si nanowires were not etched with HF and had surface oxide of 1–5 nm in thickness, PtSi formed within the Si nanowire near the center of the Si nanowire, as shown in Figure 7.23b, and from the end of the Si nanowire, as shown in Figure 7.23d, respectively. We note that in Figure 7.23a, b, the Si nanowire was across two Pt contact pads; yet in Figure 7.23c, d, the Si nanowire was an overhang from one Pt pad. As all of the phase transformations occurred in the Si nanowires, platinum atoms were the dominating diffusion species.

The different locations in forming different nano-heterostructures of PtSi/Si/PtSi, as shown in Figure 7.23a and Si/PtSi/Si, as shown in Figure 7.23b, are depicted in Figure 7.24. Figure 7.24a, c are the corresponding schematic illustrations of PtSi formation portrayed in Figure 7.23a, c. With respect to the location of nucleation where the PtSi starts, supersaturation of Pt is needed. In Figure 7.23a, due to no impedance of oxide and due to the line contact between a Pt pad and a Si nanowire, a large number of Pt atoms can quickly dissolve into the Si nanowire. Supersaturation of Pt in the Si nanowire can be reached below the line contact because of the very low equilibrium solubility of interstitial Pt atoms in Si. Therefore, the nucleation and growth of PtSi starts from the line contacts, leading to a PtSi/Si/PtSi heterostructure. On the basis of the same reason, in Figure 7.23c, PtSi forms from the contact line between the Pt pad and the overhanging Si nanowire.

Figure 7.24b, d are the corresponding schematic illustrations of PtSi formations in Figure 7.23b, d, where the Si nanowire has surface oxide. There are two reasons why nucleation cannot occur at the contact line between the Pt pad and the oxidized Si nanowire. The first is that the oxide on the Si nanowire surface limits the dissolution of Pt atoms into the Si. The second is the very rapid interstitial diffusion of Pt in Si. Thus, whenever a Pt atom dissolves into the Si nanowire, it diffuses away quickly, resulting in no supersaturation below the contact and a very low concentration gradient of Pt atoms over the entire Si nanowire. However, around the center of the Si nanowire in Figure 7.24(b), it is a meeting place of fluxes of Pt atoms coming from both sides. It is the place where the nucleation of PtSi occurs, contributing to the formation of a Si/PtSi/Si heterostructure. As for the case of Figure 7.24d, when Pt atoms diffuse interstitially to the end of the Si nanowire, they will pile up at the end because the reverse diffusion is against the concentration gradient. Again, the PtSi formation in Figure 7.24d starts at the end of the Si nanowire rather than at the contact.

In the previous discussion on point contact reaction between Ni and Si nanowires, besides interstitial diffusion of Ni in Si, surface diffusion of Ni on the oxidized Si surface could be a competing mechanism. However, surface diffusion fails to explain the silicide formation from the pad-wire contact as shown here. Also, it cannot explain the silicide formation starting from one end of Si nanowire to Pt pad, nor starting near the center between two adjacent pads. On the contrary, interstitial diffusion in a Si nanowire applies to all these conditions very well. As the silicide can be highly deficient in Pt and there are a large number of vacancies in the sublattice of Pt of the silicide, the diffusion of Pt through the silicide is very fast.

In addition to the impact on the nucleation sites, the existence of oxide affects the rate of silicide formation significantly. The rate of the PtSi growing within a Si nanowire without oxide was about 5 nm/s, while that of the PtSi growing within a Si nanowire with 2 nm thick oxide was about 0.05 nm/s. The rate difference by two orders of magnitude resulted from oxide acting as a diffusion barrier. Furthermore, a stepwise growth was observed in the oxidized Si nanowire, where an incubation time was needed for nucleation and growth with a limited number of Pt atoms to support the silicide formation. This is coherent with our previous discussion in Chapter 6.

To summarize, we have demonstrated that with or without surface oxide on the Si nanowire, different PtSi/Si/PtSi and Si/PtSi/Si nanowire heterostructures appear because of different growth mechanisms by using *in situ* TEM observations. Specifically, when a Si nanowire has no oxide, PtSi formation starts at the contact line between the Si nanowire and the Pt pads so that a nanowire heterostructure of PtSi/Si/PtSi appears. When a Si nanowire has oxide, PtSi formation starts near the center of the Si nanowire between the two Pt pads or from one end of a Si nanowire, leading to the formation of Si/PtSi/Si nanowire heterostructures. In addition, the existence of oxide seriously affects not only the growth position but also the growth rate.

7.4.3 Planar Contact Reactions

If a Si nanowire is coated with a metal film by electroplating or if we deposit a metal film over a Si nanowire, the contact in the coaxial nanowire is a surface. Furthermore, if Si diffuses out during the reaction, we can have nanotube formation as discussed in Chapter 1. The kinetic analysis of the reaction is similar to that presented in the second part of Chapter 3, hence it is not repeated here.

REFERENCES

1. Tu KN, Mayer JW, Feldman LC. *Electronic Thin Film Science*. New York: MacMillan; 1992.
2. Tu KN. *Electronic Thin Film Reliability*. Cambridge, UK: Cambridge University Press; 2011.
3. Mayer JW, Poate JM, Tu KN. Thin films and solid-phase reactions. Science 1975;190:228–234.
4. Tu KN, Mayer JW. Silicide formation Chapter 10. In: Poate JM, Tu KN, Mayer JW, editors. *Thin Films: Interdiffusion and Reactions*. New York: Wiley-Interscience; 1978.
5. Nicolet M-A, Lau SS. Formation and characterization of transition-metal silicides. In: Einspruch NG, Larrabee GB, editors. *VLSI Electronics*. Vol. 6. New York: Academic Press; 1983. For standard heat of formation of transition metal silicides and oxides, see Table A.X.
6. Chen LJ, Tu KN. Epitaxial growth of transition-metal silicides on silicon. Mater Sci Rep 1991;6:53–140.
7. Chen SH, Zheng LR, Carter CB, Mayer JW. Transmission electron microscopy studies on the lateral growth of nickel silicides. J Appl Phys 1985;57:258.
8. Kidson GV. Some aspects of the growth of diffusion layers in binary systems. J Nucl Mater 1961;3:21–29.
9. Pretorius R, Harris JM, Nicolet M-A. Reaction of thin metal films with SiO_2 substrates. Solid State Electron 1978;21:667–675.
10. Goesele U, Tu KN. Growth kinetics of planar binary diffusion couples: thin film case versus bulk cases. J Appl Phys 1982;53:3252–3260.
11. Tu KN. Selective growth of metal-rich silicide of near-noble metals. Appl Phys Lett 1975;27:221–224.
12. Foell H, Ho PS, Tu KN. Cross-sectional TEM of silicon-silicide interfaces. J Appl Phys 1981;52:250–255.
13. Tung RT. Schottky barrier formation at single crystal metal–semiconductor interfaces. Phys Rev Lett 1984;52:461–464.
14. Cui Y, Lieber CM. Functional nanoscale electronic devices assembled using silicon nanowire building blocks. Science 2001;291:851–853.
15. Patolsky F, Timko BP, Zheng G, Lieber CM. Nanowire based nanoelectronic devices in life science. MRS Bull 2007;32:142–149.
16. Lu K-C, Wu W-W, Wu H-W, Tanner CM, Chang JP, Chen LJ, Tu KN. In-situ control of atomic-scale Si layer with huge strain in the nano-heterostructure NiSi/Si/NiSi through point contact reaction. Nano Lett 2007;7(8):2389–2394.

17. Chou YC, Wu WW, Chen LJ, Tu KN. Homogeneous nucleation of epitaxial $CoSi_2$ and NiSi in Si nanowires. Nano Lett 2009;9:2337–2342.
18. Lin Y-C, Kuo-Chang L, Wen-Wei W, Bai J, Chen LJ, Tu KN, Huang Y. Single crystalline PtSi nanowires, PtSi/Si/PtSi nanowire heterostructures, and nanodevices. Nano Lett 2008;8:913–918.
19. Lu K-C, Wu W-W, Ouyang H, Lin Y-C, Huang Y, Wang C-W, Wu Z-W, Huang C-W, Chen L-J, Tu KN. The influence of surface oxide on the growth of metal/semiconductor nanowires. Nano Lett 2011;11(7):2753–2758.

PROBLEMS

7.1. What is single phase formation phenomenon in thin film contact reactions on silicon? Why is it important in silicon microelectronic technology?

7.2. What is the most important kinetic process in contact reactions on Si?

7.3. An amorphous phase of α-TiSi has been found to form as the first phase in contact reaction between Ti and Si. How can the amorphous phase form under a slow heating rather than a rapid quenching process?

7.4. Explain why M_2Si is the first phase to form when we react near-noble metals with Si wafer. What is the reaction temperature?

7.5. We have a sample of 100 nm of NiSi on a Si wafer. We deposit a thin film of Ni of 200 nm on NiSi and then anneal the Ni/NiSi/Si sample at 200 °C for one hour. What will happen?

7.6. We have a substrate of 50Si50Ge alloy. We deposit a Ni thin film of 200 nm on the SiGe substrate and anneal the sample at 200 °C for 1 h. What will happen? If you can find a ternary phase diagram of Ni–Si–Ge, it will help you. If you cannot find the ternary phase diagram, we can check the binary phase diagram of Ni–Si and Ni–Ge to have some ideas.

7.7. If we codeposit an amorphous alloy thin film of TiW on a Si wafer and anneal at 550 C for 1 h, what will happen?

7.8. On a heavily doped n-type Si and the dopant P, we deposit a Ni thin film of 200 nm on the Si and anneal at 200 °C to form Ni_2Si. The silicide formation has consumed about 100 nm of Si. What has happened to the dopant in the Si that has been consumed?

7.9. In the text, we present the case of the competing growth of two-layered compounds in which one layer exhibits diffusion-controlled growth and the other layer exhibits interface-reaction-controlled growth. Describe what happens in the following two cases when

(1) Both layers exhibit diffusion-controlled growth.

(2) Both layers exhibit interfacial-reaction-controlled growth.

7.10. A phase $A_\beta B$ ($\beta = 3$) grows between the pure A phase and the $A_\gamma B$ ($\gamma = 1.5$) phase in a diffusion couple where A is the dominant diffusing species. Given that $\tilde{D}_\beta = 1.27 \times 10^{-11}$ cm²/s. $\Delta C_\beta^{eq} = 2$ at.%. $K_\beta^{eff} = 2.79 \times 10^{-7}$ cm²/s, and Ω = atomic volume = 18×10^{-24} cm³/atom.

(1) Find the changeover thickness of the $A_\beta B$ phase.

(2) Find the $A_\beta B$ growth rate at the changeover thickness.

(3) What is the flux of A through the $A_\beta B$ phase?

7.11. Derive a pair of equations for the simultaneous growth of two intermetallic compound layers between mutually insoluble components. Find the thickness of each compound layer at a given time *t*.

CHAPTER 8

GRAIN GROWTH IN MICRO AND NANOSCALE

8.1 INTRODUCTION

The typical microstructure of a solid, such as a piece of bulk metal, is polycrystalline. The large number of grains in the microstructure has a size distribution. Because grain boundary energy is positive, grain growth occurs in the microstructure in order to reduce the grain boundary energy by reducing the grain boundary area, provided the temperature is high enough for grain boundary migration to occur. Grain growth has been studied at least since 1920, when the movement of grain boundaries was detected *in situ* by Carpenter and Elam. It is one of the fundamental subjects in materials science and it is important in the processing of most materials in order to control

Kinetics in Nanoscale Materials, First Edition. King-Ning Tu and Andriy M. Gusak.

the microstructure. The physical properties of a polycrystalline solid depend strongly on its microstructure, which is affected by grain growth. For example, the mean grain size influences the yield strength (Hall–Petch law), or grain texture influences magnetic properties.

The kinetics of grain growth has been found to be parabolic, that is, the average grain diameter has a square-root dependence on time, which tends to suggest that grain growth is a diffusion-controlled process. However, there is no long-range diffusion in grain growth. In the grain growth of a pure metal, there is no concentration gradient. Burke and Turnbull proposed that the plausible reason for the diffusion-like behavior could be the curvature of grains, which depends on the grain size or grain radius [1]. We consider a pair of neighboring grains, in which one grain is smaller than the other. As Gibbs–Thomson potential energy is inversely proportional to grain radius, the atoms in the smaller grains will have a higher Gibbs–Thomson potential energy than the atoms in the neighboring larger grain. The potential difference provides the driving force of grain growth of the larger grain. The velocity of the moving grain boundary is assumed to be a product of mobility times the driving force, and it leads to a square-root dependence on time. Hillert has given it a formal analysis [2], which is discussed later. Nevertheless, in Hillert model, the skew of the theoretical grain size distribution in the analysis is unsatisfactory.

When grain size is in nanoscale, we expect grain growth to be serious, so the thermal stability of a nanoscale microstructure is of concern. In addition, besides grain growth, grain rotation seems to occur in nanoscale microstructures. In narrow Cu interconnects on Si chips, grain growth of nanosize grains occurs at room temperature and leads to the formation of bamboo-type grains, which can be regarded as 1D microstructure (if the film is not only thin but also narrow, otherwise the structure will be 2D). In the 1D case, the migration of a vertical grain boundary in a bamboo-type microstructure causes no change in grain boundary area. In thin films of nanoscale thickness, the thickness can limit the extent of grain growth when columnar grains are already obtained. This is because there is little gain in energy in the motion of a vertical and planar grain boundary. Thin films of columnar grains can be regarded as 2D microstructure. However, abnormal grain growth occurs in Cu thin films and giant grains are observed, which is discussed in Chapter 10. In the electromigration of a stripe of Sn, grain rotation has been observed following the alignment of grain orientation to lower electrical resistance. This is because Sn is a highly anisotropic conductor.

In severely deformed plastic bulk size metals, nanoscale grains have been observed. Owing to elastic and plastic deformation, grain growth occurs by recrystallization. In binary alloy microstructures, the nanoscale grains are expected to be more stable than those in a pure metal, owing to different segregation coefficients of the alloy elements in the grain boundaries. It means that the composition at grain boundaries differs from that in the bulk of grains. This means that the movement of grains should be accompanied by the migration of segregated species. This is called *solute drag* and means the decrease of the grain boundary mobility.

Grain growth in nanoscale, the growth of the mean grain size, in principle, can be realized by the following two main mechanisms.

(1) The movement of grain boundaries and their junctions, when atoms of one grain are rearranged at the boundary, to be incorporated into the structure of another grain, where the velocity of grain boundary movement is proportional to its curvature.

(2) The merging of the neighboring grains by gradual decrease of misorientation, due to grain rotation. Angular velocity of rotation is proportional to the resulting torsion, which, in turn, is proportional to grain misorientation (at least for small misorientations).

We consider the second mechanism at the end of this chapter and find that the grain mobility with respect to rotation is inversely proportional to the fifth power of grain size. Thus, second mechanism may become realistic only for truly nanoscale grains, or in the case of external source of torsion (for example), by a strong electric current in anisotropic metallic grains. As for the first mechanism, we can divide it into normal and abnormal grain growth. Experimental observation of abnormal grain growth in Cu is presented in Chapter 10.

Grain growth in a polycrystalline body or a soap film is caused by the same thermodynamic reason as ripening, that is, the tendency to reduce surface energy, and it is realized via consuming the small grains by the large ones. While in ripening, the realization depends on long-range diffusion. Yet, in grain growth this tendency is realized without long-range diffusion (if not to mention segregation effects). Despite the absence of long-range diffusion, grain growth tends to obey a parabolic growth law, at least in very pure metals without segregation and without the corresponding solute drag. The process of grain growth seems simple, yet the three-dimensional nature of the grain boundary network makes it a difficult subject to model or analyze. In this chapter, we first cover the computer simulation of a polycrystalline microstructure, which is followed by the analysis of grain growth kinetics.

8.2 HOW TO GENERATE A POLYCRYSTALLINE MICROSTRUCTURE

We can obtain a polycrystalline microstructure from Kolmogorov–Johnson–Mehl–Avrami kinetic model of crystallization. [3–9] This model was first introduced by Soviet mathematician Kolmogorov in 1937 [4] and then, in some other form, by American researchers Johnson, Mehl, and Avrami in 1939–1941 [5–8]. Kolmogorov considered only the case of homogeneous nucleation of crystals. Avrami et al. treated more general problems, including heterogeneous nucleation. In the literature of phase transformations, the canonical equation used to describe the fraction of transformation as a function of time and temperature, or the so-called time-temperature-transformation (TTT) diagram, is represented by the Johnson–Mehl–Avrami (JMA) equation as

$$X_{\mathrm{T}} = 1 - \exp\left(-X_{\mathrm{ext}}\right) \qquad (8.1)$$

Figure 8.1 Transmission electron microscopic image of partially transformed amorphous $CoSi_2$ thin film annealed at 150 °C. The circular crystals are single crystals of the crystalline phase of $CoSi_2$.

where X_T is the fraction of volume that has been transformed and X_{ext} is the extended dimensionless volume in the transformation. A very simple derivation of the above equation given below.

Experimentally, we consider, for simplicity, 2D crystallization of an amorphous thin film, for example, the amorphous-to-crystalline transformation of $CoSi_2$ thin film. Figure 8.1 shows a transmission electron microscopic image of partially transformed amorphous $CoSi_2$ thin film annealed at 150 °C. The circular crystals are single crystals of the crystalline phase of $CoSi_2$, which are dispersed in the rather homogeneous amorphous matrix of $CoSi_{2.}$ There is no composition change in the crystallization. The nucleation is random in space and steady state in time. The growth is isotropic. The measured overall crystallization rate and the measured growth rate of individual crystal have been used to obtain the activation energy of nucleation and growth of the transformation. Below is a more general kinetic analysis of the crystallization.

Assume that the nuclei of the overcritical crystalline clusters appear homogeneously anywhere with some frequency v ($vdtdA$ is a probability of nucleation within short interval dt in the small area dA, dimension of v is s^{-1}/m^{-2}). Assume as well that the nucleated crystalline grain grows radially with a constant velocity V until meeting another grain, but even after this event of impingement, we imagine that it will grow virtually further, overlapping (virtually) with other grains. Imagine that we choose some arbitrary point in the entire thin film, and want to know what is the probability that during the time t after the beginning of the crystallization, this point will remain untransformed. In other words, what is the probability for an arbitrarily chosen point to escape being transformed, either without nucleation at this point or without being transformed because of the growth of a neighboring grain. To find this probability, we divide the whole interval (0, t) into many tiny intervals. $\left(t_i, t_i + dt\right), t_i = i \cdot dt$. To escape transformation till moment t, system should escape nucleation of "crystalline phase" during each time interval dt. In order not to reach the chosen point at time, t, no grain should nucleate within the ith time interval within the circle of radius R, where $R = V(t - t_i)$. Otherwise any nucleus within the circle nucleated at the moment t_i will grow to reach the chosen point and will transform it into the crystalline phase. The probability for such nucleus to not be born is

$$1 - \nu dt \cdot A = 1 - \nu dt \pi V^2 \left(t - t_i\right)^2 \approx \exp\left(-\nu dt \pi V^2 \left(t - t_i\right)^2\right)$$

The last step is by taking $1 - x = \exp(-x)$ when x is small. The probability of the chosen point not to be transformed is then a product of all probabilities for all the time intervals. As product of exponents is an exponent of a sum, and the sum of many very small additions is just an integral, we obtain:

$$p_0(t) \approx \prod_{i}^{N=t/dt} \exp\left(-\nu \pi V^2 \left(t - t_i\right)^2 dt\right) = \exp\left(-\int_0^t \nu \pi V^2 \left(t - t_i\right)^2 dt\right)$$
$$= \exp\left(-\frac{\pi}{3} \nu V^2 t^3\right)$$

Then the probability of being transformed, respectively, has a typical S-curve form given by

$$p_1(t) = 1 - p_0(t) \approx 1 - \exp\left(-\frac{\pi}{3} \nu V^2 t^3\right)$$

This is a form of JMA equation, Eq (8.1), where X_T is $p_1(t)$, the fraction of volume transformed (here it is area transformed), and X_{ext} is $\pi \nu V^2 t^3/3$, the extended volume (here it is the extended area). Because it is a 2D growth, the growth has a square dependence on time, but we need to add the steady state dependence of nucleation on time, so we have t^3 in the equation.

Natural logarithm of the logarithm of $p_0(t)$ (fraction of untransformed volume) depends linearly on the logarithm of time, and its tangent is called *Avrami exponent* and in the present case it is equal to 3.

We can use computer simulation to model Kolmogorov–Avrami kinetics in 2D case to obtain a simulated polycrystalline microstructure. Figure 8.2 is a simulated figure showing the result of phase transformation by nucleation and growth. At the end when the transformation is complete, we obtain a polycrystalline microstructure. We can measure the distribution of grain size in the microstructure as well as the number of the nearest neighboring grains. The network of grain boundaries in the microstructure depends on the assumed mode of the crystallization. For example, if we assume a steady state nucleation and constant isotropic growth, the grain boundaries are hyperbolic lines. If we assume that the nucleation is instant so that all nuclei occurs simultaneously and followed by constant isotropic growth, the grain boundaries will be straight lines.

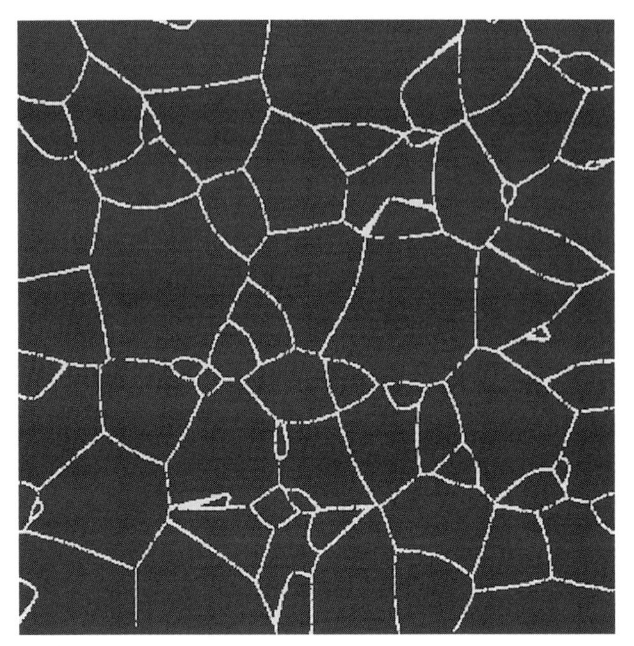

Figure 8.2 A set of simulated figures showing the progress of phase transformation by nucleation and growth followed the JMA model. At the end when the transformation is complete, we obtain a polycrystalline microstructure.

8.3 COMPUTER SIMULATION OF GRAIN GROWTH

There are two groups of models for computer simulation of grain growth: the atomistic and the phenomenological models [10–21].

8.3.1 Atomistic Simulation Based on Monte Carlo Method

Most popular is the Potts model [10], which is a generalization of the Ising model. It has been used for the analysis of the general features of second order phase transitions. In this model, atoms are immobile, but can change their state (i.e., belonging to a grain but with different lattice orientation). In ferromagnetic domain it is a change of spin orientation, and in grain growth it is a change of orientation of crystal lattice. Grain growth means migration of grain boundaries. And, grain boundary migration does not imply the long-distance migration of atoms – they may just make a little spatial shift, to be incorporated in the lattice of the coming grain: atoms just change their "passport" without substantial changing of their place.

An example from real life: Peasants, who lived in the Transcarpatian region in Europe in the twentieth century, could live all their life in the same village; however, they changed their citizenship from Austro-Hungarian empire to Czechoslovakia, then to Hungary, then to Soviet Union, and eventually to Ukraine. They stayed at one place physically, but boundaries of different countries moved over them.

In Potts model, atoms stay at their place, but one by one they change their "citizenship" – grains with different orientations. The grain boundary between atoms having different orientation is like that of an antidomain boundary. Atoms having the neighbors in different grains (or states) have higher energy. Roughly speaking, the energy of the bond between atoms belonging to different grains is equal to the sum of the bond energy (negative) between atoms of the same grain, plus the product (positive) of grain boundary tension times squared atomic size. For such atoms forming a boundary, the change of their state from the original grain to the state of neighboring grain, in grain growth, is controlled by the calculated energy change ΔE. Then the probability of state change can be found according to the normalized Boltzmann distribution:

$$p = \frac{\exp(-\Delta E/kT)}{1 + \exp(-\Delta E/kT)}$$

(and $1 - p = \frac{1}{1+\exp(-\Delta E/kT)}$ is a probability of staying in the original state). Applying such algorithm leads to the decrease of grain boundary surface and energy as well as to the decrease of the number of grains and the increase of the mean grain size.

8.3.2 Phenomenological Simulations

This kind of simulations can be divided into 3 subgroups:

(1) Numeric solution of the set of many (at least thousands) equations for grain sizes evolution in mean-field approximation or with the account of correlations.

(2) Numeric solution of viscous dynamic equations for the grain boundaries. For those cases when mobility of the triple junctions (and also the quaternary junctions in 3D case) is high enough to accommodate any grain boundary motion and can maintain mechanical equilibrium.

(3) Numeric 2D solution of viscous dynamic equations for triple junctions, in those cases when the mobility of the triple junctions is low, so that grain boundaries can be treated as straight at any moment. Force, acting on any junction, is treated as a vector sum of three grain boundary tensions. Velocity of each junction is taken as mobility times force.

For example, we consider a 2D case of phenomenological mean-field model of the first type, which is based on Hillert-like model but involves the use of some size dependence of grain size mobility [22].

We can write separate motion equations for each curved grain boundary and combine them to obtain equation for the grain evolution. This way is very complicated because of the stochastic geometry of grains and their boundaries. Therefore, more common way is to write down equations for the sizes of all the grains, relating them to the sizes of other, neighboring grains.

Let the growth/shrinking rate of grain size be determined by the equation of $V = MF$, where $V = dr/dt$ is the growth/shrinking rate, M is mobility and F is the driving force, which is, as usual, determined by chemical potential gradient: $F = -d\mu/dx$, where $d\mu$ is Gibbs–Thomson potential energy difference between a pair of neighboring grains, $d\mu = \frac{(d-1)\gamma\Omega}{r_i} - \frac{(d-1)\gamma\Omega}{r_j}$, and dx is the effective width δ of the grain boundary between the pair of grains. Factor $(d - 1)$ in Gibbs–Thomson potential is equal $3 - 1 = 2$ for 3D grain growth and $2 - 1 = 1$ – for 2D grain growth. In the mean-field approach, we consider the growth or shrinking of a grain of radius r_i against the grain of critical size, r_{cr}. Introducing renormalized mobility, $\widehat{M} = M\frac{(d-1)\gamma\Omega}{\delta}$, one obtains

$$\frac{dr_i}{dt} = \widehat{M}\left(r_i\right)\left(\frac{1}{r_{cr}} - \frac{1}{r_i}\right)$$

The expression for critical size in 2D growth can be found from the condition of area conservation:

$$\sum \pi r_i^2 = \text{const} \Rightarrow \sum r_i \frac{dr_i}{dt} = 0 \Rightarrow \sum \widehat{M}\left(r_i\right)\left(\frac{r_i}{r_{cr}} - 1\right)$$

$$= 0 \Rightarrow \frac{1}{r_{cr}} = \frac{\sum \widehat{M}\left(r_i\right)}{\sum r_i \widehat{M}\left(r_i\right)}$$

Thus, if mobilities of all grains are the same (if mobility does not depend on grain size and shape), then critical size in 2D case is just equal to average size: $r_{cr} = \frac{\sum_{i=1}^{N} r_i}{N}$.

Thus, the main viscous dynamic equation for each grain will be

$$\frac{dr_i}{dt} = \widehat{M}\left(r_i\right)\left(\frac{\sum \widehat{M}\left(r_k\right)}{\sum r_k \widehat{M}\left(r_k\right)} - \frac{1}{r_i}\right)$$

For numerical solution, such form is not very convenient because when grains shrink to zero, the term $\frac{1}{r_i}$ tends to infinity and can lead to the error of "division by zero" or "floating point operation." To escape this singularity problem, it is convenient to solve equations for the square radii, which we can obtain if we multiply the previous equation by $2r_i$:

$$\frac{dr_i^2}{dt} = 2\widehat{M}\left(r_i\right)\left(\frac{r_i \sum \widehat{M}\left(r_k\right)}{\sum r_k \widehat{M}\left(r_k\right)} - 1\right)$$

In numeric solution it means that the new value of the squared grain size after the increment of time by dt is found as

$$r_i^2\,(\text{new}) = r_i^2 + 2\widehat{M}\left(r_i\right)\left(\frac{r_i \sum \widehat{M}\left(r_k\right)}{\sum r_k \widehat{M}\left(r_k\right)} - 1\right)dt$$

If some radii become less than or equal to zero, the corresponding grain is excluded from the list of the existing grains, and the number of grains (number of summands in the sum) is decreased by 1.

After a certain number of steps, the histogram describing the grain size distribution is built and drawn. With time this distribution shifts to the right (to larger sizes), and the peak "sinks" down, due to the decreasing number of grains. Yet, if we build a distribution in the reduced size space, r/r_{cr}, and with the normalized area, then the distribution tends to some asymptotic plot.

8.4 STATISTICAL DISTRIBUTION FUNCTIONS OF GRAIN SIZE

Experimental data of grain size distributions can usually be fitted by one of three distributions: Rayleigh, Weibull, and lognormal [15–22].

Rayleigh distribution contains one parameter regulating the distribution width – dispersion σ^2:

$$f(r;\sigma) = n\frac{r}{\sigma^2}\exp\left(-\frac{r^2}{2\sigma^2}\right) = n\frac{2r}{<r^2>}\exp\left(-\frac{r^2}{<r^2>}\right), \quad \sigma = \sqrt{\frac{<r^2>}{2}} \tag{8.2}$$

Here n is the number of grains per unit volume (in 3D case) or per unit area (in 2D case), r is the characteristic linear size of the grain, $<r^2>$ is the mean-squared grain size, $f(r)\,dr$ is the number of grains (per unit volume in 3D or per unit area in 2D) within the linear sizes interval $r, r + dr$. A plot of Rayleigh distribution is shown in Figure 8.3.

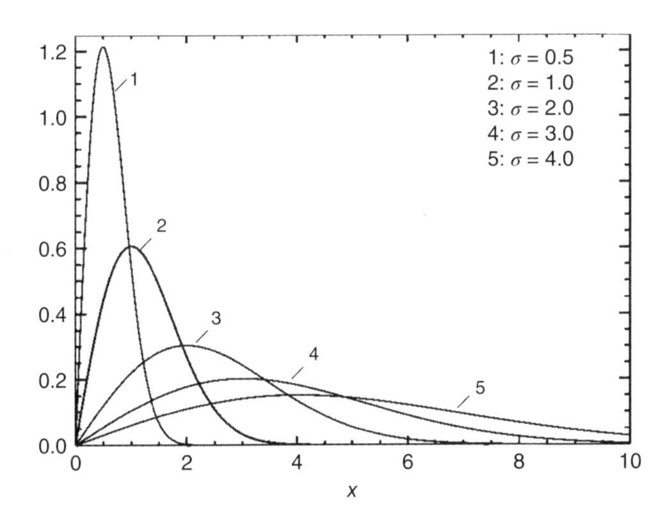

Figure 8.3 Rayleigh distribution.

The same scatter of sizes can be represented as the distribution of squared sizes:

$$f2\left(r^2;\sigma\right) = \frac{dr}{dr^2} f\left(r;\sigma\right) = \frac{1}{2r} f\left(r;\sigma\right) = n\frac{1}{2\sigma^2}\exp\left(-\frac{r^2}{2\sigma^2}\right)$$
$$= n\frac{1}{< r^2 >}\exp\left(-\frac{r^2}{< r^2 >}\right) \tag{8.3}$$

Weibull distribution is

$$f\left(r;\beta,\eta\right) = n\frac{\beta}{\eta}\left(\frac{r}{\eta}\right)^{\beta-1}\exp\left(-\left(\frac{r}{\eta}\right)^{\beta}\right) \tag{8.4}$$

with η being proportional to the mean linear size, and β determining the distribution shape. A plot of Weibull distribution is shown in Figure 8.4.

Lognormal distribution is also two-parametric, with parameters σ^2, μ to characterize the width and shape of distribution, respectively. A plot of lognormal distribution is shown in Figure 8.5.

$$f\left(r;\sigma,\mu\right) = n\frac{1}{r\sqrt{2\pi\sigma^2}}\exp\left(-\frac{(\ln r-\mu)^2}{2\sigma^2}\right)$$

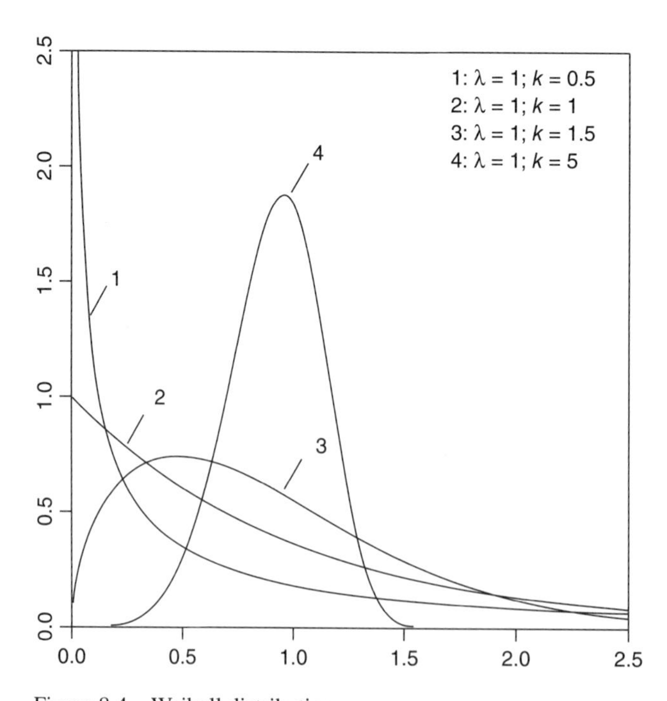

Figure 8.4 Weibull distribution.

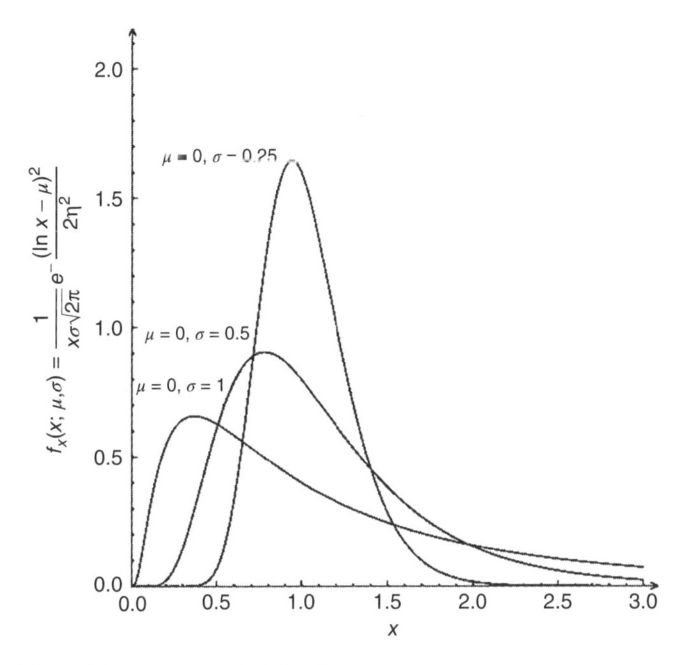

Figure 8.5 Lognormal distribution.

These three distributions differ mainly in the range of small grains, and they all give a smoothly decreasing distribution function for large grains, if the abnormal grain growth is excluded. The lognormal and Weibull distributions, though not the ideal fit, are regarded now to be the best fit among the known grain size distributions [16]. Most probably, this is because they both have two free parameters, contrary to Rayleigh distribution that has just one free parameter.

8.5 DETERMINISTIC (DYNAMIC) APPROACH TO GRAIN GROWTH

Two marginal approaches to the description of the grain growth process are known; the deterministic and the stochastic ones (and, of course, a number of combined models employing the approximations of both approaches).

In the deterministic approach (the most well known are the Burke–Turnbull model [1], Hillert's theory [2], topological model of Fradkov–Marder [17], and di Nunzio [23]), the behavior of each grain is unambiguously determined by several parameters: size (radius, cross-sectional area, or volume), number of nearest neighbors, etc., [13, 15, 24, 25].

In stochastic approximation, first suggested by Louat [18, 19] and later modified by other authors [20, 21], the grains are treated as "drunk sailors" who randomly walk within a semi-infinite size space with boundary zero point working just as a sink (without possibility of nucleation of new grains). As a nonzero possibility of getting to zero point (and disappearing) always exists, the number of "travelers" (grains) diminishes while the average size of the grains increases.

Deterministic approach seems to be more valid physically because it deals with the explicit form of the driving force of grain growth process, namely the reduction of the surface energy of grains. However, this approach in the standard form of Hillert's model leads to the wrong prediction of size distribution of grains.

In the deterministic approach, all grains of the 3D polycrystalline sample can be roughly represented as spheres of various effective radii, which correspond to the grain volumes $V_i = (4/3)\,\pi r_i^3$. Each sphere is under Laplace pressure of $2\gamma/r_i$, so atoms of these grains have the additional Laplace term $2\gamma\Omega/r_i$ in chemical potential. So, atoms belonging to the adjacent (neighboring) grains of different size will have a difference of Gibbs–Thomson chemical potentials as

$$\Delta\mu = \frac{2\gamma\Omega}{r_i} - \frac{2\gamma\Omega}{r_j} \tag{8.5}$$

Thus, there is a direct thermodynamic force for the transition of atoms from a smaller grain (higher chemical potential) to a larger grain (lower chemical potential). In the tradition of nonequilibrium thermodynamics (irreversible processes), the difference of chemical potentials should lead to the flux of atoms proportional to the thermodynamic force. The force is

$$X_{ij} = -\frac{\Delta\mu}{\delta} = \frac{2\gamma\Omega}{\delta}\cdot\left(\frac{1}{r_j} - \frac{1}{r_i}\right) \tag{8.6}$$

where δ is the effective width of grain boundary between grain r_i and r_j. Atomic flux is, by definition, equal to the boundary velocity, V_{ij}, times atomic density, $1/\Omega$, because $J = C<V>$. We note that for a pure metal, $C\Omega = 1$, where C is concentration and Ω is atomic volume. We have

$$J_{ij} = \frac{V_{ij}}{\Omega}$$

Thus,

$$\Omega J_{ij} = V_{ij} = LX_{ij}$$

$$V_{ij} = L\frac{2\gamma\Omega}{\delta}\left(\frac{1}{r_j} - \frac{1}{r_j}\right) \tag{8.7}$$

Taking as usual, for a pure metal or element, $L = D/kT$ (D is the diffusivity of atoms across the grain boundary), we obtain

$$V_{ij} = \frac{D_{\mathrm{GB}}^{\mathrm{across}}}{kT}\,\frac{2\gamma\Omega}{\delta}\left(\frac{1}{r_j} - \frac{1}{r_i}\right) \tag{8.8}$$

Taking *i*th grain as a central one, and *j*th grain as the representative of **mean-field**, we have

$$\frac{dr_i}{dt} = \frac{D_{\mathrm{GB}}^{\mathrm{across}}}{kT}\frac{2\gamma\Omega}{\delta}\left(\text{mean-field inverse radius} - \frac{1}{r_i}\right) \tag{8.9}$$

To derive the expression of mean-field inverse radius (mfir), we consider the constraint of constant volume:

$$\sum r_i^3 = \text{const} \Rightarrow \sum r_i^2 \frac{dr_i}{dt} = 0 \tag{8.10}$$

Substituting the rate equation of dr_i/dt of Eq (8.9) into Eq (8.10) and dropping the parameters, we obtain,

$$\sum r_i^2 \frac{dr_i}{dt} = \sum r_i^2\left(\text{mfir} - \frac{1}{r_i}\right) = 0 \Rightarrow \text{mfir}\sum r_i^2 - \sum r_i = 0$$

$$\sum r_i = \text{mfir} \cdot \sum r_i^2 \Rightarrow \text{mfir} = \frac{\sum r_i}{\sum r_i^2} = \frac{<r>}{<r^2>} \tag{8.11}$$

We note that in the last equation, we have simultaneously divided the nominator and the denominator by the total number of grains, to obtain mean values. Thus

$$\frac{dr_i}{dt} = \frac{D_{\mathrm{GB}}^{\mathrm{across}}}{kT}\frac{2\gamma\Omega}{\delta}\left(\frac{<r>}{<r^2>} - \frac{1}{r_i}\right) \tag{8.12}$$

In 2D case, for grain growth of columnar grains in a thin film, the analogous considerations, with the constraint of constant area ($\sum r_i^2 = \text{const} \Rightarrow \sum r_i^1 \frac{dr_i}{dt} = 0$) and with twice lower Laplace pressure, give,

$$\frac{dr_i}{dt} = \frac{D_{\mathrm{GB}}^{\mathrm{across}}}{kT}\frac{\gamma\Omega}{\delta}\left(\frac{1}{<r>} - \frac{1}{r_i}\right) \tag{8.13}$$

Both of these (3D and 2D) equations can be written in unified form, as it was first performed by Hillert [2].

$$\frac{dr_i}{dt} = (d-1)M\gamma\left(\frac{1}{r_{\mathrm{cr}}(t)} - \frac{1}{r_i}\right) \tag{8.14}$$

where d is a dimension (3 or 2), $r_{\mathrm{cr}} = \frac{<r^2>}{<r>}$ for 3D case and $r_{\mathrm{cr}} = <r>$ for 2D case. Critical size, r_{cr}, means approximately the same as in nucleation theory, where grains larger than the critical will grow and grains lesser than the critical will shrink. However, in Ostwald ripening, the critical size itself grows with time, so it will "catch up" practically all larger grains, one by one, transforming them from growing to shrinking. [26, 27] Finally, only one grain should survive.

Let us take an average of both parts of the previous equation and multiply them by the average size and by 2. It gives

$$\frac{d(<r>)^2}{dt} = (d-1)M\gamma\left(\frac{<r>}{r_{\mathrm{cr}}} - <r>\left\langle\frac{1}{r}\right\rangle\right) \quad (8.15)$$

For example, we obtain in the 2D case,

$$\frac{d(<r>)^2}{dt} = M\gamma\left(1 - <r>\left\langle\frac{1}{r}\right\rangle\right) \quad (8.15')$$

We note that the product $<r> \langle\frac{1}{r}\rangle$ is constant in time, so that all right-hand side $M\gamma(1- <r> \langle\frac{1}{r}\rangle)$ is constant in time and, evidently, positive because grain growth is always an energetic process favoring the growth of mean size. Thus, the time derivative of the square mean grain size is constant.

$$\frac{d(<r>)^2}{dt} = \mathrm{const}$$

Therefore, the mean size itself should satisfy the parabolic law. Actually, it coincides with the main idea of Burke and Turnbull that the growth rate of the mean size is proportional to the inverse mean size [1], and this immediately leads to parabolic growth law.

Last equation can be written simultaneously for all grains of the polycrystalline sample and form the self-consistent set of equations for the self-regulated system. A simple program for numerical solution of such set (with initial number of grains, say, about 10,000) can give, in a rather short computation time, the time dependence of the mean size, as well as the distribution histogram. On the other hand, the analytical approaches to such problems has been first invented by Lifshitz and Slezov for description of Ostwald ripening [26, 27], and later used by Hillert for description of grain growth [2]. The analytic approach produces the following results:

1. The time law for critical size obeys parabolic law,

$$r_{\mathrm{cr}}(t) = \sqrt{\frac{d-1}{2}M\gamma t} \quad (8.16)$$

where $d=2$ or 3.

2. The size distribution is for nondimensional size $u \equiv r/r_{\mathrm{cr}}$

$$f(u) = n \cdot (2e)^d \frac{d \cdot u}{(2-u)^{2+d}} \exp\left(-\frac{2d}{2-u}\right) \quad (8.17)$$

To find this distribution function, Hillert had to solve the equation for size distribution in the size space:

$$\frac{\partial f(t,r)}{\partial t} = -\mathrm{div}\left(V\left(r, r_{\mathrm{c}}(t)\right) f(t,r)\right) = -(d-1)M\gamma\frac{\partial}{\partial r}\left(\left(\frac{1}{r_{\mathrm{c}}(t)} - \frac{1}{r}\right) f(t,r)\right) \quad (8.18)$$

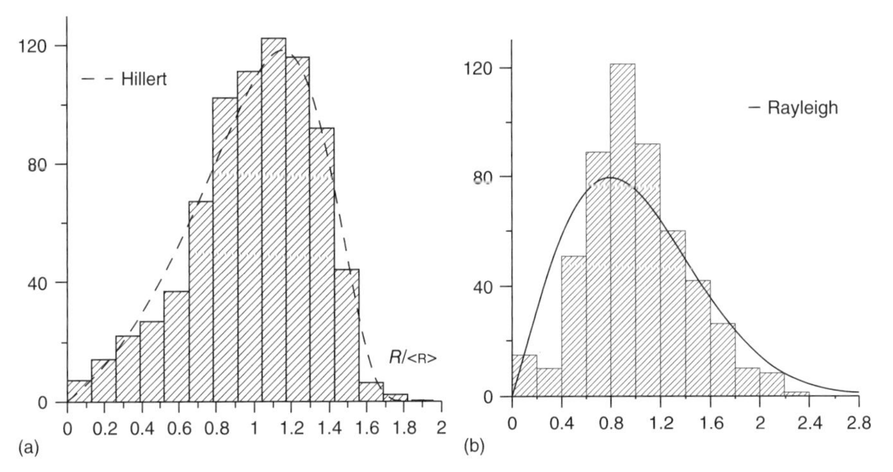

Figure 8.6 (a) Hillert distribution and (b) experimental distribution of grain growth and fitted with Rayleigh distribution.

Unfortunately, such an elegant theory disagrees with the experimental data on grain size distribution. The main difference is the *skewness*. It is a measure of the asymmetry of any probability distribution.

$$\text{skewness} = \frac{\left\langle (r - <r>)^3 \right\rangle}{\left(\sqrt{\left\langle (r - <r>)^2 \right\rangle}\right)^3} = \frac{\int_0^\infty (u - <u>)^3 f(u)\, du}{\left(\sqrt{\int_0^\infty (u - <u>)^2 f(u)\, du}\right)^3} \tag{8.19}$$

Same as the size distribution in Lifshiz–Slezov–Wagner (LSW) theory on ripening, Hillert distribution has **negative** skewness (right-hand), indicating a wide distribution of smaller grains; yet the real size distribution of grains has **positive** skewness (left-hand), indicating a wide distribution of larger grains as shown by Rayleigh distribution in Figure 8.3. Figure 8.6 shows the discrepancy between Hillert distribution and experimental distribution.

8.6 COUPLING BETWEEN GRAIN GROWTH OF A CENTRAL GRAIN AND THE REST OF GRAINS

The main reason of inadequacy of the existing deterministic grain growth theories is the lack of proper account of cooperative movement of grain boundary network in the polycrystalline microstructure. Any boundary in a grain boundary network cannot change its length (in 2D model) or area (in 3D model) without changing the length or area of the other boundaries. To make this coupling clear, we consider a system of N grains, consisting of an arbitrary grain of size r (we define it as the "central" grain) and the rest of $N-1$ grains with a mean size $\bar{r}$ (we define them as the "reservoir").

So far we have not specified the concrete form of mean value (in principle, it can be $<r>, (\langle r^2 \rangle)^{1/2}, (\langle r^3 \rangle)^{1/3}$, etc.). The reservoir can be regarded as a "mean-field" for the central grain. Considering all grains as spheres for simplicity, we have the constraint of constant volume in the form of

$$V^{\text{total}} = \frac{4}{3}\pi \left(r^3 + (N-1)\bar{r}^3 \right) = \text{const} \tag{8.19a}$$

In turn, we have (if all grains in the reservoir change equally)

$$d\bar{r} = -dr \left(\frac{r}{\bar{r}} \right)^2 \frac{1}{N-1} \tag{8.19b}$$

It means that a change of size of the "central" grain leads to a change of all of the other grains owing to the constraint of total volume. While the latter change is very small for each grain in the reservoir, it gives a nonnegligible effect on the change of the total grain boundary area,

$$dS^{\text{total}} = \frac{1}{2} 4\pi d \left(r^2 + (N-1)\bar{r}^2 \right) = 4\pi \left(\frac{1}{r} - \frac{1}{\bar{r}} \right) r^2 dr \tag{8.20}$$

where the factor of 1/2 accounts for the sharing of a grain boundary by two neighboring grains. Thus, the driving force of grain growth of the central grain is

$$-\frac{\partial F}{\partial r} = -\gamma \frac{\partial S^{\text{total}}}{\partial r} = 4\pi r^2 \gamma \left(\frac{1}{\bar{r}} - \frac{1}{r} \right) \tag{8.21}$$

where $F = \gamma S^{\text{total}}$ is a free energy of the whole system, and γ is surface energy per unit area. We note that the driving force contains comparable inputs from both the central grain and the reservoir. Such coupling at thermodynamic and kinetic levels may lead to cross-term effects between a grain and its surrounding. Therefore, the analysis is rather complicated and has obstructed the progress of the theory of grain growth in the past. We demonstrate below that this basic difficulty can be circumvented by using the normalized size space. In the normalized size space, we can decouple them, so that we can use the simple model [1] of Burke–Turnbull of grain growth of a central grain.

8.7 DECOUPLING THE GRAIN GROWTH OF A CENTRAL GRAIN FROM THE REST OF GRAINS IN THE NORMALIZED SIZE SPACE

Let $R(t)$ be some average characteristics of the grains, such as a unit length, being proportional to all kinds of average of the same dimension:

$$R \propto <r> \propto \sqrt{<r>} \propto \sqrt{\frac{<r^3>}{<r>}} \propto \cdots$$

The proper choice of R will be made below to satisfy the constraint of constant total volume. We introduce a nondimensional space (or the normalized size space) with R being a unit length in the space, and furthermore we consider grain growth in this space.

In real space, the free energy, related to grain boundaries, is given as $F = \frac{1}{2}\sum_{i=1}^{N}\gamma S_i$, where S_i is then surface area of ith grain, $S_i = q_3 r_i^2, r_i \equiv (\frac{3V_i}{4\pi})^{1/3}$, q_3 is a geometrical factor of surface area (equal to 4π for a sphere), which is constant under the assumption of fixed shape, and V_i is the volume of the ith grain. In real space, the dimension of free energy is kg·m^2/s^2. In the normalized space the free energy is represented by $\widetilde{F} = F/R^2$, with a dimension of kg/s^2, as

$$\widetilde{F} = \frac{1}{2}\gamma q_3 \sum_{i=1}^{N} \widetilde{r}_i^2, \quad \widetilde{r}_i = \frac{r_i}{R} \tag{8.22}$$

In the framework of mean-field approach, we consider an arbitrary grain 1 as the "central" grain, and all the others as the reservoir:

$$\widetilde{F} = \frac{1}{2}\gamma q_3 \left(\widetilde{r}^2 + \sum_{i=2}^{N} \widetilde{r}_i^2\right) = \frac{1}{2}\gamma q_3 \left(\widetilde{r}^2 + (N-1)\left\langle \widetilde{r}^2 \right\rangle\right) \tag{8.23}$$

Because R is proportional to average size by definition, the ratio $\frac{\langle r^2\rangle}{R^2} = \langle \widetilde{r}^2 \rangle$ is constant. The number of grains N can be treated as constant when the change of size is infinitesimal. Thus, the second term in Eq (8.23) is constant, and the change of normalized free energy of the central grain (defined in normalized size space) is independent of the reservoir.

We note that we have decoupled the central grain from the reservoir in the normalized size space. The influence of reservoir will be present only after the transition back to the real space: $dF = R^2 d\widetilde{F} + \widetilde{F} dR^2$.

Thus, in the normalized size space we can follow the original Burke–Turnbull approach by considering only the energy change of the "central" grain alone, with the correct sign for interrelation among pressure, mobility and velocity:

$$\widetilde{p} = -\frac{\partial \widetilde{F}}{\partial \widetilde{V}} = -\frac{\partial\left(\frac{\gamma q_3}{2}\widetilde{r}^2\right)}{\partial\left(\frac{4\pi}{3}\widetilde{r}^3\right)} = -\frac{\gamma q_3}{4\pi}\frac{1}{\widetilde{r}} \tag{8.24}$$

The dimension of $\widetilde{p}$ is (joule/m^2)/(m^3/m^3) = joule/m^2.

$$\frac{d\widetilde{r}}{dt} = \widetilde{M}\widetilde{p} = -\widetilde{M}\frac{\gamma q_3}{4\pi}\frac{1}{\widetilde{r}} \tag{8.25}$$

Here $\widetilde{M}$ is mobility in the normalized size space. The dimension of $\widetilde{M}$ (from Eq (8.25)) is (1/s)/(joule/m^2). The dimension of M is (m/s)/(joule/m^3) = m^4/s joule. So the dimension of the ratio is $M/\widetilde{M} = m^2$. [22]

The "minus" sign in Eq (8.25) means that in the normalized size space it is thermodynamically favorable to decrease the size of any grain. In real space it translates to mean that even if some grain is growing, its growth rate is less than the growth rate of the mean size grain. (One can check that this characteristics is valid in the case of LSW ripening for all sizes [26, 27], except the maximum, $r_{max} = \frac{3}{2} r_{crit}$, for which $\frac{d\tilde{r}}{dt} = 0$).

Thus, we have

$$\widetilde{M} = \frac{M}{R^2} \tag{8.26}$$

where M is mobility in real space. Following Burke–Turnbull's approach, we take M to be constant. Actually, this is a major assumption of most grain growth theories. Substituting Eq (8.26) into Eq (8.25), we obtain the rate equation for grain sizes in the normalized space:

$$\frac{d\tilde{r}^2}{dt} = -\frac{k_3}{R^2(t)} \tag{8.27}$$

$$k_3 = M\frac{\gamma q_3}{2\pi} \tag{8.28}$$

To relate R to average values, we use the constraint of constant total volume in real space:

$$\begin{aligned} 0 &= \sum r_i \frac{dr_i^2}{dt} = \sum R\tilde{r}_i \frac{d\left(R^2\tilde{r}_i^2\right)}{dt} = R\sum \tilde{r}_i\left(R^2\left(-\frac{k_3}{R^2}\right) + \tilde{r}_i^2\frac{dR^2}{dt}\right) \\ &= RN\left(-k_3 <\tilde{r}> + \frac{dR^2}{dt} <\tilde{r}^3>\right) \end{aligned} \tag{8.29}$$

Thus,

$$\frac{dR^2}{dt} = k_3\frac{<\tilde{r}>}{<\tilde{r}^3>} = R^2\frac{<r>}{<r^3>} \tag{8.30}$$

so that

$$\frac{d\ln R^2}{dt} = k_3\frac{<r>}{<r^3>} \tag{8.31}$$

Substituting the condition of Eq (8.30) or Eq (8.31) into Eq (8.27), we have the growth/shrinkage equation for grain sizes in real space:

$$\frac{dr^2}{dt} = k_3\left(\frac{r^2<r>}{<r^3>} - 1\right) \tag{8.32a}$$

or

$$\frac{dr}{dt} = \frac{1}{2}k_3\left(\frac{r<r>}{<r^3>} - \frac{1}{r}\right) \tag{8.32b}$$

Substituting Eq (8.32b) into the continuity equation in size space (or Eqs (8.27) and (8.30) into the continuity equation in the normalized size space), we obtain

$$\frac{\partial f}{\partial t} = -\frac{k_3}{2}\frac{\partial}{\partial r}\left(f\left(\frac{r<r>}{<r^3>} - \frac{1}{r}\right)\right) \tag{8.33}$$

where $f(t, r)$ is a size distribution. By using a standard mathematical procedure of separation of variables, the following asymptotical size distribution function is obtained, which practically coincides with Rayleigh distribution, and it fits experimental observations much better than Hillert distribution;

$$f(t,r) = \frac{\text{const}}{\left(\frac{2}{3}k_3 t\right)^2}\frac{r}{\left(\frac{2}{3}k_3 t\right)^{1/2}}\exp\left(-\frac{r^2}{\frac{2}{3}k_3 t}\right) \tag{8.34}$$

$$<r> = \frac{\sqrt{\pi}}{2}\left(\frac{2}{3}k_3 t\right)^{1/2}, <r^2> = \frac{2}{3}k_3 t,$$

$$<r^3> = \frac{3\sqrt{\pi}}{4}\left(\frac{2}{3}k_3 t\right)^{3/2}$$

$$\frac{<r^3>}{<r>} = k_3 t\frac{d\ln R^2}{d\ln t} = 1 \tag{8.35}$$

8.8 GRAIN GROWTH IN 2D CASE IN THE NORMALIZED SIZE SPACE

As the basic idea remains the same, we can streamline the consideration and derivations of 2D case, indicating only the key equations. The total free energy of grain boundaries for the system, consisting of a "central" grain and a "reservoir" of all the other grains, is

$$F = \frac{1}{2}\gamma\left(l + \sum_{i=1}^{N} l_i\right) = \frac{1}{2}\gamma q_2\left(r + \sum_{i=1}^{N} r_i\right) \tag{8.36}$$

where γ is the grain boundary free energy per unit length of grain boundaries in an ordinary 2D space, l_i is the perimeter length of ith grain, $l_i = q_2 r_i, r_i \equiv \sqrt{S_i/\pi}$, S_i is the grain's area, and q_2 is a geometrical factor, equal to 2π for circular grains.

The free energy in the normalized space is

$$\widetilde{F} = \frac{1}{2}\gamma q_2 \widetilde{r} + \text{const} \tag{8.37}$$

The pressure has the same dimension as in the 3D case and is equal to

$$\widetilde{p} = -\frac{\partial \widetilde{F}}{\partial \widetilde{S}} = -\frac{\partial\left(\frac{1}{2}\gamma q_2 \widetilde{r}\right)}{\partial\left(\pi \widetilde{r}^2\right)} = -\frac{\gamma q_2}{4\pi}\frac{1}{\widetilde{r}} \tag{8.38}$$

The grain velocity in the normalized size space is

$$\frac{d\widetilde{r}}{dt} = \widetilde{M}\widetilde{p} = -\widetilde{M}\frac{\gamma q_2}{4\pi}\frac{1}{\widetilde{r}} \tag{8.39}$$

As velocity and pressure in the normalized 2D case have the same dimensions as in the normalized 3D case, we have for mobility, $\widetilde{M} = \frac{M}{R^2}$, and

$$\frac{d\widetilde{r}^2}{dt} = -\frac{k_2}{R^2} \tag{8.40}$$

with

$$k_2 = M\frac{\gamma q_2}{2\pi} \tag{8.41}$$

The constraint of conservation for the total area, $\sum r_i \frac{dr_i}{dt} = 0$, leads to the condition for R:

$$\frac{d \ln R^2}{dt} = \frac{k_2}{<r^2>} \quad \text{or} \quad \frac{dR^2}{dt} = \frac{k_2}{<\widetilde{r}^2>} \tag{8.42}$$

so that the growth/shrinkage law in the ordinary 2D case is

$$\frac{dr^2}{dt} = k_2\left(\frac{r^2}{<r^2>} - 1\right) \tag{8.43}$$

or

$$\frac{dr}{dt} = \frac{1}{2}k_2\left(\frac{r}{<r^2>} - \frac{1}{r}\right) \tag{8.44}$$

The corresponding size distribution coincides well with Rayleigh distribution and fits not ideally but reasonably with most parts of experimental observations:

$$f(t,r) = \frac{\text{const}}{\left(k_2 t\right)^{3/2}}\frac{r}{\left(k_2 t\right)^{1/2}}\exp\left(-\frac{r^2}{k_2 t}\right) \tag{8.45}$$

$$<r^2> = k_2 t \tag{8.46}$$

Thus, we obtained very reasonable size distributions and parabolic time dependence for normal grain growth in both 3D and 2D cases in the frame of deterministic approach. The main idea of the above analysis is based on the decoupling of a "central" grain from a "reservoir" consisting of all the other grains (a mean-field approach), by means of transition to a normalized size space.

Inclusion of a stochastic term into the above equations for distribution in the normalized space can lead to two-parametric distribution, which will fit even better.

By the way, the deterministic mean-field grain growth might be formulated directly in real space by taking into account the cooperative movement of the grain boundary network. In this case the mobility ceases to be constant. For example, the discussed Hillert's model can lead to a reasonable size distribution, provided that

1. Average size $R(t)$ is the mean-squared radius $\sqrt{<r^2>}$ in 2D case, and $\sqrt{\frac{<r^3>}{<r>}}$ in 3D case
2. The mobility is modified from a constant M (in Hillert's theory) to

$$M' = M\left(1 + \frac{r}{R}\right) \tag{8.47}$$

so that

$$\frac{dr}{dt} = M'\gamma\left(\frac{1}{R} - \frac{1}{r}\right) = M\gamma\left(1 + \frac{r}{R}\right)\left(\frac{1}{R} - \frac{1}{r}\right) = M\gamma\left(\frac{r}{R^2} - \frac{1}{r}\right) \tag{8.48}$$

that practically coincides with Eqs (8.24b) in 3D case and (8.44) in 2D case, which we have derived.

8.9 GRAIN ROTATION

The concept of grain rotation was used first only for subgrains of one grain, with small misorientation between each other but with large misorientation with other grains. Later it was concluded that when grains are less than 10 μm they can rotate under stress, providing the superductility. We see below that the mobility of grain rotation is inversely proportional to the fifth power of size. It means that the size plays the key role in mobility. In bulk nanoalloys and especially in nanofilms, the grain rotation might be a key player in morphology evolution, including grain growth, by merging grains.

Thermodynamic reason for the grain rotation in the absence of external torques is a dependence of grain boundary tension on misorientation [28]. In the ideal case, grain rotation should lead to the formation of large single crystal grain as a result of the merging of all grains. Until it happens, the resulting torque (force x arm), M, acting on one grain in a thin film of thickness h, with vertical bamboo structure, is equal to,

$$M = -\frac{\partial W^{\text{surf}}}{\partial \vartheta} = -\frac{\partial}{\partial \vartheta} h \sum_{j=1}^{n} l_j \gamma_j \tag{8.49}$$

Here γ_j is a tension of grain boundary with jth neighboring grain, l_j is a length of grain boundary with this grain.

For each neighbor, we can change the derivative of grain boundary tension over the orientation of "central" grain by the derivative (with opposite sign) over the orientation of neighboring grain, or over the misorientation ϑ_j

$$\frac{\partial \gamma_j}{\partial \vartheta} = -\frac{\partial \gamma_j}{\partial \vartheta_j} \tag{8.50}$$

Then we obtain for torque:

$$M = h\sum_{j=1}^{n} l_j \frac{\partial \gamma_j}{\partial \vartheta_j} \tag{8.51}$$

Of course, dissipation of energy in such slow process as grain rotation proceeds much faster than the process itself. Therefore, it is reasonable to assume that the torque is proportional not to the angular acceleration but instead to the angular velocity (velocity of rotation):

$$\omega = \frac{d\vartheta}{dt} = L \cdot M \tag{8.52}$$

where L is a mobility in respect to grain rotation.

In fact, mobility depends both on grain boundary and on bulk diffusivities, helping to adjust the grain formation to each other during rotation. If grain boundary diffusion is dominant ($D^{GB}\frac{\delta}{R} \gg D^{bulk}$),rather complicated considerations show that the mobility has the following explicit form:

$$L \approx 10^3 \frac{\Omega \delta D^{GB}}{kT} \frac{1}{R^4 d} \tag{8.53}$$

Because grain size in thin films is typically of the same order as the thickness, we can rewrite the last expression as

$$L \approx 10^3 \frac{\Omega \delta D^{GB}}{kT} \frac{1}{R^5} \tag{8.54}$$

When bulk diffusion is the main mechanism, mobility is inversely proportional to the fourth power of size instead of the fifth power for the case of grain boundary diffusion. It means that rotation as a mechanism of morphology evolution indeed should play an important role for nanograined alloys. If rotation is the only mechanism of grain coarsening, the time dependence of the grain size would be $t^{1/(p-1)}$, where p is, respectively, 5 and 4 for grain boundary and for bulk diffusion mechanism of rotation.

8.9.1 Grain Rotation in Anisotropic Thin Films Under Electromigration

Misorientation is not a single possible reason of grain rotation. Actually, any torque should induce grain rotation if grains are not large and the time of torque action is long enough. We show it here for the case of electromigration in pure Sn thin films with free surface [29].

White tin (or β-Sn) has a body-center tetragonal crystal lattice (with lattice parameters $a=b=0.583$ nm and $c=0.318$ nm), for which essentially anisotropic properties are typical. In particular, specific resistivities along the a (b) and c directions are 13.25×10^{-8} (Ωm) and 20.27×10^{-8} (Ωm), respectively. First, the reduction of resistivity was noticed in tin line under electromigration during 1 day at constant direct current. Apparently, some structural changes take place in tin. On the one hand, the grains with lower resistivity may consume (relative to the fixed current direction) those with higher resistivity. On the other hand, a change of grain orientation (grain rotation) from c to a or b in parallel to the current direction may take place resulting in the reduction of resistivity.

Grain rotation requires a torque consisting of a force and a moment arm to rotate an object. We consider possible forces and arms below.

Electromigration-induced vacancy fluxes directed from cathode to anode have different values in neighboring grains with different orientations. This difference (singular flux divergence) leads to supersaturation with vacancies of the boundary at one side of the grain and undersaturation at the other side. Because the top surface of tin (Sn) line is stress-free and can be considered to be a good sink/source of vacancies, it must have the equilibrium vacancy concentration. That is why the vacancy gradients and fluxes should be built up in vertical directions. Vertical vacancy gradients imply vertical stress gradients along the grain boundaries. Thus, they indicate transversal forces (up and down) acting on the atom at the two grain boundaries of the grain in rotation. This pair of forces is the origin of the torque.

Many studies have focused on electromigration in metal interconnects. However, these studies are based on electrically isotropic materials, such as aluminum or copper. The vacancy flux in such materials causes the inverse gradient of mechanical stress (so-called back-stress) mainly in the direction parallel to the current flow, not in the direction normal to it, and hence there is no grain rotation.

We consider a simple and geometrically ideal situation of three grains in Figure 8.7: a "bad" grain, the grain 2 in the middle, with its c-axis directed along the electron flow direction, and it is situated between two "good" grains, grain 1 and 3 at the left and the right, respectively, with their a-axis directed also along the electron flow direction. The electrons flow from left to right. As both resistivity and diffusivity of grains along a- and c-axes are different, the electron wind effect and the corresponding vacancy fluxes in the bad and good grains are different:

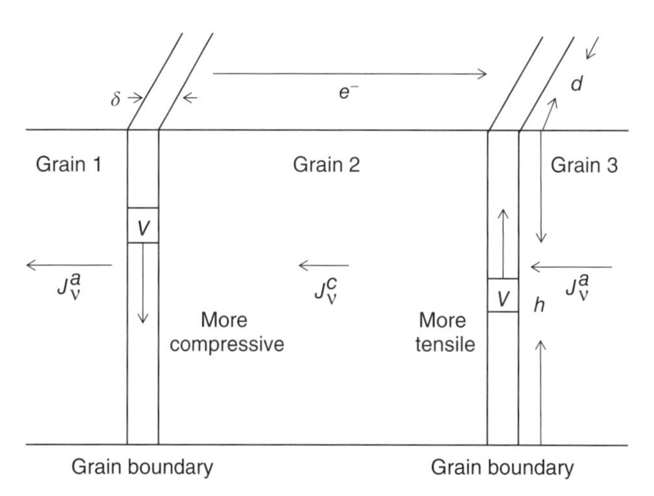

Figure 8.7 Schematic diagram of grain 2 with "bad" orientation between grains 1 and 3, having "good" orientation. The height of the grain is "h," the width is "d," and the width of grain boundaries is "δ." Inside the grain boundaries, "v" stands for vacancies and the arrows indicate the direction of vacancy flows.

$$J_{\mathrm{v}}^{c} = \frac{C_{\mathrm{v}}^{\mathrm{bulk}} D_{\mathrm{v}}^{c,\mathrm{bulk}}}{kT} Z^* e \rho^c j$$

$$J_{\mathrm{v}}^{a} = \frac{C_{\mathrm{v}}^{\mathrm{bulk}} D_{\mathrm{v}}^{a,\mathrm{bulk}}}{kT} Z^* e \rho^a j \tag{8.55}$$

J_{v}^{c} and J_{v}^{a} are the vacancy fluxes along a- and c-axes, respectively. We have the reference data below for the diffusivity and resistivity of tin atoms along these two directions: $D_c = 5.0 \times 10^{-17}\,\mathrm{m}^2/\mathrm{s}$, $\rho_c = 20.33 \times 10^{-8}\Omega m$; $D_a = 1.33 \times 10^{-16}\mathrm{m}^2/\mathrm{s}$, $\rho_a = 13.33 \times 10^{-8}\Omega m$.

The atomic flux under electromigration should be in the same direction of electron flow. Therefore, a counterflux of vacancies flows from right to left. The effective charge, Z^*, is assumed to be the same in both directions. As $D_c \rho_c < D_a \rho_a$, we find that a larger vacancy flux reaches grain boundary II from grain 3 to grain 2; yet a smaller vacancy flux leaves grain boundary II going into grain 2. This smaller vacancy flux then goes through grain 2 and reaches grain boundary I. Correspondingly, a larger vacancy flux leaves the grain boundary I and goes into the grain 1. In grain 2, depletion (undersaturation) of vacancies occurs at the grain boundary I and corresponds to compressive stresses near the boundary. On the other hand, supersaturation of vacancies and corresponding tensile stresses occur at grain boundary II.

In this model, the divergence of vacancies at the grain boundaries occurs in the manner that the vertical gradient of stresses acts as the vertical forces at each atom in the grain boundaries. It acts downward (atomic flux moves down) at grain boundary II, but upward at grain boundary I. Therefore, this pair of forces generates a clockwise torque and leads to the rotation of the bad grain of grain 2.

At slow (quasi-steady state) process, the vertical gradient of vacancy concentration can be found from the flux balance condition: the number of excess vacancies entering the boundary following the difference in horizontal fluxes equals the number of vacancies arriving at free surface,

$$\delta D_{\mathrm{v}}^{\mathrm{GB}} \frac{\Delta C_{\mathrm{v}}}{h} \cong \frac{C_{\mathrm{v}}^{\mathrm{bulk}} Z^* e j}{kT} \left(D_{\mathrm{v}}^{a,\mathrm{bulk}} \rho^a - D_{\mathrm{v}}^{c,\mathrm{bulk}} \rho^c \right) \times h \tag{8.56}$$

As the vertical gradient is what makes the torque to rotate the grain, it is important to correlate the vacancy concentration gradient to the stress gradient. Because the stress is in equilibrium with vacancy concentration, $C_{\mathrm{v}} = C_{\mathrm{v}0} \cdot \exp\left(\frac{\sigma\Omega}{kT}\right)$, the stress gradient along the grain boundary is

$$\frac{\partial \sigma}{\partial h} \cong \frac{kT}{\Omega C_{\mathrm{v}}} \frac{\partial C_{\mathrm{v}}}{\partial h} \cong \pm \frac{Z^* e j}{\Omega \delta D_{\mathrm{v}}^{\mathrm{GB}}} \left(D_{\mathrm{v}}^{a,\mathrm{bulk}} \rho^a - D_{\mathrm{v}}^{c,\mathrm{bulk}} \rho^c \right) \times h \tag{8.57}$$

By multiplying the stress gradient with the atomic volume, we obtain the potential gradient, which is the force acting on an atom along the grain boundary,

$$F_y' = \Omega \times \frac{\partial \sigma}{\partial h} \cong \pm \frac{Z^* e j}{\delta D_{\mathrm{v}}^{\mathrm{GB}}} \left(D_{\mathrm{v}}^{a,\mathrm{bulk}} \rho^a - D_{\mathrm{v}}^{c,\mathrm{bulk}} \rho^c \right) \times h \tag{8.58}$$

To sum up the effect from every atom, the total force that acts on all atoms in the grain boundary becomes

$$F_y = \frac{d \times h \times \delta}{\Omega} F'_y \cong \pm \frac{Z^* ej}{\Omega D_{\mathrm{v}}^{\mathrm{GB}}} \left(D_{\mathrm{v}}^{a,\mathrm{bulk}} \rho^a - D_{\mathrm{v}}^{c,\mathrm{bulk}} \rho^c \right) \times h^2 \times d \tag{8.59}$$

For the full torque (from both grain boundaries) that acts on the unit width of the grain and causes the grain to rotate will be

$$M = 2 \times F_y \times \frac{h}{2} \times \frac{1}{d} \cong \pm \frac{Z^* ej}{\Omega D_{\mathrm{v}}^{\mathrm{GB}}} \left(D_{\mathrm{v}}^{a,\mathrm{bulk}} \rho^a - D_{v}^{c,\mathrm{bulk}} \rho^c \right) \times h^3 \tag{8.60}$$

For a standard equation for angular velocity of grain rotation,

$$\omega = \frac{d\theta}{dt} = LM$$

with a mobility

$$L = \frac{4096 \Omega \delta D^{\mathrm{GB}}}{kT} \frac{1}{d^5} \tag{8.61}$$

where δ is the width of the grain boundary, $D_{\mathrm{v}}^{\mathrm{GB}}$ is the diffusivity at grain boundary, and Ω is the atomic volume. Inserting M and L from the above equations, the angular velocity becomes

$$\omega = \frac{d\theta}{dt} = \frac{4096 \delta D^{\mathrm{GB}}}{kT} \frac{1}{h^2} \frac{Z^* ej}{\Omega D_{\mathrm{v}}^{\mathrm{GB}}} \left(D_{\mathrm{v}}^{a,\mathrm{bulk}} \rho^a - D_{\mathrm{v}}^{c,\mathrm{bulk}} \rho^c \right) \tag{8.62}$$

As the diffusivity of vacancy and atomic coefficient of self-diffusion can be correlated both in the bulk and at the surface, $D^{\mathrm{bulk}} = C_{\mathrm{v}}^{\mathrm{bulk}} D_{\mathrm{v}}^{\mathrm{bulk}}$, $D^{\mathrm{GB}} = C_{\mathrm{v}}^{\mathrm{GB}} D_{\mathrm{v}}^{\mathrm{GB}}$, we have in this case

$$\omega = \frac{d\theta}{dt} = \frac{C_{\mathrm{v}}^{\mathrm{GB}}}{C_{\mathrm{v}}^{\mathrm{bulk}}} \frac{4096 \delta}{kTh^2} Z^* ej \left(D^{a,\mathrm{bulk}} \rho^a - D^{c,\mathrm{bulk}} \rho^c \right) \tag{8.63}$$

In a general case where the c-axis of the "bad" grain is oriented at an arbitrary angle u to the current direction, the magnitudes of vacancy flux density along the c- and a-axes will be equal to

$$J_{\mathrm{v}}^c = \frac{C_{\mathrm{v}}^{\mathrm{bulk}} D_{\mathrm{v}}^{c,\mathrm{bulk}}}{kT} Z^* e \rho^c j \cos\theta, \quad J_{\mathrm{v}}^a = \frac{C_{\mathrm{v}}^{\mathrm{bulk}} D_{\mathrm{v}}^{a,\mathrm{bulk}}}{kT} Z^* e \rho^a j \sin\theta \tag{8.64}$$

Considering the electron flow direction is along x-axis, only the x direction of the overall vacancy flux would contribute to the torque.

$$J_{\mathrm{v}x}^{\mathrm{bulk}} = -J_{\mathrm{v}}^c \cos\theta - J_{\mathrm{v}}^a \sin\theta$$

By following the same logic, the angular velocity would be

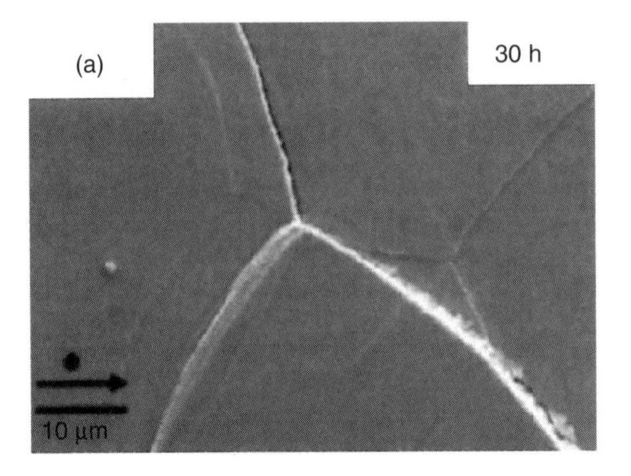

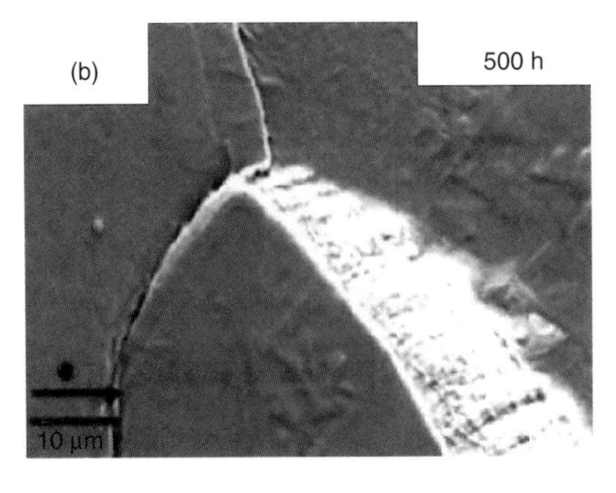

Figure 8.8 (a) Sn strip under electromigration at the current density of $2 \times 10^4\,\mathrm{A/cm^2}$ at 100 °C for 30 h. (b) Morphology of the strip shows grains rotation after the electromigration for 500 h.

$$\omega = \frac{d\theta}{dt} = K\cos^2\theta$$

where

$$K = \frac{4096\delta D^{\mathrm{GB}}}{kT}\frac{1}{h^2}\frac{Z^*ej}{D_{\mathrm{V}}^{\mathrm{GB}}}\left(D_{\mathrm{V}}^{a,\mathrm{bulk}}\rho^a - D_{\mathrm{V}}^{c,\mathrm{bulk}}\rho^c\right)$$

Thus

$$\omega = \frac{d\theta}{dt} = \frac{C_{\mathrm{V}}^{\mathrm{GB}}}{C_{\mathrm{V}}^{\mathrm{bulk}}}\frac{4096\delta}{kTh^2}Z^*ej\left(D^{a,\mathrm{bulk}}\rho^a - D^{c,\mathrm{bulk}}\rho^c\right)\cos^2\theta \tag{8.65}$$

For a real tin strip sample, as can be seen in Figure 8.8, the sample was tested at the current density of $2 \times 10^4\,\mathrm{A/m^2}$ at 100 °C. The grain size is around 50 μm, and the grain boundary width is about 0.5 nm. The effective charge number of tin is 17. The D^{GB} at 100 °C is $2 \times 10^{-7}\,\mathrm{cm^2/s}$. The only unknown parameter is the ratio of vacancy concentration in the grain boundary and in the bulk. At 100 °C, it is reasonable to take this ratio to be 100. The calculated time for the grain to rotate is 447 h. The experiment shown in Figure 8.8 was carried out for 500 h. So, it seems reasonable.

Thus, a mechanism of grain rotation can be proposed to depict the microstructure evolution in anisotropic conducting materials in electromigration: divergence of the vacancy fluxes from two neighboring but differently oriented grains induces a vacancy flux for an atomic flux in the reverse direction along the grain boundary between them. The flux divergence exists because of the anisotropic conduction. The grain boundary diffusion or creep will result in a rotation of the grain.

REFERENCES

1. Burke JE, Turnbull D. Recrystallization and grain growth. Progr Met Phys 1952;3:220–292.
2. Hillert M. On the theory of normal and abnormal grain growth. Acta Metall 1956;13(3):227–238.
3. http://en.wikipedia.org/wiki/Avrami_equation. Accessed 2014 Jan 24.
4. Kolmogorov A. H. About statistical theory of metals crystallization. Isvestiya Akdemii Nauk, mathematics, 1937;1(3):355–359.
5. Avrami M. Kinetics of phase change. I. General theory. J Chem Phys 1939;7(12):1103–1112.
6. Avrami M. Kinetics of phase change. II. Transformation-time relations for random distribution of nuclei. J Chem Phys 1940;8(2):212–224.
7. Avrami M. Kinetics of phase change. III. Granulation, phase change, and microstructure. J Chem Phys 1941;9(2):177–184.
8. Johnson WA, Mehl RF. Reaction kinetics in processes of nucleation and growth. Trans AIME 1939;135:416.
9. Cahn JW. Transformation kinetics during continuous cooling. Acta Metall 1956;4(6):572–575.
10. Holm EA, Glazier JA, Srolovitz DJ, Grest GS. Effects of lattice anisotropy and temperature on domain growth in the two dimensional potts model. Phys Rev A 1991;43:2662–2668.
11. Mullins WW. Two dimensional motion of idealized grain boundaries. J Appl Phys 1956;27:900–904.
12. Stavans J. The evolution of cellular structures. Rep Prog Phys 1993;56:733–789.
13. Thompson CV. Grain growth in thin films. Annu Rev of Mater Sci 1990;20:245–268.
14. Fayad W, Thompson CV, Frost HJ. Steady state grain size distributions resulting from grain growth in two dimensions. Scr Mater 1999;40:1199–1205.
15. Thompson CV. Grain growth and evolution of other cellular structures. Solid State Phys(Academic Press) 2000;55:269–314.
16. Carpenter DT, Codner JR, Barmak K, Rickman JM. Issues associated with the analysis and acquisition of thin film grain size data. Mater Lett 1999;41:296–302.
17. Marder M. Soap-bubble growth. Phys Rev A 1987;36:438–440.
18. Luoat NP. On the theory of normal grain growth. Acta Metall 1974;22:721–724.
19. Louat NP, Duesbery MS, Sadananda K. On the role of random walk in normal grain growth. Mater Sci Forum 1992;94–96:67–76.
20. Pande CS. On a stochastic theory of grain growth. Acta Metall 1987;35:2671–2678.
21. Pande CS. Stochastic theory of grain growth. Mater Sci Forum 1992;94–96:351–360.
22. Gusak AM, Tu KN. Theory of normal grain growth in normalized size space. Acta Materialia 2003;51:3895–3904.
23. Di Nunzio PE. A discrete approach to grain growth based on pair interactions. Acta Mater 2001;49:3635–3643.
24. Lucke K, Abruzzese G, Heckelmann I. Statistical theory of 2-D grain growth based on first principles and its topological foundation. Mater Sci Forum 1992;94–96:3–16.
25. Mullins WW. Grain growth of uniform boundaries with Scaling. Acta Mater 1998;46:6219–6226.
26. Lifshiz IM, Slezov VV. The kinetics of diffusive decomposition of oversaturated solid solutions. Sov JETP 1958;35:479–492.
27. Wagner C. Theorie der altrung von niederschlagen durch umblosen (Ostwald-Reinfund). Z Electrochem 1961;65:581–591.
28. Moldavan D, Wolf D, Philipot SR. A discrete approach to grain growth based on pair interactions. Acta Mater 2001;49:3521–3535.
29. Wu AT, Gusak AM, Tu KN, Kao CR. Electromigration-induced grain rotation in anisotropic conducting beta tin. Appl Phys Lett 2005;86:241902–241903.

PROBLEMS

8.1. What is the difference among recovery, recrystallization, and grain growth?

8.2. In amorphous-to-crystalline transformation, when two crystalline grains join and form a grain boundary, why can the grain boundary plane be straight or hyperbolic?

8.3. Consider physical vapor deposition of a "thick" film on the substrate (more than one grain along the height of film). Predict and explain how the mean size of grains in the film should depend on the (i) substrate temperature and (ii) incident flux density. Is it possible to obtain amorphous film by deposition if we consider the grain size in an amorphous film as atomic size?

8.4. Typically, in reactive diffusion, the layer of the growing intermediate phase initially consists of very small (nanosized) grains. Later, in the process of layer growth, the mean grain size $R(t)$ grows. Let this growth obey the power law $R(t) \approx kt^{1/2}$. At median temperatures the main diffusion flux through the phase layer goes along the grain boundaries, so that effective diffusivity of the phase is $D_{\mathrm{ef}} \approx D^{\mathrm{GB}} \frac{\delta}{R}$. ($\delta$ is a width of grain boundary). What will be the time law of phase layer growth?
Answer $\Delta X \propto t^{1/4}$.

8.5. Solve the previous problem for the case when time law of grain growth is not fixed but instead the mean grain size is proportional to the phase layer thickness: $R \propto \Delta X$.
Answer $\Delta X \propto t^{1/3}$.

8.6. In a powder mixture of Cu and Ni at 300 °C, the bulk diffusion is totally frozen so that even Fisher model of grain boundary penetration with continuous leakage into grains by bulk diffusion cannot be realized because the characteristic depth of bulk diffusion during reasonable time is less than a nanometer. Nevertheless, experiment shows that sintering of powder mixture at 300 °C gives partial homogenization (proved by X-ray analysis). At that, some grains indeed are transformed into solid solution, and some remain practically pure Ni or Cu. Explain!

8.7. In standard grain growth models it is usually assumed that the triple joints of grain boundaries are sufficiently mobile, so that the grain growth kinetics is fully determined by the mobility of grain boundaries, and their joints just follow them. Describe the possible picture of the grain growth if joints are slow (e.g., due to segregation of impurities), so that boundaries follow the joints.

8.8. Consider the grain structure of binary nanoalloy. Owing to the scatter of sizes, there will be a scatter of Laplace tensions inside grains. It should lead to a scatter of compositions. Estimate the scatter of compositions by assuming that the mean deviation of sizes is known. Here, neglect segregation at the grain boundaries and take into account redistribution only among grains. Gibbs potential per atom as a function of concentration C as well as the mean surface tension is known.

8.9. Consider grain boundary as a layer of certain thickness δ with tension, linearly dependent on concentration in this layer: $\gamma = C_{\mathrm{B}}^{\mathrm{GB}} \gamma_{\mathrm{B}} + C_{\mathrm{A}}^{\mathrm{GB}} \gamma_{\mathrm{A}}$. Gibbs potential per atom as a function of concentration C is known. Predict the difference between concentrations in the bulk of the grain and at the grain boundary.

8.10. Can you imagine a microstructure in which the grain boundaries become energetically favorable?

8.11. It is known that in thin film deposition, grain size increases with film thickness. It means grain growth accompanies film deposition. The grain growth occurs in both lateral and vertical direction with respect to the normal of the substrate. In the lateral growth, grain boundary must be consumed. This will generate tensile stress in the film, the intrinsic stress. Calculate the intrinsic stress in Ni thin films of 30 nm thick.

8.12. Consider a bilayer thin film of Au/Ag. Both are polycrystalline with the average grain size of 30 nm. When it is annealed at a low temperature of 200 °C, we assume that no lattice diffusion occurs, yet grain boundary diffusion can take place. It has been found that Au will diffuse in the grain boundaries of Ag. Furthermore, the grain boundary of Ag will lead to migration, paving way for the formation of Au/Ag alloy. This phenomenon is called *diffusion induced grain boundary migration* or DIGM. Explain the driving force of DIGM. What will be the grain growth rate?

CHAPTER 9

SELF-SUSTAINED REACTIONS IN NANOSCALE MULTILAYERED THIN FILMS

9.1 INTRODUCTION

Self-sustained high-temperature synthesis (SHS) reactions were realized initially in powder mixtures. In the 1990s, SHS reactions in multilayered thin films of nanothickness, for example, Al/Ni/Al/Ni, around 40 nm were found [1–10]. It was found that a small amount of energy in the form of electrical spark, laser pulse, or mechanical impact at room temperature will ignite a localized reaction by initiating interdiffusion and reaction in the multilayered structure at the point of spark or impact. The heat release from the point of reaction is intensive enough to lead to heat transfer by thermal diffusion laterally along the multilayered structure and to heat up the neighboring region to a much higher temperature for a very fast atomic interdiffusion and reaction. As the reaction proceeds, more heat is generated and released. Consequently, a self-sustained reaction or explosive reaction occurs. For example, to ignite the reaction mechanically at room temperature, the dropping of a pointed pencil on the surface of a multilayered Al/Ni thin film will cause an explosive reaction. Accompanying the

Kinetics in Nanoscale Materials, First Edition. King-Ning Tu and Andriy M. Gusak.

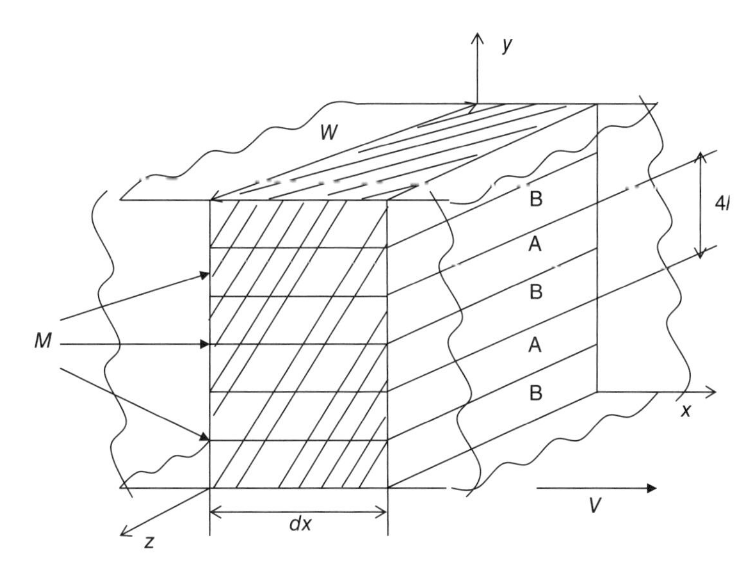

Figure 9.1 A schematic diagram of a multilayered thin film.

explosive reaction, a flash light of orange color can be seen. A schematic diagram of the multilayered thin films is depicted in Figure 9.1.

To control the main characteristics of SHS reactions, that is, the flame velocity or the reaction front velocity of the lateral reaction and the maximum reaction front temperature, we have to consider several factors, such as the composition of the bilayer, their thickness and period, minimum number of periods, microstructure of the as-prepared multilayer (crystalline or amorphous), stability of the multilayered structure stored at room temperature, and ignition parameters (a localized temperature rise by an electrical pulse or a localized pressure increase by a shock impact).

On composition of the bilayer, several pairs of A/B systems have been tried. Somehow, Al/Ni turns out to be a very good choice and it remains up to now to be the most studied system; it is also commercially available for localized heating applications. In the next section, we discuss the criteria of selection of bilayers for SHS reactions. In selecting a bilayer, enough care should be taken to choose the one that possesses a low melting point. This is because the low melting point will enhance the interdiffusion and reaction needed in the self-sustained reaction. The other one should be selected on the basis of its room temperature stability with the first one.

What is the optimal thickness of A/B and what is the minimum number of period needed for self-sustained reaction are less critical. Owing to heat dissipation from the film surfaces, it is obvious that a single bilayer will not have the self-sustained reaction. If the thin film surfaces can be made adiabatic, we need less periods. In actual samples, it was found experimentally that about 20 periods of bilayered thin films is sufficient for the self-sustained reaction to occur. On the thickness of each metal film in the period, if they are too thin, the multilayered structure is unstable at room temperature. If they are too thick, it takes too long to react thoroughly, so the rate of heat dissipation will be too slow to keep the reaction self-sustained. Again, experimentally, it is found that a period of 20–40 nm seems to have worked very well in the case of Al/Ni samples.

On flame velocity, it is typically of the order of m/s and the maximum temperature of the reaction front in the case of Al/Ni is assumed to be near or above the melting point of Al, which means that the Al in the multilayered structure may melt during the explosive reaction.

Mathematically, the control of SHS reactions means the solution of an inverse problem, that is, finding out the magnitudes of the abovementioned parameters that would provide the required flame velocity and the maximum temperature. Moreover, mathematical modeling of SHS in multilayered thin films seems more transparent and predictable in comparison with powder mixtures, because in the thin film case we can use one-dimensional diffusion equations with known diffusion parameters (activation energy and preexponential factor), contrary to the case of powder where we have to use three-dimensional equations with fitting coefficients for the reaction rate.

Generally speaking, SHS reactions are in nonstationary conditions because SHS systems normally have complex phase diagrams with several intermediate phases. For example, the phase diagram of Al–Ni is shown in Figure 9.2. There are several intermetallic compound (IMC) phases, and they can grow either sequentially or simultaneously during the SHS reaction, so that the SHS process must be described as a competition between exothermic and endothermic reactions with varying local temperatures. All this depends on Gibbs free energies of phase formation and diffusivities. Variation of the reaction temperature is affected by changes in the phase composition of products and different modes of heat dissipation. The process is strongly nonlinear, so that small variations may lead to bifurcation of regimes or

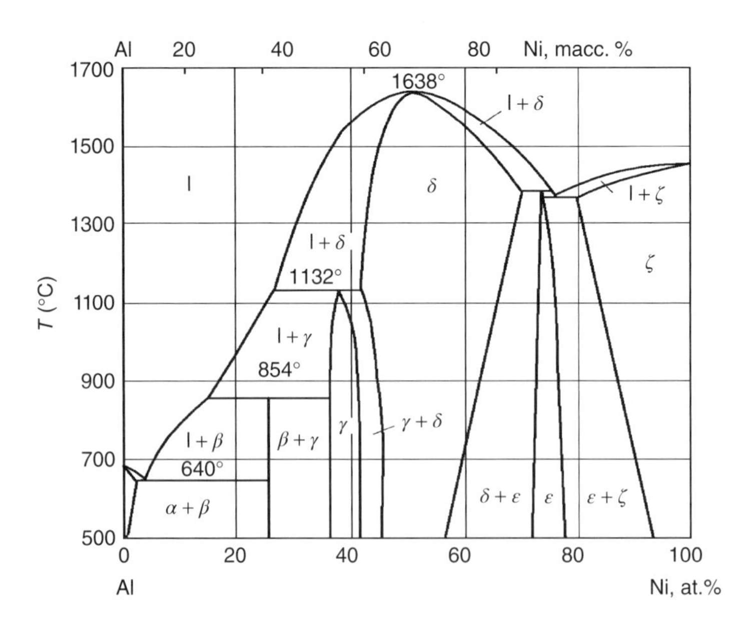

Figure 9.2 Equilibrium binary phase diagram of Al–Ni.

uncertainty in phase evolution. All this makes mathematical modeling of SHS very difficult, thus in this chapter simple assumptions have to be made so that SHS in multilayers can be treated as a sequence of diffusion-controlled phase transformations (which is nearly transversal to temperature gradient), which is discussed in Section 9.3.

A very simple assumption is to assume that SHS reaction yields a single-phase product that has the highest free energy of formation during the explosive reaction so that we can ignore the existence of all the other phases (actually, this is the case when the growing phase has won the competition over other phases), and it propagates over the entire multilayer in a steady state mode.

In the following, we first present a general rule of selection of a binary system to form the multilayered structure for SHS. A simple kinetic model is given to analyze the interfacial reaction in the multilayered structure. Then the frame velocity is estimated and the reaction product is analyzed. Finally, SHS in silicidation reactions is briefly reviewed.

9.2 THE SELECTION OF A PAIR OF METALLIC THIN FILMS FOR SHS

The criteria of selection of an A/B pair of metallic thin films for SHS reaction are presented below.

(1) The first criterion is that the multilayered thin film structure must be relatively stable at room temperature storage. It will not react without an act to trigger the reaction.

(2) The second criterion is that after the reaction is triggered by an electrical pulse or mechanical impact, the reaction will be self-sustained and will lead to explosive propagation of the reaction front to the entire multilayered thin film.

(3) The third criterion is that in order to sustain the reaction without external heat source, the pair of metals or elements must react to form a high melting point IMC so that the gain in the free energy of formation or the heat released from the reaction is as large as possible.

(4) Because IMC formation is required, we do not consider bimetallic systems, such as Cu–Ni, that form solid solutions, which belong to interdiffusion in man-made superlattices. Nevertheless, the interfacial reactions to form IMC in the multilayered structure begin with interdiffusion. To facilitate the interdiffusion, one of the metals in the pair should be a low melting point metal so that the other metal can diffuse rapidly into the low melting point one at the reaction temperature.

(5) A low melting point metal could melt at the reaction temperature, which enhances the interdiffusion and reaction for self-propagation. But, it could also affect the stability at room temperature storage. We have three low melting point metals for consideration, and they are Sn, Pb, and Al, which have a melting point of 232, 327, and 660 °C, respectively. However, it is known that noble metals (Cu, Ag, Au) and near-noble metals (Ni, Pd, Pt) diffuse interstitially into group four elements (Si, Ge, Sn, Pb), so it has been found that noble and near-noble metals react with Sn and Pb at room temperature to form IMCs. For example, a pair of Cu/Sn or Ni/Sn is unstable at room temperature and they form Cu_6Sn_5 and Ni_3Sn_4, respectively. Thus we avoid the use of Sn and Pb for SHS. Besides, Pb is toxic.

(6) Thus, Al becomes the first metal to be selected.

(7) To select the second metal, we require that it must diffuse rapidly into Al and form a high melting point IMC with Al. On diffusion in the self-sustained reaction, we may simplify the picture by assuming that the reaction temperature is near the melting point of Al, that is, the front temperature of propagation of the self-sustained reaction is near the melting point of Al. We know for face-centered cubic metals near their melting point, the equilibrium vacancy concentration is about 10^{-4} and the diffusivity is about 10^{-8} cm^2/s. The high concentration of vacancy will enable the other metal to diffuse rapidly into Al and react to form IMC quickly. This assumption has enabled us to estimate the flame velocity of the reaction to be of the order of 1 m/s, as given in Section 1.11. If melting occurs, the diffusivity will increase to 10^{-5} cm^2/cm, and in turn it will increase the flame velocity by more than one order of magnitude.

(8) We can select the second metal from noble and near-noble metals, for example, Cu or Ni. If we compare the binary phase diagrams of Al–Cu and Al–Ni, the latter has a very high melting point IMC, the δ-phase, which has a melting point at 1638 °C, as shown in Figure 9.2. Besides, because of the high melting point of Ni, the diffusion of Al into Ni is negligible and the multilayer Al/Ni/Al/Ni is quite stable at room temperature as well as during the deposition of the multilayered thin films. The latter is a concern for Al/Cu multilayered thin films, as it was found to have formed interfacial IMC during room temperature deposition. Thus, if we choose to use Al/Cu, a thicker period will be needed. On the other hand, we can deposit the multilayered thin film at liquid nitrogen temperature to avoid interfacial reactions, but it is unpractical.

(9) Therefore, Al/Ni was the first pair of metals selected for SHS and it remains to be the pair most widely used for SHS, including commercial samples.

(10) We do not consider transition metals and refractory metals to pair with Al, this is because of their tendency of oxidation in nanothickness.

(11) Because of the latent heat of crystallization, amorphous Si has been studied for SHS. It was found that Ni will not react with Si wafer to form silicide until the temperature is close to 200 °C. Thus, multilayered thin films of α-Si/Ni have been studied for SHS as well as other metals in explosive silicidation reactions, which is discussed in Section 9.5.

9.3 A SIMPLE MODEL OF SINGLE-PHASE GROWTH IN SELF-SUSTAINED REACTION

We consider binary multilayered thin films consisting of M periods of A and B. For simplicity, we take atomic volumes of A and B to be equal and the product phase A_1B_1 as stoichiometric, but it may have some (narrow but nonzero) homogeneity range.

Each layer of both A and B has thickness $2l$ and width W, and the period in the multilayered structure is $4l$, as shown in Figure 9.1 or Figure 1.18. As a rule, at each interface the premixing thin layer of product phase with thickness Δy_0 may exist. To model the SHS reaction, the following assumptions are taken.

(1) The reaction front is considered as a steady state one, propagating along x-axis with a constant velocity, V.

(2) The reaction front width L along y-axis is much wider than the multilayer period $4l$. Therefore, all interdiffusion and reaction fluxes are directed practically normal to the x-axis and the temperature gradient.

(3) Temperature variation along diffusion flux direction (along y-axis) is neglected.

(4) The last three assumptions are valid if $a^2(L/V) \gg l^2$ (a^2 is thermal diffusivity).

(5) The stoichiometric phase A_1B_1 (δ-phase in the binary Al–Ni phase diagram) with a narrow homogeneity range $c_{\text{left}} < c < c_{\text{right}}$ ($\Delta c \equiv c_{\text{right}} - c_{\text{left}} \ll 1$) is the first phase to grow and simultaneously the final product phase. Phase competition during the SHS reaction will not be considered here.

(6) All concentrations at all interphase boundaries are regarded as quasi-equilibrium. The composition at both sides of each boundary corresponds to common tangent rule at the local temperature.

(7) Diffusion through the forming phase layer is treated as a steady state process so that the velocity of a moving interface of the forming phase between Al and Ni is constant.

(8) All diffusivities are proportional to vacancy concentration, which is considered as quasi-equilibrium. It means the sources and sinks of vacancies are fully operative, so vacancy concentration corresponds to the equilibrium value at the local temperature. Thus, in the Al/Ni multilayered structure, at a given reaction temperature near the melting point of Al, the vacancy concentration in Al is near 10^{-4}, which is many orders of magnitude larger than that in Ni. The interdiffusion occurs by the diffusion of Ni into Al and we ignore the diffusion of Al into Ni.

(9) The heat is released only at the moving interphase boundary but almost instantly and uniformly distributed within each period transversal to the propagation direction.

(10) Side walls of the multilayered structure is assumed to be adiabatic so the outflux of heat is ignored.

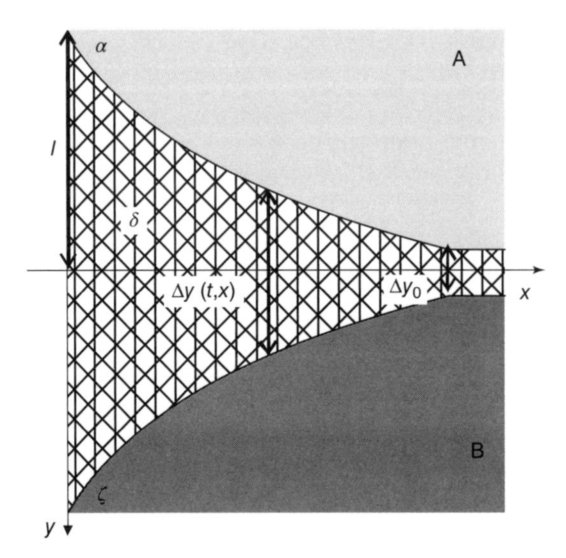

Figure 9.3 A schematic morphology of the flame front.

(11) A typical morphology of the flame front is presented in Figure 9.3. It is similar to grain boundary diffusion that the penetration along the grain boundary is faster than that in the bulk of the two grains on the two sides of the grain boundary. However, as SHS is a high-temperature reaction, we ignore the effect of penetration of IMC along the interfaces. Besides, the depth of penetration or the length of the "tongue" along the interface is not much more than that of the reaction front due to the very high velocity of flame propagation.

On the basis of energy balance, we obtain the following heat release density per unit volume per unit time of $q_\delta(x)$ in units of heat/cm^3 s, averaged over the cross-section of the multilayered thin films at a local layer thickness of $\Delta y_\delta(t,x)$. In the balance, we ignore the energy of interfaces and grain boundaries.

$$q_\delta(x) = \frac{\Delta g_{\delta(\alpha,\zeta)} d\Delta y_\delta(x) \cdot dx \cdot W/\Omega/dt}{2l \cdot dx \cdot W} = \frac{\Delta g_{\delta(\alpha,\zeta)} d\Delta y_\delta(x)}{2l\Omega dt} \tag{9.1}$$

where $\Delta g_{\delta(\alpha,\zeta)}$ is the driving force of reaction per atom in $A + B \rightarrow A_1B_1$, or it is the formation energy of δ-phase per atom by the reaction of Al (α-phase) and Ni (ζ-phase) as shown in the Al–Ni phase diagram. Ω is atomic volume. The other dimensional parameters are shown in Figure 9.1.

To derive Eq. (9.1), we take $dy_\delta(x)dxW$ to be the volume transformed by moving reaction interface within a space interval of dx during a time interval of dt. Thus, $dy_\delta(x)dxW/\Omega$ is the number of atoms involved in transformation during this time interval. Then, $\Delta g_{\delta(\alpha,\xi)} dy_\delta(x)dxW/\Omega$ is the amount of heat generated or to be released in this process. This heat is assumed to be distributed uniformly among all atoms in a half-period thickness of the multilayered structure within the interval dx. The reason for considering a half-period is shown in Figure 1.18, Section 1.9. It is because we consider the diffusion of Ni into Al from both the upper and the lower interfaces (so a single period contains two interfaces), which means interdiffusion occurs over a volume of $2 \cdot dl \cdot dx \cdot W$, where dl is the distance of interdiffusion. If during the time interval of dt, the diffusion of Ni can reach the middle of the Al layer so the reaction is complete, we have $dl = 2l$ or $\Delta y_\delta(x) = 2l$. These considerations give us a heat release rate per unit volume, as given in Eq. (9.1).

Mathematically, we develop the model to give self-consistent profiles of temperature and phase formation by the simultaneous solution of the following equations for temperature evolution under heat release by the phase formation. Temperature is taken to change only along propagation direction by assuming fast equalizing so that there is no temperature gradient in any cross-section perpendicular to this direction. The one-dimensional temperature profile changes with time according to heat transfer equation, taking into account the heat release in reactions, as shown in the right-hand side of the following equation.

$$\frac{\partial T}{\partial t} - a^2 \frac{\partial^2 T}{\partial x^2} = \begin{cases} 0, \quad x < Vt, \quad \Delta y\,(x) = 2l - \text{beyond the front} \\ q_\delta(x)/c_p\rho, \quad Vt < x, \quad \Delta y_0 < \Delta y(x) < 2l - \text{in front} \end{cases} \tag{9.2}$$

where c_p is the specific heat capacity with unit of heat per °C/mol, and ρ is the density with unit of mol/cm^3. The term of $q_\delta(x)/c_p\rho$ is the source/sink of heat in the heat transfer equation. The parameter, a^2, is defined as thermal diffusivity with a unit of cm^2/s. We note that it is not the same as thermal conductivity, κ, as in the heat flux equation where $J_q = -\kappa(dT/dx)$. It is known that $a^2 = \kappa/C_p\rho$, or $\kappa = a^2 C_p \rho$, where C_p is heat capacity per unit mass and its unit is joule/K-m_0 and $\rho = m_0/\Omega$ is mass density, where m_0 is atomic mass and Ω is atomic volume.

For single-phase growth in nonisothermal conditions, we do not use Fick's second law of diffusion equation, which is similar to Eq. (9.2), to describe the reaction. Instead, we can write the atomic flux balancing equations for the velocity of two interphase interfaces (with coordinates $y_{\alpha\delta}$ and $y_{\delta\zeta}$) moving across the multilayers (in y-direction or opposite direction) by using Fick's first law, and we follow Kidson's analysis of single layer growth as presented in Section 7.2.1. This is because it is much simpler.

$$\begin{cases} (c_\delta - 0) \cdot \dfrac{dy_{\alpha\delta}}{dt} = -\dfrac{D_\delta \Delta c_\delta}{y_{\delta\zeta} - y_{\alpha\delta}} \\ (1 - c_\delta) \cdot \dfrac{dy_{\delta\zeta}}{dt} = +\dfrac{D_\delta \Delta c_\delta}{y_{\delta\zeta} - y_{\alpha\delta}} \end{cases} \tag{9.3}$$

where c_δ is the average molar fraction of B component in δ-phase, D_δ is the averaged interdiffusion coefficient in δ-phase, Δc_δ is the homogeneity interval of δ-phase.

After a simple algebra to combine the two equations in Eq. (9.3), we obtain the growth rate of the δ-phase in the y-direction.

$$\frac{d\Delta y_\delta^2(t,x)}{dt} = \frac{2\Delta c_\delta}{c_\delta(1 - c_\delta)} D_\delta \tag{9.4}$$

where $y_{\delta\zeta} - y_{\alpha\delta} = \Delta y_\delta(t,x)$. We recall that Eq. (9.4) is the same as Eq. (7.12).

When integrating Eq. (9.4), we could take into account that typically, before the start of SHS reaction, the multilayered structure may already contain preexisting phase layer of thickness of Δy_0 formed during film deposition and during storage.

$$\Delta y_\delta^2(t,x) - \Delta y_0^2 = \frac{2}{c_\delta(1-c_\delta)} \int_{-\infty}^{t} \left\{ D_\delta\left[T\left(t',x\right)\right] - D_0 \right\} \Delta c_\delta dt' \qquad (9.5)$$

The main purpose of the integration of the above solution is to obtain the time when the maximum amount of reaction occurs at $y_\delta = 2l$. We need to know the time because it is the coupling link between the atomic diffusion and the heat transfer. However, to do so, we need to know the interdiffusivity, which can be measured separately in the thin film couple of Al/Ni at a certain temperature range where the single δ-phase forms. On the other hand, using the Darken's and Wagner's expression for interdiffusivity, we can express the product of $D_\delta \Delta C_\delta$ in terms of intrinsic diffusivities and the driving force of reaction, Δg_δ, as presented in Section 7.2.5.

$$\begin{aligned} \Delta c_\delta D_\delta &= \Delta c_\delta [c_\delta D_\delta(A) + (1-c_\delta) D_\delta(B)] \\ &= D_{0\delta} \exp\left[-\frac{Q_\delta}{kT\,(t,x)}\right] \frac{\Delta g_\delta}{kT} \end{aligned} \qquad (9.6)$$

where $D_\delta(A)$ and $D_\delta(B)$ are intrinsic diffusivity of A and B in the IMC of AB. $D_{0\delta}^*$ and Q_δ are respectively preexponential factor and activation energy of the interdiffusivity.

Substituting Eq. (9.6) into Eq. (9.5), we can reformulate the condition of full reaction in some moment t^* at some section x of the multilayered foil with period $4l$ and initial thickness Δy_0 of resulting phase as

$$(2l)^2 - (\Delta y_0)^2 = \frac{2\overline{D_{\delta 0}^*}}{c_\delta(1-c_\delta)} \int_{-\infty}^{t*} \left(\exp\left(-\frac{Q_\delta}{kT\,(t,x)}\right) - \exp\left(-\frac{Q_\delta}{kT_0}\right) \right) \frac{\Delta g_\delta}{kT(t,x)} dt \qquad (9.7)$$

Change the argument in the last integral from time t to local temperature T, so that $dt = dT/(\partial T/\partial t)$:

$$(2l)^2 - (\Delta y_0)^2 = \frac{2\overline{D_{\delta 0}^*}}{c_\delta(1-c_\delta)} \int_{T_0}^{T_{\max}} \left(\exp\left(-\frac{Q_\delta}{kT}\right) - \exp\left(-\frac{Q_\delta}{kT_0}\right) \right) \frac{\Delta g_\delta}{kT(t,x)} \frac{1}{\partial T/\partial t} dT \qquad (9.8)$$

For steady state propagation of SHS with constant velocity V, we can take (see Section 9.4),

$$\frac{\partial T}{\partial t} = -V \frac{\partial T}{\partial x} \approx \frac{V^2}{a^2} T \qquad (9.9)$$

Then

$$(2l)^2-(\Delta y_0)^2 = \frac{2\overline{D^*_{\delta 0}}}{c_\delta(1-c_\delta)}\int_{T_0}^{T_{\max}}\left(\exp\left(-\frac{Q_\delta}{kT}\right)-\exp\left(-\frac{Q_\delta}{kT_0}\right)\right)\frac{\Delta g_\delta}{kT(t,x)}\frac{1}{\partial T/\partial t}dT$$

$$= \frac{a^2}{V^2}\frac{2\overline{D^*_{\delta 0}}}{c_\delta(1-c_\delta)Q_\delta}\int_{T_0}^{T_{\max}}\Delta g_\delta\left(\exp\left(-\frac{Q_\delta}{kT}\right)-\exp\left(-\frac{Q_\delta}{kT_0}\right)\right)d\left(-\frac{Q_\delta}{kT}\right) \tag{9.10}$$

In all real situations at room temperature, the reaction is practically frozen: $\exp(-Q_\delta/kT_0) \approx 0$. Taking the driving force Δg_δ as approximately constant, we obtain

$$(2l)^2-(\Delta y_0)^2 = \frac{a^2}{V^2}\frac{2\overline{D^*_{\delta 0}}}{c_\delta(1-c_\delta)}\frac{\Delta g_\delta}{Q_\delta}\exp\left(-\frac{Q_\delta}{kT_{\max}}\right) \tag{9.11}$$

This gives us the steady state propagation velocity of SHS,

$$V = \frac{1}{l}\sqrt{\frac{a^2}{2(1-((\Delta y_0)^2/4l^2))}\frac{\overline{D^*_{\delta 0}}}{c_\delta(1-c_\delta)}\frac{\Delta g_\delta}{Q_\delta}\exp\left(-\frac{Q_\delta}{kT_{\max}}\right)} \tag{9.12}$$

To calculate the maximum temperature, if we neglect the transversal heat outflux, and take Dulong–Petit law for heat capacity ($3k_B$ per atom), then the maximum temperature in the SHS front can be found from energy conservation. As part of the thin film had reacted before the initiation of SHS, the heating is only from the remaining unreacted films. The energy released in the reaction per atom is

$$\Delta g_i\frac{2l-\Delta y_0}{2l} = \Delta g_i\left(1-\frac{\Delta y_0}{2l}\right)$$

In the absence of the outflux of heat, all this energy goes for heating;

$$\Delta g_i\left(1-\frac{\Delta y_0}{2l}\right) = 3k_B(T_{\max}-T_0)$$

$$T_{\max} = T_0 + \frac{\Delta g_\delta}{3k_B}\left(1-\frac{\Delta y_0}{2l}\right) \tag{9.13}$$

For AlNi IMC, $\Delta g_\delta = 6.6\times 10^{-20}$joule/atom. Thus, with $\Delta y_0 = 0$, we obtain $T_{\max} - T_0 = (6.6\times 10^{-20}/3\times 1.38\times 10^{-23}) \approx 1600$ K.

9.4 A SIMPLE ESTIMATE OF FLAME VELOCITY IN STEADY STATE HEAT TRANSFER

In the calculation of flame velocity, we can greatly simplify the mathematical analysis in the above by making the following assumption that the time interval dt can be estimated from the simple equation of $l^2 = D(dt)$. Near the melting point of Al, we can take the diffusivity $D = 10^{-8}\ \mathrm{cm^2/s}$. The distance of diffusion is a quarter of the period, so $l = 10$ nm or 10^{-6} cm. Then, dt is about 10^{-4} s. Knowing the time, we can estimate dx traveled by the heat from the heat transfer equation. Then we can obtain the flame velocity, $V = dx/dt$. In this very simple picture, we have ignored the growth of the δ-phase. Taking the following relation that

$$dt = \frac{l^2}{D} = \frac{(dx)^2}{a^2} \tag{9.14}$$

By knowing the magnitude of thermal diffusivity, $a^2 \approx 1-10\ \mathrm{cm^2/s}$, we obtain

$$V = \frac{dx}{dt} = \frac{\sqrt{a^2D}}{l} \cong 100\frac{\mathrm{cm}}{s} \cong 1\frac{\mathrm{m}}{s} \tag{9.15}$$

On the other hand, when melting occurs, the atomic diffusivity in a melt is about $10^{-5}\ \mathrm{cm^2/s}$. Using this value, V will be able to reach 100 m/s. Experimentally, the velocity of explosive reaction in Al/Ni multilayered thin films has been measured to be about 4 m/s by using a high-speed camera shooting at 6000 frames/s. Typically, it is of the order of m/s as we have estimated.

Furthermore, we can also simplify the heat transfer equation to obtain an estimation of flame velocity V. In steady state approximation, $\partial T/\partial t = -V(\partial T/\partial x)$, the heat transfer equation has the form of

$$-V\frac{\partial T}{\partial x} - a^2\frac{\partial^2 T}{\partial x^2} = 0$$

with a solution as

$$\frac{\partial T}{\partial x} = \left(\frac{\partial T}{\partial x}\right)_0 \exp\left(-\frac{V}{a^2}x\right) \approx \exp\left(-\frac{x}{W}\right) \tag{9.16}$$

where $W = a^2/V$ is the distance of heat propagation in time dt, and we note that $W = dx$ as in the last section. Thus

$$dt = \frac{W}{V} = \frac{a^2}{V^2} = \frac{a^2}{(dx/dt)^2}$$

and we have the same as given in Eq. (9.14)

$$dt = \frac{(dx)^2}{a^2}$$

We note that the solution of $W = a^2/V$ is expected because of the dimension of thermal diffusivity, which is cm^2/s. Again it shows that time is the link between atomic diffusion and heat transfer in the process of SHS.

9.5 COMPARISON IN PHASE FORMATION BY ANNEALING AND BY EXPLOSIVE REACTION IN AL/NI

For comparison, multilayered Al/Ni thin films of thickness ratio of 3Al:1Ni with a period of 25 nm were annealed from RT to 1000 K in calorimeter. The formation of the Al_3Ni phase was found to start from 500 K and completed about 550 K. The Al_3Ni phase is stable and is the only phase to form up to 1000 K. This finding is in agreement with that found in bilayer Al/Ni thin film annealed in a furnace that Al_3Ni is the first phase to form. The fact that Al_3Ni is the first phase to form, which is an Al-rich phase, indicates that Ni diffuses into Al during its formation.

In self-explosive reactions of the 3Al:1Ni multilayered thin films, besides Al_3Ni, small amount of Al_3Ni_2 was found with some unreacted Al. When it was followed by an annealing at 750 K, these phases were completely converted to Al_3Ni. It indicated an incomplete reaction during the explosive reaction. The phase diagram of Al–Ni, as shown in Figure 9.2, shows that Al_3Ni is not a congruently melting phase, but rather melts by a peritectic reaction at a temperature of 1127 K, above which a wide two-phase region of Al_3Ni_2 plus an Al-rich liquid phase exists. Thus the results indicate that the temperature in the explosive reaction front is at or higher than 1127 K, which we note is above the melting point of Al at 933 K. Thus we can conclude that the Al film was partially melted during the explosive reaction. Assuming adiabatic heating, the heat of formation of Al_3Ni (−36 kJ/mol), which would be released during the explosive reaction, is more than sufficient to compensate for the heat required to heat the Al/Ni film to 1127 K (22 kJ/mol). Furthermore, the additional evidence that a significant heating has occurred is the emission of visible orange light during the explosive reaction. Figure 9.4 shows transmission electron micrographs of 3Al:1Ni multilayered thin films; (a) annealed to 1000 K, and (b) explosively reacted. The latter shows a much larger grain size.

On the effect of film thickness, it was found that explosive reactions were observed to occur at room temperature only in multilayers with a period less than 50 nm. For samples heated to 470–570 K, explosive reactions were observed in some of the multilayers having a period of 50–70 nm. No explosive reactions were observed in multilayers with a period of 150 and 300 nm.

9.6 SELF-EXPLOSIVE SILICIDATION REACTIONS

In silicide formation by reacting an amorphous Si film and a crystalline Rh metal film, an amorphous Rh–Si alloy thin film was formed [11]. It was unexpected that an amorphous alloy thin film can be formed by slow heating. This is because, generally speaking, rapid quench is required in amorphous alloy formation. In order to use this slow heating reaction to form a thick or bulk amorphous alloy of Rh–Si, multilayered

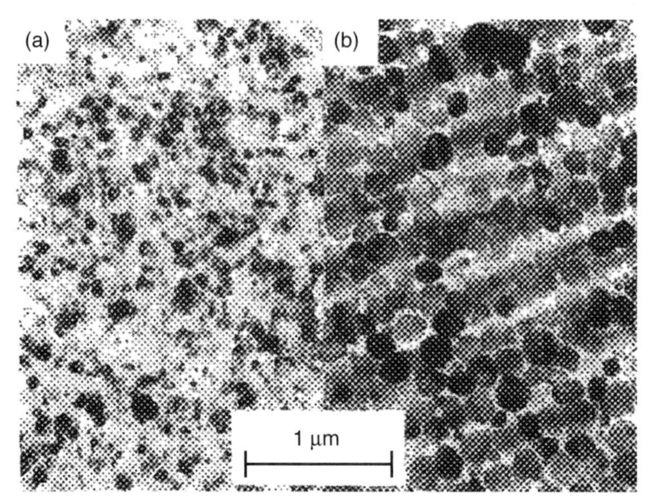

Figure 9.4 Plan-view of transmission electron micrographs of 3Al:1Ni multilayered thin films. (a) Annealed to 1000 K and (b) explosively reacted. The latter shows a much larger grain size. *Source*: Reproduced with kind permission from [1].

Rh/Si thin films were prepared. Then, self-explosive reaction was observed to occur in the multilayered thin films. It has led to a series of study of explosive silicidation in amorphous Si/Ni, amorphous Si/V, amorphous Si/Zr, and amorphous Si/Ti thin films. As these are high melting point elements, interdiffusion coefficients are low at room temperature. It seems that the latent heat of the transition from amorphous Si to crystalline Si may have assisted the self-explosive reactions. Indeed, self-explosive crystallization of amorphous Sb, Ge, and CdTe was reported.

In the study of explosive silicidation of amorphous Si/Ni, the multilayered thin films were prepared on microscopic slides for reaction front temperature and velocity measurements. The periods of 14, 20.8, and 28 nm were prepared, and in each period the composition or atomic ratio of Si and Ni were controlled to be 2 Ni atoms to 1 Si atoms. Two different total thicknesses of 0.73 and 1.62 μm of the multilayered thin films were made.

The experimental setup for high-speed photography and temperature measurements is schematically depicted in Figure 9.5. The reaction front velocities were measured using a high-speed camera shooting at 6000 frames/s. The reaction front temperature was measured with a pyrometer operating at narrow near-infrared wavelength bands centered at 0.81 and 0.95 μm. A digital oscilloscope and a specially designed amplification system were used to obtain temperature data at the sampling rate of one measurement every 500 ns.

Figure 9.6 shows the high-speed temperature measurement for the explosive silicidation occurring in amorphous Si/Ni. At zero time, as the reaction front has just come into the field of view of the pyrometer, the temperature reading was around 1580 K, which is slightly above the melting point of Ni_2Si at 1565 K. The temperature decreases slightly as the front moves away.

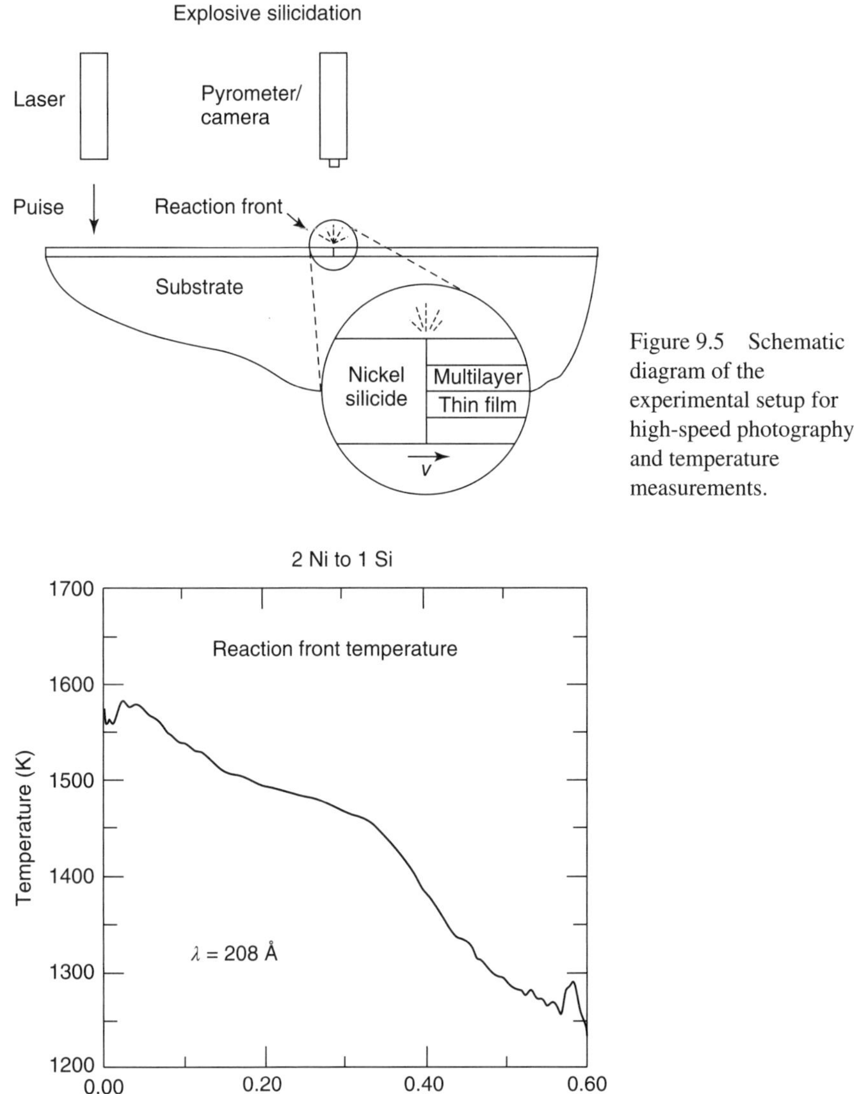

Figure 9.5 Schematic diagram of the experimental setup for high-speed photography and temperature measurements.

Figure 9.6 Temperature measurement for the explosive silicidation occurring in amorphous Si/Ni. At zero time, as the reaction front has just come into the field of view of the pyrometer, the temperature reading was around 1580 K and it decreases slightly as the front moves.

Figure 9.7 shows the high-speed video output showing the bright band of flame of the explosive reaction front at 12 different times, with 1/12 ms apart. The first output is shown at the bottom in Figure 9.7 as the front moves from left to right. The front velocity is measured to be 23 m/s.

Figure 9.8 shows the explosive reaction front velocity versus modulation period and the two total film thicknesses of amorphous Si/Ni.

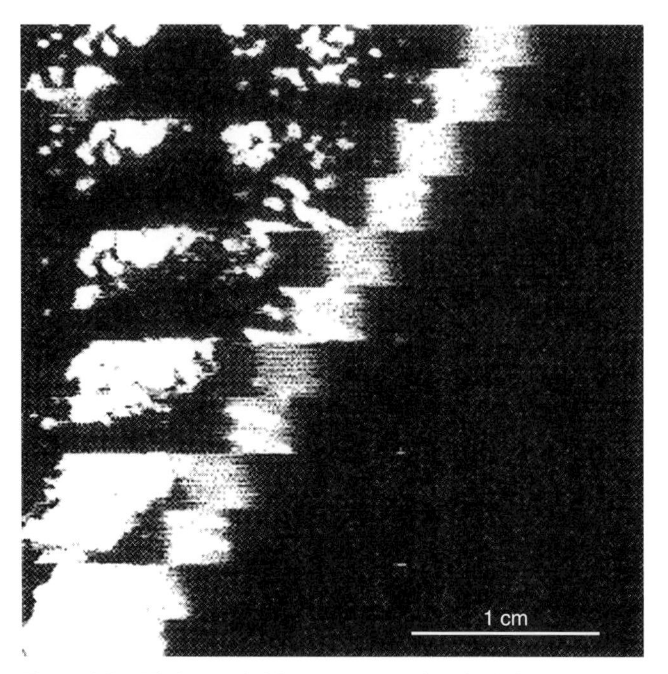

Figure 9.7 High-speed video output showing the bright band of the explosive reaction front at 12 different times, with 1/12 ms apart. The first time is shown at the bottom as the front moves from left to right. The front velocity is measured to be 23 m/s. *Source*: Reproduced with kind permission from [3].

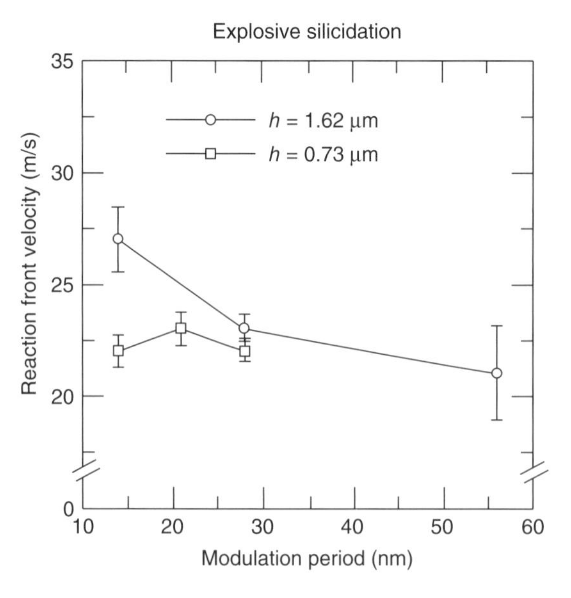

Figure 9.8 A plot of the explosive reaction front velocity versus modulation period and total film thickness of amorphous Si/Ni. *Source*: Reproduced with kind permission from [3].

(a)

(b)

Figure 9.9 Plan-view transmission electron micrographs of amorphous Si/Ni multilayered thin films having an atomic ratio of 1Si:2Ni, a period of 14 nm, and a total thickness of 70 nm. (a) The thin film microstructure after explosive reaction and (b) after conventional annealing. *Source*: Reproduced with kind permission from [3].

Figure 9.9 shows plan-view transmission electron micrographs of amorphous Si/Ni multilayered thin films having an atomic ratio of 1Si:2Ni, a period of 14 nm, and a total thickness of 70 nm. Figure 9.9a is the thin film microstructure after explosive reaction, and Figure 9.9b is that after conventional annealing.

REFERENCES

1. Ma E, Thompson CV, Clevenger LA, Tu KN. Self-propagating explosive reaction in Al/Ni multilayer thin films. Appl Phys Lett 1990;57:1262–1264.
2. Floro JA. Propagation of explosive crystallization in thin Rh–Si multilayer films. J Vac Sci Technol A 1986;4:631–636.

3. Clevenger LA, Thompson CV, Tu KN. Explosive silicidation in nickel/amorphous-silicon multilayer thin films. J Appl Phys 1990;67:2894–2898.
4. Wickersham CE, Poole JE. Explosive crystallization in Zr/Si multilayers. J Vac Sci Technol A 1988;6:1699–1702.
5. De Avillez RR, Clevenger LA, Thompson CV, Tu KN. Quantitative investigation of Ti/amorphous-Si multilayer thin film reactions. J Mater Res 1990;5:593–600.
6. Weihs TP. *Handbooks of Thin Film Process Technology*. Bristol, UK: Institute of Physics; 1998.
7. Gavens AJ, Heerden DV, Mann AB, Reiss ME, Weihs TP. Effect of intermixing on self-propagating exothermic reactions in Al/Ni nanolaminate foils. J Appl Phys 2000;87:1255–1263.
8. Wang J, Besnoin E, Knio OM, Weihs TP. Investigating the effect of applied pressure on reactive multilayer foil joining. Acta Mater 2004;52:5265–5274.
9. Tong M, Sturgess D, Tu KN, Yang J-M. Explosively reacting nanolayers as localized heat sources in solder joints – a microstructural analysis. Appl Phys Lett 2008;92:144101–144103.
10. Colgan EG, Nastasi M, Mayer JW. Initial phase formation and dissociation in the thin-film Ni/Al system. J Appl Phys 1985;58:4125–4129.
11. Herd S, Tu KN, Ahn KY. Formation of an amorphous Rh–Si alloy by interfacial reaction between amorphous Si and crystalline Rh thin films. Appl Phys Lett 1983;42:597–599.

PROBLEMS

9.1. Design a new pair of metals, A/B, which is better than Al/Ni for explosive reaction. Explain why your design is better.

9.2. If we take multilayered nanothickness A/B thin films and roll it into a long rod, will the rod show explosive reaction upon ignition?

9.3. What will be the maximum frame speed that can be obtained? How to design the multilayered thin film structure to achieve it?

9.4. Estimate the **width** of the SHS front in Ni–Al multilayer.

9.5. Consider SHS reaction in multilayer Ni/Al with equal amount of Ni and Al atoms the phases appear sequentially, as in usual thin film reactions – at first Ni1Al3, then, after consumption of aluminum, Ni1Al1.

- **(1)** Try to draw the snapshot of reaction morphology in this case.
- **(2)** Try to draw the snapshot for the case when both phases appear and grow simultaneously.

9.6. SHS reaction can be clearly suppressed by sufficient big heat outflux via the marginal layers in the case of good thermal contact with some plates with high heat conductivity and heat capacity. Estimate the ratio of the heat loss via side outflux PER ATOM during the flame front passing to the heat production (PER ATOM) in reaction during the same period. Introduce any parameters you need.

9.7. Experiment shows that the dependence of SHS front velocity NONMONOTONOUSLY depends on the preexisting thickness Δy_0 of reacted material. Try to explain it (a) qualitatively and (b) using Eq. (9.12).

9.8. Computer experiments, carried out with the method of Molecular Dynamics with Ni–Al multilayer, show that even under isothermal annealing with temperature WELL below the melting temperature of aluminum, a system Ni–Al shows rapid growth of disordered (liquid or amorphous) solution in the first nanoseconds, before the appearance of any IMCs. How can it be? Actually, aluminum has eutectics with Ni1Al3, but at temperature very close to the melting of pure aluminum.

9.9. Sometimes the movement of the SHS front is not steady state but instead the oscillatory one. Try to explain the possible reasons for oscillatory regime.

CHAPTER 10

FORMATION AND TRANSFORMATIONS OF NANOTWINS IN COPPER

10.1 INTRODUCTION

Copper is the most important metallic conductor. Besides the use as cables for long-distance transport of electricity, Cu has been widely used as multilayered thin film interconnects in microelectronic technology. On a Si chip of the size of our fingernail, the total length of Cu interconnect is as long as a kilometer. Thus,

Kinetics in Nanoscale Materials, First Edition. King-Ning Tu and Andriy M. Gusak.

any improvement in the physical properties of Cu is of keen interest. Recently, a high density of nanotwins, with twin plane spacing of the order of 10 nm, has been produced in bulk pieces of Cu by pulse electroplating [1, 2]. What is very attractive of the nanotwinned Cu is that it has superb mechanical properties as compared to that of nontwinned Cu. The yield strength is 10 times higher and yet the ductility remains as good as that of ordinary or nontwinned Cu. In addition, the electrical conductivity of nanotwinned Cu decreases only very slightly from that of ordinary Cu. A large number of publications on nanotwins in Cu are available in the literature. High-resolution transmission electron microscopic images of the nanotwin microstructure and mechanical and electrical properties of the nanotwinned Cu are published [1].

The very large increase in yield strength is indeed a rare accomplishment because high mechanical strength has been a major challenge in classic metallurgical applications. For a long time, there have been only four techniques to increase mechanical strength; they are solution hardening, precipitation or dispersion hardening, work or strain hardening, and grain size reduction (Hall–Petch effect). Now, the fifth one is nanotwin.

After the introduction of nanotwin in Cu, it has been extended to some other metals, for example, Au. Pure Au is very soft. For this reason, pure Au cannot be used to hold diamond or other precious stones in jewelry. Thus solution hardening of Au by alloying with Ag and Cu to form 14 K or 18 K gold was invented. But the color of them is not as attractive as pure Au. Now, nanotwinned Au is being developed in jewelry industry to hold diamonds and precious stones. In other words, if solution hardening in Au can be replaced by nanotwin hardening, it will be an interesting application of nanoscale metals.

In Section 10.2, we review why a high density of nanotwin can be formed and what the energy of formation of nanotwinned Cu is. Then, in Section 10.3, we discuss how to control the growth direction of nanotwins so that they have an oriented growth, for example, all the twin planes are parallel to the substrate surface. In other words, we achieve an oriented growth of nanotwins so that the normal of all of them is parallel to the normal of the substrate surface. It means that we can control the growth of the nanoscale microstructure, and we obtain a highly oriented and high density of nanotwin in Cu.

With the formation of the oriented nanotwins, we can perform two interesting transformations. First, we can have a unidirectional growth of Cu_6Sn_5 intermetallic compound (IMC) during the wetting reactions of molten solder on the oriented nanotwinned Cu. Second, we can transform the [111] oriented nanotwinned Cu to a very large single crystal of (100) Cu having no twins, by an extremely anisotropic grain growth of Cu.

Finally, in Section 10.4, the application of nanotwinned Cu to interconnect technology in reducing electromigration is presented. Also, the application of the oriented Cu_6Sn_5 and the (100) Cu single crystal to the manufacturing of a large number of microbumps in three-dimensional integrated circuits (3D IC) packaging technology is discussed briefly.

10.2 FORMATION OF NANOTWINS IN Cu

About the microstructure of twin, we recall that the crystal structure of a face-centered-cubic (FCC) metal can be represented by the stacking of (111) planes in the repeating sequence of ABCABC. When an error of stacking occurs and changes the stacking to ABCACBA, in which the middle plane of A can be regarded as a mirror of the atomic planes on both sides, a twin is formed. The middle plane of A is called the twin plane. The stacking error increases the internal energy of the crystal only slightly because each atom in the twin plane still has 12 nearest neighbors as in the regular FCC crystal; thus twin is considered as a coherent plane defect of low energy. The energy increase is of the same order of a stacking fault. FCC metals such as Cu and Ag have low stacking fault energy and low twin energy, so ordinarily they tend to have microtwins, meaning the twin plane spacing is of the order of microns. FCC metals such as Al and Ni have high stacking fault energy, so they seldom show twins.

Even though the twin in Cu is a low energy defect, a high density of them will require a certain amount of energy to be compensated. The question is why does high density of nanotwins form? It has been proposed that the strain energy stored in Cu during electroplating is traded to form the nanotwins [3]. Then we ask why does electroplating of Cu induce strain? We recall that in thin film deposition, it is known that even when a thin film is deposited at room temperature and kept at room temperature, intrinsic stress or strain is found. For example, Ni thin films tend to have a very high intrinsic tensile stress and tend to peel off easily from their substrates when the deposited Ni thin film is more than 300 nm thick. It is mainly because during a high rate of deposition at a low temperature, most of the deposited atoms are jammed into nonequilibrium positions. If there is not enough time for them to relax to the equilibrium positions of the lowest energy, strain energy builds up. Intrinsic stress in thin films is a subject covered in books on thin film processing and is not discussed here.

A first principle calculation will be presented next to show that indeed nanotwins can be formed in Cu with just a reasonable amount of strain energy [3]. It is very unlikely to have nanotwins in Al because its twin energy is very high, so the amount of strain energy needed will be too high to compensate for nanotwin formation.

10.2.1 First Principle Calculation of Energy of Formation of Nanotwins

The formation energy of twin plane has been studied by first principle total energy calculations using the Vienna ab initio simulation package (VASP) with the local density approximation (LDA). Periodic nanospaced twinning structure has been constructed and strain has been introduced to estimate the evolution of the total energy with increasing strain. To simplify the calculation, a biaxial stress state was used. It assumes the condition of a Cu thin film on a substrate so that it is stress-free along the normal direction to the film surface.

Figure 10.1 Schematic diagram of the periodic nanotwinned structure by repetition of the chosen supercell. *Source*: Reproduced with kind permission from [5].

Supercells having the fcc structure and periodic nanotwins are constructed and relaxed using a conjugated-gradient algorithm. Figure 10.1 shows the periodic nanotwinned structure by repetition of the chosen supercell. The fcc Cu is constructed by having three (111) layers with atoms in each layer occupying A, B, and C positions. Each supercell of periodic nanotwinned model contains 12 (111) atomic planes in a stacking sequence of ... ABCAB[C]BACBA[C]ABCAB[C]BACBA[C] ..., so that in the 24 atomic planes shown, there are two twin planes in each of the two supercells, as indicated by the four [C] planes. There are 6 (111) planes in between 2 twin planes, and these planes can be increased to 9 or 12 planes or more in order to increase the twin spacing or to reduce the twin density. The twin fault energy can be calculated by using the following equation,

$$\gamma_{\text{twin}} = \frac{E' - NE_{\text{fcc}}}{2A} \tag{10.1}$$

where E' is the total energy of the supercell nanotwin system, N is the number of atoms in the nanotwinned supercell, E_{fcc} is the energy per atom in an ideal fcc Cu, A is the surface area of each twin plane in the supercell, and the factor of 2 accounts for the presence of two twin planes in a supercell.

To study the influence of stress/strain on twin formation in Cu thin films, isotropic biaxial stress was introduced and the supercells with strains in the three-dimensional structure were created using the Poisson's ratio of 0.34 of bulk Cu. Two types of biaxial stresses were studied as depicted in Figure 10.2: biaxial stresses in the {111} plane with x-axis//[110] and y-axis//[112] and biaxial stresses in the {112} plane with x-axis//[110] and z-axis//[111].

The calculated twin fault energy is close to those from other theoretical and experimental studies. The twin fault energy is very small, only about two-thousandths of 1 eV/atom. We recall that interatomic pair potential energy is typically of 0.5 eV/atom.

Figure 10.3a shows the evolution of the total energies for nontwinned fcc Cu (the lower curve) and nanotwinned Cu (the upper curve) when they respond to biaxial stresses in the {111} plane. Figure 10.3b shows the results when the biaxial stresses

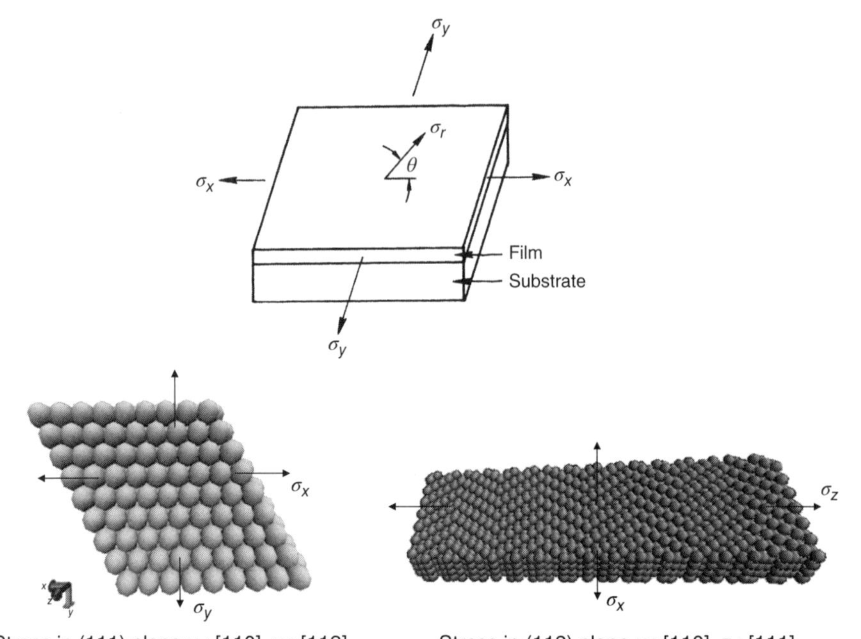

Figure 10.2 Biaxial stress in the model of calculation. *Source*: Reproduced with kind permission from [5].

are in the {112} plane. The pair of curves in each of these figures indicates that when the strain in the nontwinned Cu thin film is larger than a certain value, the total energy of the strained fcc Cu system becomes higher than the total energy of the strain-free Cu nanotwin system. To see it, we can draw a horizontal straight line tangent to the bottom of the upper curve in Figure 10.3a. The horizontal line will cut the lower curve at 0.008 on the left-hand side (the compressive strain side) and at 0.011 at the right-hand side (the tensile strain side). This means that if we strain the nontwin Cu by a compressive strain of more than 0.8%, it will be higher than the energy of the strain-free Cu having the nanotwins, so the strain energy can be traded to form nanotwins of twin spacing of 6 {111} atomic planes to reduce energy. Or if we strain nontwin Cu by a tensile strain more than 1.1%, it can be traded to form nanotwins of twin spacing of 6 {111} atomic planes to reduce energy. Figure 10.3a also shows that for nanotwins of larger twin spacing, such as 9 or 12 (111) layers, the strain energy needed will be even less. The results in Figure 10.3b are similar. The conclusion is that a highly strained nontwin Cu is energetically less stable than a strain-free nanotwinned Cu.

It is obvious that the nanotwinned Cu may not be completely strain-free. This is because under a given strain, if the rate of nucleation and growth of nanotwins, or the density of nanotwin formed, is insufficient to compensate the entire strain, certain amount of residual strain will remain in the nanotwin structure. This is discussed in the next section.

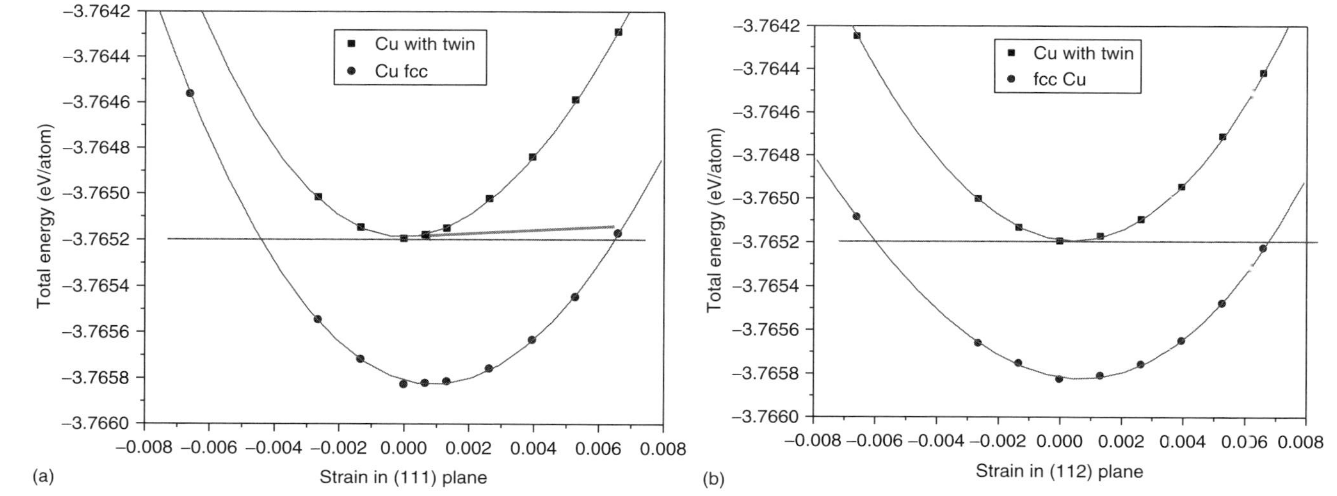

Figure 10.3 (a) Calculated curves show the evolution of the total energies for fcc Cu (the lower curve) and nanotwinned Cu (the upper curve) when they response to strains in the {111} plane. (b) Calculated curves of the results when the biaxial stresses are in the {112} plane. The pair of curves indicates that when the strain in the thin film is larger than a certain value, as indicated by the cut of the horizontal line with the upper curve, the total energy of the strained fcc Cu system becomes higher than the total energy of the strain-free Cu nanotwin system.

10.2.2 *In Situ* Measurement of Stress Evolution for Nanotwin Formation During Pulse Electrodeposition of Cu

In situ stress measurements by bending beam method were performed during high frequency pulse electroplating of nanotwinned Cu thin films [4]. It shows repeating cyclic events of stress build-up and relaxation in the pulse plating. Nanotwin formation in the plated Cu films has been measured by transmission electron microscopy and the measured twin density is in agreement with the first principle calculation.

Figure 10.4 shows the *in situ* stress measurement system. It consists of an electrochemical cell, an optical system that can detect with high sensitivity the bending of the cantilever beam on which the Cu film is being plated, and a signal processing unit. The laser reflection response was recorded every 1 ms and it is fast enough to record the bending during every pulse deposition. When the bending or the radius of the cantilever beam was measured, the stress was calculated on the basis of Stoney's equation.

$$\sigma_{\mathrm{f}} = \left(\frac{Y}{1-\nu}\right)_{\mathrm{S}} \frac{t_{\mathrm{S}}^2}{6rt_{\mathrm{f}}} \tag{10.2}$$

where σ_{f} is the biaxial stress in the film, Y and ν are the Young's modulus and Poisson's ratio of the cantilever beam substrate, t_{S} is the substrate thickness, t_{f} is

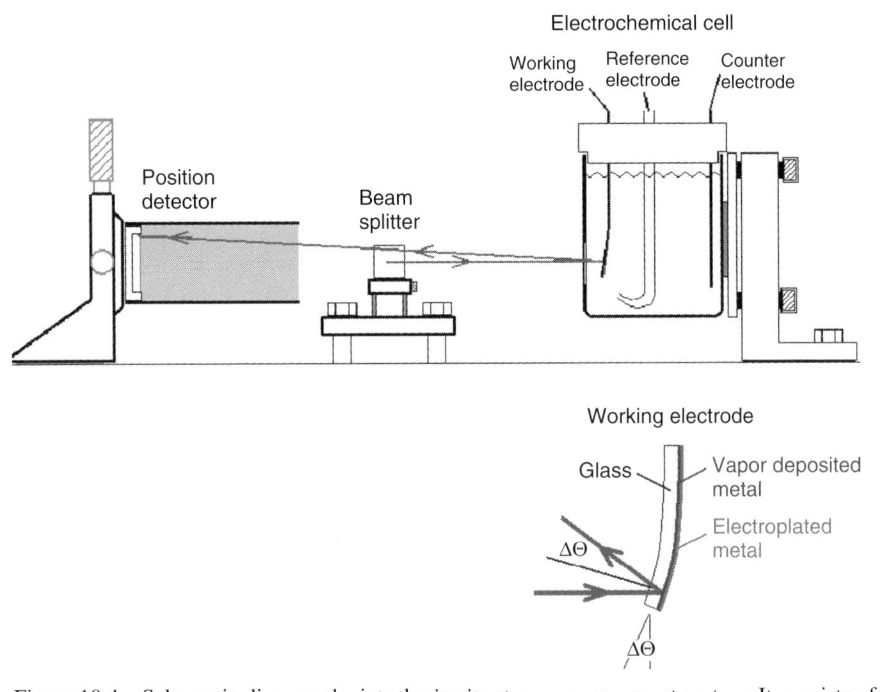

Figure 10.4 Schematic diagram depicts the *in situ* stress measurement system. It consists of an electrochemical cell, an optical system that can detect with high sensitivity the bending of the cantilever beam on which the Cu film is being deposited, and a signal processing unit.

the deposited film thickness, and r is the radius of the cantilever beam. The equation shows that it requires knowing the film thickness in order to calculate the biaxial stress.

Typically, the film thickness was measured after deposition. However, in the pulse electroplating, we cannot measure *in situ* the deposited film thickness during each pulse. As we can measure the *in situ* radius during each pulse, we move the film thickness of t_f from the right-hand side to the left-hand side in Stoney's equation, Eq (10.2), so it is the product of $(\sigma_f \times t_f)$ that will be measured during each pulse deposition when the radius is obtained.

Figure 10.5a shows the evolution of the product of stress times film thickness as a function of pulse deposition time. The deposition potential was – 0.70 V versus Cu, resulting in a pulse current density of about 0.22 A/cm^2. The pulse deposition time consists of 0.1 s pulse-on time and 9.9 s pulse-off time, so the duty cycle was 1%. In Figure 10.5a, the product evolves with a periodicity that is consistent with the pulse cycle. During the pulse-on time period, the stress-thickness product moves in the tensile direction, whereas during the pulse-off period, approximately 60% of this tensile stress is relaxed.

To calculate the stress, we need to estimate the nominal film thickness produced by a single pulse. It was found to be about 8.2 nm on the basis of the deposition charge. The single pulse thickness can be checked at the end of the deposition by dividing

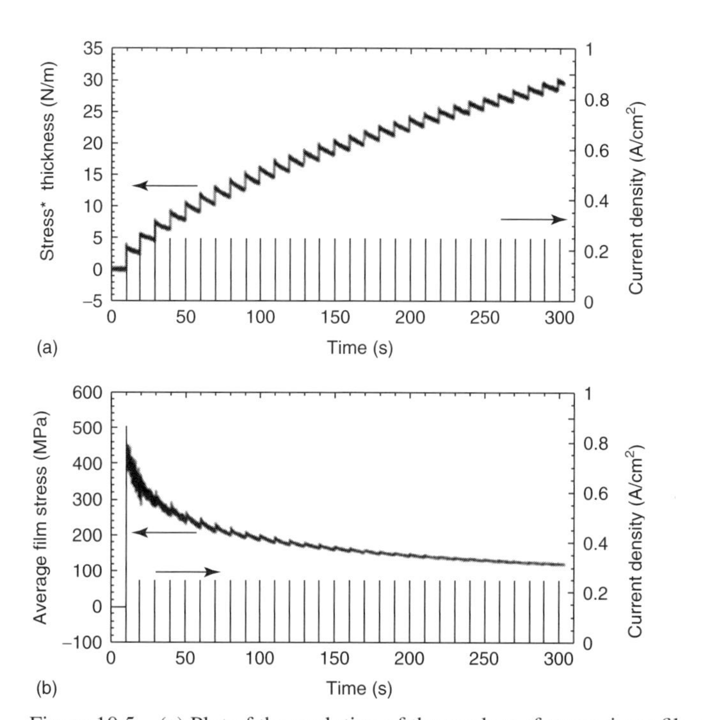

Figure 10.5 (a) Plot of the evolution of the product of stress times film thickness as a function of pulse deposition time. (b) Plot of the evolution of average film stress as a function of pulse deposition time.

the total film thickness by the number of pulse. Figure 10.5b shows the evolution of average film stress as a function of pulse deposition time. The maximum tensile stress recorded was 400 MPa at a film thickness of 10 nm. As shown in Figure 10.5b, while stress relaxation is observed during each pulse-off period, the overall stress decreases with increasing deposition time, which leads to an interesting question. Why? One possible explanation is that the film is getting hotter, so atomic relaxation becomes faster. When the stress decreases, we expect that the twin density will decrease or the twin spacing will increase with the film thickness on the basis of the first principle calculation. Furthermore, the rate of decrease slows down with increasing deposition time. This is reasonable because the atoms deposited in the earlier period have more time to relax, so the average stress decreases.

If the duty cycle is changed, it was found, a longer pulse-on time results in a higher tensile stress, suggesting that the relaxation in pulse-off cycle is linked to the amount of films deposited during the pulse-on cycle, which in turn affects nanotwin formation in the Cu film. It was found that the tensile stress and the extent of stress relaxation depend on both the pulse-on and pulse-off periods, as well as the potential of the pulse. More detailed study is needed in order to determine quantitatively the effect of pulse electroplating condition on twin density.

The nanotwins formed by the pulse electroplating were examined by X-ray diffraction and scanning and transmission electron microscopy. The films obtained were polycrystalline Cu having no strong preferred grain orientation. In turn, the (111) twin planes were formed more or less in random orientation with respect to the normal of the substrate surface. In a later section, we discuss the formation of oriented nanotwins with all the <111> axis that is normal to the twin plane to be parallel to the normal of the substrate surface.

Figure 10.6 shows the histogram of distribution of the measured twin spacing in the films. The average twin spacing is about 20 nm. Knowing the measured average tensile stress, the twin spacing can be predicted by first principle calculation to be about 28 nm, which is consistent with observed twin spacing by TEM characterization.

10.2.3 Formation of Nanotwin Cu in Through-Silicon Vias

In order to plate Cu into high-aspect ratio through-Si-vias (TSVs) of varying opening size (15–100 μm) and depth (200–400 μm), aspect-ratio-dependent (ARD) pulsed electroplating technique was used [5]. After the arrays of through-holes in silicon wafer were prepared by deep reactive ion etching, SiO_2 and Si_3N_4 layer were deposited on the surface and sidewall of the holes as insulation layer. An electrical contact substrate with a seed layer (Cr/Au) was temporarily bonded with the etched wafer using photo-resist. Then the controlled ARD electroplating to deposit Cu from the bottom of the holes to the top of the holes was performed. In ARD technique, electroplating parameters were continually varied with respect to the change of the unfilled depth during plating in order to maintain the uniform current distribution inside the through-holes. In the beginning of electroplating, a low forward current density (10 mA/cm^2) and a high reverse current density (60 mA/cm^2) were used to achieve void-free fine Cu grains on the bottom of the through-holes. In the

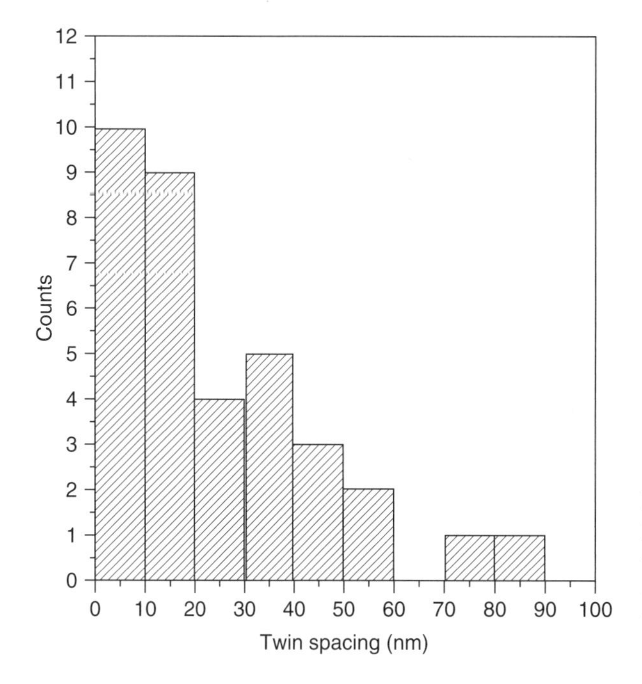

Figure 10.6 The histogram of distribution of the measured twin spacing in the films. The average twin spacing is about 20 nm.

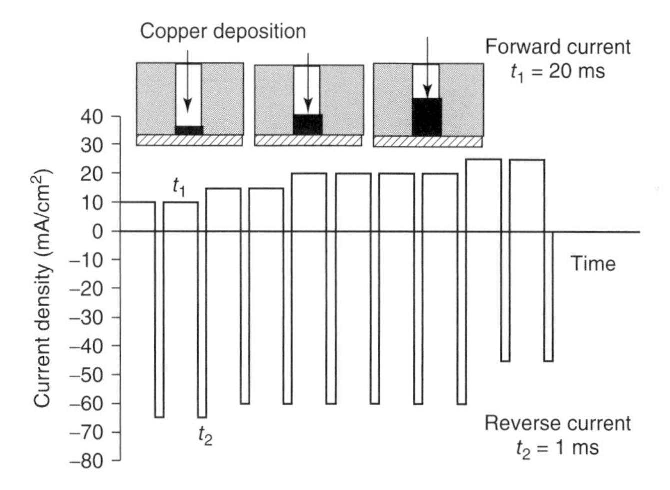

Figure 10.7 In the ARD deposition, as the electroplating continuous, the forward current density was gradually increased in steps while the reverse current density was reduced.

reverse cycle, polarity of the electrodes reverses and some of the deposited Cu is removed. The alternative cycles of deposition and removal result in a more uniform Cu deposition and void-free. As the electroplating continues, the forward current density is gradually increased in steps while the reverse current density is reduced, as shown in Figure 10.7.

Figure 10.8 The electroplated through-wafer Cu columns. (a) The cross-section view without etching the Si. (b) The freestanding Cu columns after the Si was dissolved by KOH solution. The diameter of each Cu column is 30 μm, the height is 225 μm, and the pitch is 80 μm. *Source*: Reproduced with kind permission from JAP 2007 [4].

The electroplated Cu columns in TSV are shown in Figure 10.8. Figure 10.8b exhibit the freestanding Cu columns after the Si was dissolved by KOH solution. The diameter of each Cu column is 30 μm, the height is 225 μm, and the pitch is 80 μm. Figure 10.8a shows the cross-section view without etching the Si. Nanoindentation on one of the cross-sections of Cu was performed. The measured hardness has a value of about 1.7 GPa near the top and about 2.5 GPa near the bottom of the Cu column. It indicates that the hardness of the earlier deposited Cu (bottom of the hole) is higher than the later deposited (top of the hole). The range of hardness values is similar to that published in the literature for nanotwinned Cu.

The variation of hardness of Cu column from top to bottom is related to the nanotwin density. The nanotwin density of the electroplated Cu column at three different locations was measured by transmission electron microscopy; with the highest twin density at the bottom, it decreases at the middle, and further decreases at the top. Figure 10.9 shows TEM images of the nanotwins. Most likely, the nanotwin formation is due to the strain in the Cu column, which is confined within the Si via.

Figure 10.9 TEM images of nanotwins in the Cu column. *Source*: Reproduced with kind permission from JAP 2007.

The change of twin density is due to the change in strain in the Cu column, as discussed in the previous section. The strain change, in turn, may be due to the change in relaxation because of the temperature increase with increasing deposition and the change in ARD pulse plating. The trend is similar to that discussed in the last section.

10.3 FORMATION AND TRANSFORMATION OF ORIENTED NANOTWINS IN Cu

Because the twin plane is (111) plane of Cu and if we can have this plane to be parallel to the substrate surface during electroplating, we have a unidirectional growth of (111) oriented Cu film or we have a very highly textured Cu film on a substrate. To do so, we need to control the nucleation and growth of nanotwins. We may use a seeding layer to enhance the nucleation of <111> oriented Cu nuclei on a given substrate, and after the oriented nucleation, we may further enhance the oriented growth of <111> twins, which is discussed in Section 10.3.1.

The purpose of obtaining the <111> oriented Cu having a high density of nanotwins is that we can perform subsequently phase transformations to control the formation of unique nanoscale microstructures.

First, if we wet the <111> oriented twin surface of Cu by molten solder, it forms oriented IMC of Cu_6Sn_5, which is similar to the formation of oriented Cu_6Sn_5 on <111> oriented single crystal of Cu. In other words, we translate the unidirectional growth of nanotwinned Cu to the unidirectional growth of the IMC, which finds an important application in microbumps of 3D IC technology in the stacking of Si chips, which is discussed in Section 10.3.2.

Second, we can transform the <111> oriented and nanotwinned Cu to a very large grain size of (100) oriented single crystals of Cu (without twins) on a given substrate. The substrate can be a Si wafer.

10.3.1 Formation of Oriented Nanotwins in Cu

To have a unidirectional growth of <111> oriented Cu film or a highly textured <111> Cu film on a substrate, we might say that it is the second choice after the first choice of a bulk single crystal of <111> oriented Cu. Indeed, the <111> oriented nanotwin Cu thin films have been obtained by sputtering [6, 7]. In electroplating, we can also do so by using a seeding layer to enhance the nucleation of <111> oriented Cu nuclei on a given substrate, for example, on a Si wafer. Then the oriented growth of <111> nanotwins can be enhanced by a high rate of rotation of the Si wafer or the rotation of electrolyte during plating. The axis of rotation is the surface normal of the wafer and it is also the normal of the <111> twin plane.

The beneficial effect of the rotation is that it cools the wafer and keeps the surface temperature of the wafer constant. In addition, because it produces a shear force between the rotating wafer and the plating solution, it will not only remove any gas or bubble formation on the film surface, but also create instability of Cu adatoms on the Cu film surface, the (111) plane, so that the probability of twin nucleation due to stacking error is enhanced.

Figure 10.10a shows a focused-ion-beam (FIB) image that has both the top view and cross-sectional view of an oriented and nanotwinned Cu film. The cross-section was prepared by FIB that had rendered a very clear view of the top surface as well as the cross-sectional interface. On the cross-section, columnar Cu grains are shown and in each grain there are layered parallel planes. The layered images are the nanotwins and they are parallel to the film surface or the wafer surface. As the twin plane is (111), we have the oriented <111> nanotwins. Figure 10.10b shows the X-ray diffraction spectrum of the sample; only a very strong <111> reflection and <222> reflection of Cu are detected. On the top surface as shown in Figure 10.10a, there are cones on each columnar grain, which are formed by the stacking of multilayers of (111) planes having smaller diameter going from the base of each cone to the top of the cone. It suggests that the nanotwins nucleated in the center of the cone and grew sidewise.

10.3.2 Unidirectional Growth of Cu–Sn Intermetallic Compound on Oriented and Nanotwinned Cu

As the trend of miniaturization of large-scale-integration of circuits on Si chip technology is approaching the limit of Moore's law, microelectronic industry has been looking for ways to overcome or to extend the limit. One of the ways is to turn to 3D IC by combining chip technology and packaging technology together. This is because when the concept of Moore's law is applied to packaging technology, it has not reached its limit yet. We can examine its limit by considering the diameter of solder joints. It is about 100 μm in today's flip chip technology, and the diameter of

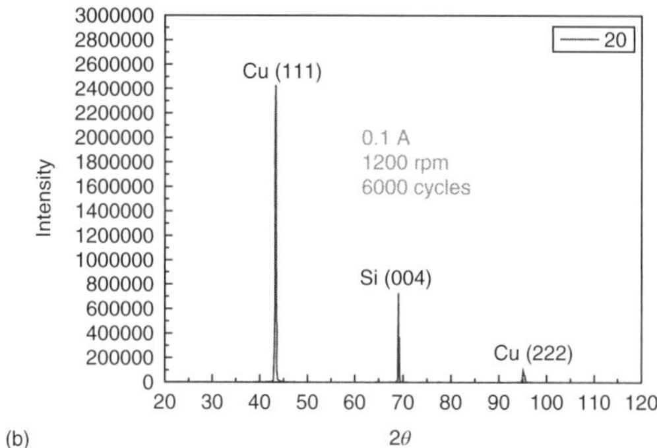

Figure 10.10 (a) SEM images of both the top view and cross-sectional view of an oriented and nanotwinned Cu film. (b) X-ray diffraction spectrum of the oriented and nanotwinned Cu sample; only a very strong <111> and <222> reflections of Cu were detected.

microbumps in 3D IC is now about 10 μm. It is quite possible that the microbump diameter can be reduced to 1 μm; thus there is ample room in the near future for the packaging technology to advance. This is because the density of solder joint per unit area can be increased by four orders of magnitude. As the Moore's law has projected that it will take about 15–20 years for the density of transistors on a Si chip to increase three orders of magnitude, and if we apply this rate to solder joints, the microelectronic industry will have another 15–20 years to grow when the chip technology is combined with the packaging technology.

When the microbump diameter is reduced by 10 times, from 100 μm to 10 μm, the volume is reduced by 1000 times. Thus, if we consider an extreme case in which the grain size is assumed to be 10 μm, the microbump may have only one grain, but the flip chip solder joint of 100 μm may have 1000 grains. In the latter case, we can assume the properties of the solder joint to be isotropic because of the large number of grains, but in the former case, we cannot do so. This has a serious consequence on isotropic or anisotropic properties of microbumps because there are thousands of them on a chip,

As Sn has a body-centered-tetragonal crystal structure, it has anisotropic properties; for example, its conductivity is anisotropic and its deformation is dominated by twinning. It is also known that the diffusion of Cu and Ni in Sn is anisotropic, the diffusivity along the *c*-axis is about two to three orders of magnitude faster than that along the *a*- and *b*-axis [8–10]. Thus the formation of IMC on a *c*-axis oriented Sn grain will be much faster than that on an *a*-axis or *b*-axis oriented Sn grain. In transforming Sn to Cu–Sn or Ni–Sn IMC, the IMC has a noncubic or an anisotropic crystal structure too. This is a concern because it means that among a large number of microbumps on a TSV chip, the properties of each of the microbumps can be different because it may have only a few grains or just one grain. From the point of view of reliability, it could mean a wide distribution in statistical failure, so some microbumps could result in early failure. Thus, it is important to control the microstructure of the grains of every microbump on a TSV chip in order to have a uniform microstructure or a fixed grain orientation in all of the microbumps. However, there are hundreds or thousands of microbumps on a TSV chip and therefore how to do so is challenging.

This situation has led to the development of a new paradigm in electronic packaging technology that precision control of microstructure is urgently needed in manufacturing. We recall that the past trend of miniaturization of Si chip technology, which is the foundation of Moore's law, is based on precision control of the step-by-step reduction of critical feature sizes such as the gate width in Si devices. The feature size has to be the same among hundred millions of transistors on a chip so that the transistors behave the same. Today, we need to extend the precision control of microstructure from chip technology to packaging technology.

In the following, we concentrate on IMC formation in the microbumps and we consider how to control IMC formation in order to have a unidirectional growth and in turn to obtain a uniform microstructure of all the oriented IMC grains. We start from the control of grain orientation in Cu under-bump-metallization (UBM) by forming the oriented nanotwins in the Cu UBM. Then, we follow by the formation of oriented Cu–Sn IMC on the oriented nanotwin Cu in every microbumps on a TSV chip.

We recall that when (100) or (111) oriented bulk single crystal Cu was wetted with molten solder for the study of IMC formation, the oriented Cu_6Sn_5 with roof-top shape or prism-type was observed on both (100) and (111) oriented single crystal Cu [11, 12]. On (100) oriented single crystal Cu, the oriented Cu_6Sn_5 grains were elongated along two perpendicular directions. On (111) oriented single crystal of Cu, the oriented Cu_6Sn_5 grains were elongated along three preferential directions with 120° separation. The orientation relationship between Cu_6Sn_5 and (100) Cu and (111) Cu were determined and published [11, 12]. They are not going to be repeated here. Now, on the (111) oriented nanotwinned Cu, we can have a similar oriented growth of Cu_6Sn_5 as that on (111) oriented single crystal of Cu [13].

Figure 10.11a shows an optical image of the cross-section of a microbump. Figure 10.11b shows the FIB cross-sectional view of a microbump formed with the oriented nanotwin Cu as UBM on both the top and bottom sides. The layered microstructure in both the top and the bottom UBM are the oriented nanotwins. The scallops in between are the Cu_6Sn_5 IMC. When the scallops from the upper and the lower UBM touched each other, they joined to form columnar grains of IMC, and, surprisingly, there does not seem to be a grain boundary between the upper and the lower Cu_6Sn_5 grains. It is possible that a rapid ripening or grain growth has

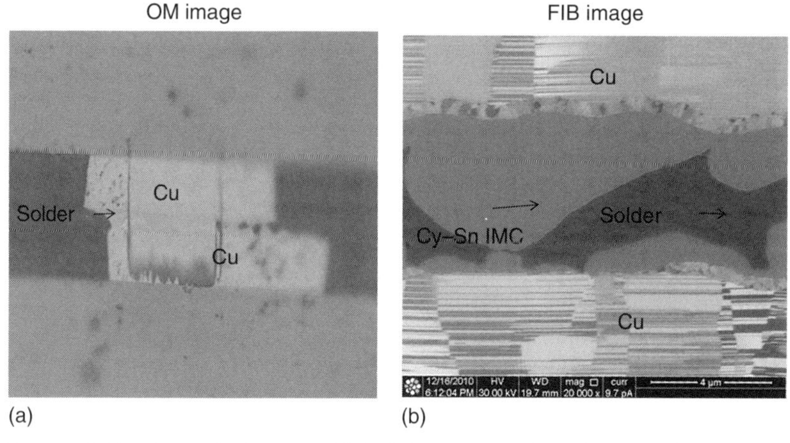

Figure 10.11 (a) Optical image (OM) of the cross-section of a Cu-to-Cu microbump joined by Pb-free solder. (b) FIB image of the cross-section of oriented and nanotwinned Cu on top and bottom UBM in the microbump.

occurred between them. This observation has also been reported in the literature [12]. Between the Cu_6Sn_5 scallops and the nanotwins, there is a layer of Cu_3Sn.

What is of interest in Figure 10.11b is to compare the microstructure of Cu_6Sn_5 to that of Cu_3Sn. While the latter is thinner and has a large number of small nanosize grains, its polycrystalline microstructure is very clear. On the other hand, the Cu_6Sn_5 has much larger grains, and each scallop seems to be a single crystal grain of Cu_6Sn_5. In particular, the image of their microstructure looks rather uniform and no grain boundary can be seen; furthermore, all the scallops seem to look the same, at least from their physical appearance. In addition, when the upper Cu_6Sn_5 joined the lower Cu_6Sn_5, they seem to have merged into one grain quickly.

Figure 10.12a shows the FIB cross-sectional view of another microbump with 20 μm diameter nanotwinned Cu as UBM on the top and the bottom sides. Figure 10.12b, c shows, respectively, the electron backscatter diffraction (EBSD) images of Cu_6Sn_5 scallops and unreacted Pb-free solder in the microbump. Figure 10.12d shows EBSD images of the joining of Cu_6Sn_5 grains to form columnar grains in the microbump after 260 °C for 5 min. The three triangular sections in the lower right corner show the distribution of orientations of grains in the EBSD images.

When Cu_3Sn forms between Cu_6Sn_5 and Cu, Kirkendall void formation is reported. This is because when one molecule of Cu_6Sn_5 is converted to two molecules of Cu_3Sn, the three Sn atoms left behind will attract nine Cu atoms in order to form three more molecules of Cu_3Sn. The diffusion of nine Cu atoms from the Cu UBM will require a reverse diffusion of nine vacancies, which tends to form voids at the interface between Cu_3Sn and Cu as well as within Cu_3Sn. Initially, voids nucleated and formed at the interface. When Cu_3Sn grows thicker, the voids are embedded into the Cu_3Sn as the interface moves into the Cu.

Crystal structure of Cu_3Sn is such that the Cu (the majority) sublattice is percolating, so that copper atoms can diffuse via its sublattice without violating the

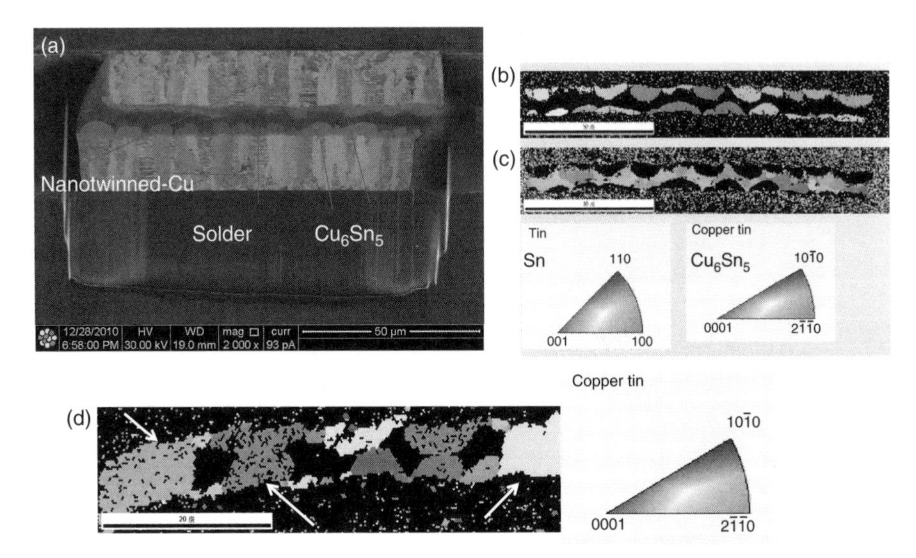

Figure 10.12 (a) FIB image of the cross-section of a microbump. (b) and (c) are EBSD cross-sectional images of the Cu_6Sn_5 scallops and the unreacted Pb-free solder, respectively. (d) EBSD images of the formation of single crystal columnar grains of Cu_6Sn_5 in the joint.

long-range order. On the other hand, Sn (the minority) atoms are surrounded by the sites of Cu atoms, so the substitutional diffusion of Sn atoms cannot occur without violating the order. Therefore, the diffusion of Cu is much faster than that of Sn in Cu_3Sn, and it leads to vacancy flux toward the Cu substrate. If vacancy sinks are not working properly, voids will form.

However, when Cu_3Sn forms on nanotwinned Cu, very few or no Kirkendall voids are observed. Figure 10.13a, b shows the cross-sectional FIB images of Cu_3Sn formation between nanotwin Cu and Cu_6Sn_5: there are no Kirkendall voids. This is a significant finding from the point of view of microbump reliability. It was proposed that the steps and kinks on the interface between Cu_3Sn and a (111) twin plane of Cu can serve as sinks to absorb vacancies [13]. Figure 10.13c is a high-resolution TEM image of the interface between nanotwin Cu and Cu_3Sn. As Cu_3Sn grows by consuming the (111) planes of the nanotwin Cu, it will soon meet a twin plane. When the Cu_3Sn is in contact with a twin plane, the interface has to adjust to the change of the (111) stacking error. It increases the probability of step and kink formation to serve as vacancy sinks. A high density of nanotwins will provide a high density of vacancy sinks. When no supersaturation of vacancies occurs at the interface between the Cu_3Sn and nanotwinned Cu, there is no nucleation and growth of voids.

10.3.3 Transformation of <111> Oriented and Nanotwinned Cu to <100> Oriented Single Crystal of Cu

We recall Figure 1.19 in Chapter 1, which shows a SEM image of a sample of <111> oriented and nanotwinned Cu annealed at 450 °C for 5–10 min. In the center of the

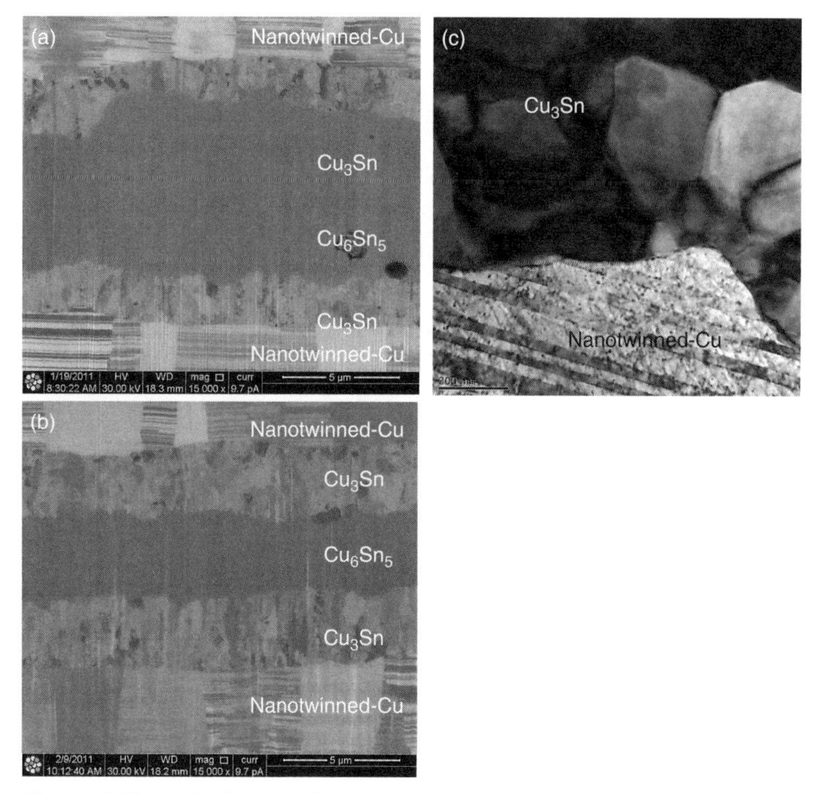

Figure 10.13 (a, b) Cross-sectional images of IMC microbump. No Kirkendall voids were observed in the Cu_3Sn. (c) A high-resolution TEM images of the interface between the Cu nanotwin and the Cu_3Sn grains.

image, there is a very large grain surrounded by a large number of much smaller size grains. The abnormal large grain was found to have <100> normal to the surface. The surrounding smaller grains are <111> nanotwinned grains.

As the grain size of the central grain is much larger than those grains in the matrix of Cu annealed at 450 °C for a few minutes, it seems to be an abnormal grain growth. Abnormal grain growth is known to occur in Cu, and it leads to a bimodal distribution of grain size, in which one set of grains is exceptionally large. However, what is shown in Figure 1.19 is an extreme case of anisotropic grain growth. After a prolonged annealing of several hours at 450 °C, all the <111> oriented and nanotwinned Cu grains transformed into many very large <100> oriented grains. There is no bimodal distribution of grain size. The average diameter of the <100> grains is large than 100 μm, which we note is larger than that of a microbump.

The phenomenon of the extremely anisotropic grain growth suggests that we have a unique phase transformation. As we can form <111> oriented and nanotwinned Cu as UBM in every microbump by electroplating, we can transform every one of them into (100) oriented single crystal Cu UBM. In other words, we grow an array of a large number of (100) oriented single crystals of Cu on a Si wafer

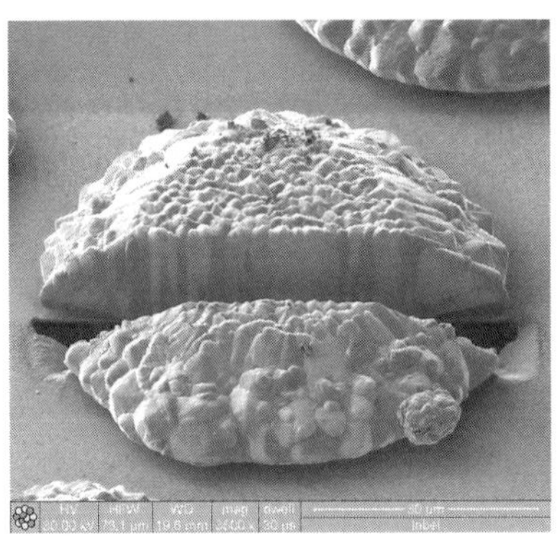

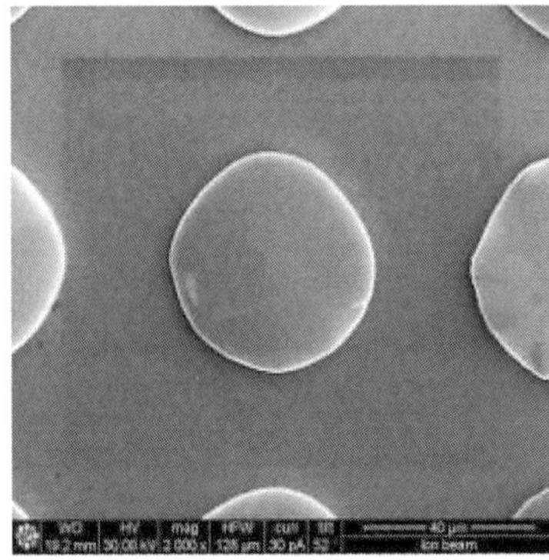

Figure 10.14 SEM images of an oriented <111> nanotwinned Cu of 20 μm diameter before and after annealing at 500 °C for 60 min. Before the annealing, the cross-section by FIB shows columnar grains of Cu having the <111> oriented nanotwins. After the annealing, the Cu has transformed to a <100> oriented single crystal. *Source*: Reproduced with kind permission from Chih Chen, National Chiao Tung University, Taiwan.

surface. It means that we have a precision control of the microstructure of Cu UBM on the basis of the oriented nanotwins. Figure 10.14 shows SEM images of a Cu UBM of 20 μm in diameter and 10 μm in thickness, before and after annealing at 500 °C for 60 min. It became a (100) single crystal of Cu after the annealing. The transformation from <111> oriented nanotwin to <100> oriented single crystal of Cu can occur in every microbump on a TSV chip. We obtain a mono-size distribution of (100) single crystal of Cu on the chip.

10.4 POTENTIAL APPLICATIONS OF NANOTWINNED Cu

We review briefly here the potential applications of nanotwinned Cu to interconnect technology and 3D IC packaging technology.

10.4.1 To Reduce Electromigration in Interconnect Technology

Electromigration has been the most persistent reliability issue in Al and Cu interconnect technology. This is because of the trend in device miniaturization. The interconnect dimension, especially its cross-section, has to be reduced together with the critical feature size. Consequently, the current density increases. According to Huntington's model, the driving force of electromigration is directly proportional to current density, hence electromigration is becoming more serious in smaller interconnects. To overcome electromigration-induced failures in Al interconnect technology, it was found that by adding about 1 at.% of Cu into Al, electromigration is retarded. Also, at the same time, the electrical conductivity is not much increased. Thus, in Cu interconnect technology, the question is what can we add to

Cu to improve its resistance to electromigration. The answer is to add nanotwins to Cu [14].

There is a major difference between the electromigration in Cu and that in Al at the device operation temperature of 100 °C. In Al, it occurs by grain boundary diffusion, but in Cu it occurs by surface diffusion. This is not only because of the difference in their melting point, and in turn the difference in their activation energy of diffusion, but also because Al has a protective native oxide that has prevented electromigration to occur along the interface between Al and its native oxide. On the other hand, Cu does not have a protective native oxide. Furthermore, in building the multilayered Cu interconnects, after the dual-damascene process of electroplating Cu into the trenches and vias, the surface of Cu has to be chemical-mechanically planarized before the deposition of a dielectric layer for repeating the dual-damascene process. The interface between the Cu and the deposited dielectric is incoherent and becomes a freeway for rapid electromigration. This is because of the poor adhesion between the Cu and the dielectric; electromigration of Cu along the interface is as fast as that of electromigration of Cu on a free surface of Cu.

Why nanotwin can affect surface electromigration in Cu? This is due to the fact that when a pair of twin planes intersects a free surface (or a grain boundary), a zigzag-type surface structure is created by a pair of triple points. Surface electromigration slows down in going over a triple point. It was found that when a surface step approaches a triple point, its motion is temporarily stopped for a while before it starts moving again after passing the triple point. As the step passed a triple point, it has to move on a new surface, say from (111) to (422) or vice versa. Then the nucleation of a new step is needed. The delay time is most likely due to the incubation time of nucleating the new step [14].

10.4.2 To Eliminate Kirkendall Voids in Microbump Packaging Technology

As we have mentioned in Section 10.3.2, the trend in microelectronic industry is moving from two-dimensional to three-dimensional integrated circuits (3D IC) by combining chip technology and packaging technology. In 3D stacking of semiconductor chips, the technology depends on vertical interconnects by using TSV and micro solder joints. While the joint becomes small, about 10 μm in diameter, the processing temperature and time remains the same. The entire solder joint can be converted into IMC of Cu_6Sn_5 and Cu_3Sn by Cu-Sn reaction. The formation of Cu_3Sn is accompanied by Kirkendall voids, which tends to weaken the mechanical properties of the joint. As most IMCs are brittle, the void formation will make it worse. If we can eliminate Kirkendall void formation, we may be able to use IMC joint instead of solder joint.

To use IMC joint, we must consider its fracture property. So, Kirkendall void formation is undesirable. However, it was reported that on nanotwinned Cu, the formation of Cu_6Sn_5 and the transformation of Cu_6Sn_5 to Cu_3Sn does not introduce Kirkendall voids (see Figure 10.13) [13].

REFERENCES

1. Lu L, Shen Y, Chen X, Qian L, Lu K. Ultrahigh strength and high electrical conductivity in copper. Science 2004;304:422–426.
2. Lu L, Chen X, Huang X, Lu K. Revealing the maximum strength in nanotwinned copper. Science 2009;323:607–610.
3. Xu D, Kwan WL, Chen K, Zhang X, Ozolins V, Tu KN. Nanotwin formation in copper thin films by stress/strain relaxation in pulse electrodeposition. Appl Phys Lett 2007;91:254105.
4. Xu D, Sriram V, Ozolins V, Yang J-M, Tu KN, Stafford GR, Beauchamp C. In situ measurements of stress evolution for nanotwin formation during pulse electrodeposition of copper. J Appl Phys 2009;105:023521–023526.
5. Xu L, Dixit P, Miao J, Pang JHL, Zhang X, Tu KN. Through-wafer electroplated copper interconnect with ultrafine grains and high density of nanotwins. Appl Phys Lett 2007;90:033111.
6. Anderoglu O, Misra A, Wang H, Ronning F, Hundley MF, Zhang X. Epitaxial nanotwinned Cu films with high strength and high conductivity. Appl Phys Lett 2008;93:083108.
7. Anderoglu O, Misra A, Wang H, Zhang X. Thermal stability of sputtered Cu films with nanoscale growth twins. J Appl Phys 2008;103:094322.
8. Minhua L, Shih D-Y, Lauro P, Goldsmith C, Henderson DW. Effect of Sn grain orientation on electromigration degradation mechanism in high Sn-based Pb-free solders. Appl Phys Lett 2008;92 (211909).
9. Yeh DC, Huntington HB. Extremely fast diffusion system – nickel in single crystal tin. Phys Rev Lett 1984;53:1469–1472.
10. Dyson BF, Anthony TR, Turnbull D. Interstitial diffusion of Cu in tin. J Appl Phys 1967;38:3408.
11. Suh J-O, Tu KN, Tamura N. Dramatic morphological change of scallop-type Cu_6Sn_5 formed on (001) single crystal copper in reaction between molten SnPb solder and Cu. Appl Phys Lett 2007;91:051907.
12. Zou HF, Yang HJ, Zhang ZF. Morphologies, orientation relationships and evolution of Cu_6Sn_5 grains formed between molten Sn and Cu single crystals. Acta Mater 2008;56:2649–2662.
13. Hsiao H-Y, Liu C-M, Lin H-W, Liu T-C, Lu C-L, Huang Y-S, Chen C, Tu KN. Unidirectional growth of microbumps on (111)-oriented and nanotwinned copper. Science 2012;336:1007–1010.
14. Chen K-C, Wu W-W, Liao C-N, Chen L-J, Tu KN. Observation of atomic diffusion at twin-modified grain boundaries in copper. Science 2008;321:1066–1069.

PROBLEMS

10.1. What are the four classical methods of strengthening a metal? Give an example for each of them.

10.2. What is the difference between a twin plane and a stacking fault in terms of energy and structure inface-centered cubic metals?

10.3. Can silicon have twin plane? If so, describe the twin plane structure in silicon.

10.4. What is the difference between a coherent twin plane and an incoherent twin plane?

10.5. When a twin having two twin planes meets a free surface, a ridge and a valley are formed. Discuss the configuration of the ridge. Can it be an equilibrium configuration?

10.6. Gold foils can be hammered down to nanometer thickness and have been widely used as coating materials for decoration. What is the microstructure of the ultrathin Au foils? Why does it have such good ductility?

10.7. A 300 nm oxide thin film is deposited on a 500 μm thick bare Si wafer that has a radius of curvature of 300 m before the deposition. After deposition the radius of curvature has changed to 300 m. A 600 nm nitride film is subsequently deposited on the oxide and the radius of curvature is 240 m. Calculate the dual film stress and the stress of the nitride film. ($\nu_{Si} = 0.272$, $Y_{Si} = 1.9 \times 10^{12}\,\text{dyne/cm}^2$)

10.8. Stress in a film on a wafer can be determined by the amount of bow of the wafer. The bow can be measured by a surface profileometer with an 18 cm scan length.

(1) Show that the Stoney equation is equivalent to the equation below

$$\sigma = \left(\frac{\delta}{3\rho^2}\right)\left(\frac{Y}{1-\nu}\right)\left(\frac{t_S^2}{t_f}\right)$$

where δ is the maximum bow height of the profileometer scan and ρ is half the scan length.

(2) Given a scan length of 5 cm and a bow of 2000 nm, calculate the stress for an unknown film of 2 μm thick. The wafer is a 200 μm thick (100) Si wafer and we have $Y/(1-\nu) = 1.8 \times 10^{11}\,\text{N/m}^2$ for the Si.

APPENDIX A

LAPLACE PRESSURE IN NONSPHERICAL NANOPARTICLE

We consider a cubic particle, having an edge length of h, which remains cubic upon growth or shrinkage (due to anisotropy of crystal). There is no definite curvature for all surfaces: for flat facets the curvature is zero (radius tends to infinity), but at the edges (junctions of facets) the curvature tends to infinity (radius formally tends to zero, and actually to atomic size). Nevertheless, we can introduce effective Laplace pressure and surface input into the chemical potential of the cubic particle. Indeed, let the atom of atomic volume Ω "smashed" over the whole surface of the cube while retaining the cubic shape of the precipitate: $\Omega = \Delta(h^3) = 3h^2\Delta h$, so that $\Delta h = \Omega/3h^2$. The change of surface energy will be $\Delta E^{\text{surf}} = \gamma \cdot \Delta(6h^2) = 12\gamma h\Delta h = 12\gamma h \cdot (\Omega/3h^2) = 4\gamma\Omega/h = 2\gamma\Omega/(h/2)$. So we can treat the ratio $\Delta E^{\text{surf}}/\Omega = 2\gamma/(h/2)$ as a Laplace pressure. In other words, $p^{\text{Laplace}} = \partial E^{\text{surf}}/\partial V = \partial(\gamma \cdot 6h^2)/\partial(h^3) = 12\gamma hdh/3h^2dh = 2\gamma/(h/2)$.

Next, we consider a more complicated example: a circular disk of radius R and height h, with two different surface tensions – $\gamma_{//}$ for the top and bottom and γ_{side} for the sidewall. We introduce the shape parameter, the aspect ratio $\phi \equiv h/R$. We assume that the kinetics of surface diffusion is fast enough to provide the minimization of surface energy at each given volume. $E^{\text{surf}} = \gamma_{//} \cdot 2 \cdot \pi R^2 + \gamma_{\text{side}} \cdot 2\pi Rh$. At that, the volume $V = \pi R^2 h = \phi \cdot \pi R^3 \Rightarrow R = (V/\pi)^{1/3} \cdot \phi^{-1/3}$, $h = \phi R = (V/\pi)^{1/3} \cdot \phi^{2/3}$. Thus, the total surface energy, as a function of volume and of shape parameter, $E^{\text{surf}} = \gamma_{//} \cdot 2 \cdot \pi R^2 + \gamma_{\text{side}} \cdot 2\pi Rh = 2\pi(\gamma_{//}(V/\pi)^{2/3} \cdot \phi^{-2/3} + \gamma_{\text{side}}(V/\pi)^{2/3} \cdot \phi^{1/3})$, can be optimized over the shape at each fixed volume: $\partial E^{\text{surf}}/\partial\phi = 0 = 2\pi(V/\pi)^{2/3}(-(2/3)\gamma_{//} \cdot \phi^{-5/3} + (1/3)\gamma_{\text{side}} \cdot \phi^{-2/3}) \Rightarrow \phi^{\text{opt}} = 2\gamma_{//}/\gamma_{\text{side}}$ (a particular case of the Wulff rule). Thus, the optimized surface energy at fixed volume is equal to $E^{\text{surf}}_{\text{optimal}} = 2\pi(V/\pi)^{2/3}(\gamma_{//} \cdot (\gamma_{\text{side}}/2\gamma_{//})^{2/3} + \gamma_{\text{side}} \cdot (2\gamma_{//}/\gamma_{\text{side}})^{1/3}) = 3(2\pi\gamma_{//}\gamma^2_{\text{side}})^{1/3}V^{2/3}$.

Thus, in the case of unrestricted shape optimization, the chemical potential should have the following Gibbs–Thomson input: $\Delta\mu^{\text{GT}} = \Omega(\partial E^{\text{surf}}_{\text{optim}}/\partial V) = 2(2\pi\gamma_{//}\gamma^2_{\text{side}}/V)^{1/3}\Omega$, with $\partial E^{\text{surf}}_{\text{optim}}/\partial V = 2(2\pi\gamma_{//}\gamma^2_{\text{side}}/V)^{1/3}$ being the Laplace pressure. It is very interesting that when we express volume V in terms of optimal shape parameter and optimal radius, $V = \phi^{\text{opt}} \cdot \pi R^3_{\text{opt}} = (2\gamma_{//}/\gamma_{\text{side}})\pi R^3_{\text{opt}}$, we obtain $p^{\text{Laplace}} = 2(2\pi\gamma_{//}\gamma^2_{\text{side}}/V)^{1/3} = 2((2\pi\gamma_{//}\gamma^2_{\text{side}})/((2\gamma_{//}/\gamma_{\text{side}})\pi R^3_{\text{opt}}))^{1/3} = 2\gamma_{\text{side}}/R_{\text{opt}}$.

Kinetics in Nanoscale Materials, First Edition. King-Ning Tu and Andriy M. Gusak.

Finally, if we assume the shape is not optimized fast enough, and the facets are still competing between each other, we can introduce the chemical potential for each facet. It should contain the total increase of surface energy due to smashing of an additional atom only at this very facet. In such a case, we should take into account that the elementary movement of each facet generates the change of area not only of this very facet, but also the area change of neighboring facets. To simplify the situation, we consider the previous example of disk but without automatic shape optimization. If an additional atom is smashed, say, just over the top surface, the top surface area does not change (because of rectangular disc); instead the sidewall area increases by $2\pi R dh = 2\pi R(\Omega/\pi R^2) = (2\Omega/R)$ and surface energy increases, respectively, by $\gamma_{\text{side}}(2\Omega/R)$; $\Delta\mu_{\text{top}} = \gamma_{\text{side}}(2\Omega/R)$. We emphasize that the surface input to the chemical potential of top atoms (and all the atoms) is determined by the sidewall surface tension. On the other hand, if we smash an additional atom only over the sidewall, $\Omega = h \cdot 2\pi R dR \Rightarrow dR = \Omega/2\pi hR$, it leads to an increase in both the sidewall area ($2\pi h dR = 2\pi h(\Omega/2\pi Rh) = \Omega/R$) as well as energy ($\gamma_{side}(\Omega/R)$) and the top and bottom areas ($2 \cdot 2\pi R dR = 4\pi R(\Omega/2\pi Rh) = 2\Omega/h$) as well as their energy ($\gamma_{//}(2\Omega/h)$). Thus, the input of surface to the chemical potential $\Delta\mu_{\text{side}} = \gamma_{\text{side}}(\Omega/R) + \gamma_{//}(2\Omega/h)$. We can check whether this result is reasonable. Indeed, at the equilibrium shape, the abovementioned chemical potentials should be equal: $\gamma_{\text{side}}(2\Omega/R) = \gamma_{\text{side}}(\Omega/R) + \gamma_{//}(2\Omega/h)$. This gives $\gamma_{\text{side}}(\Omega/R) = \gamma_{//}(2\Omega/h) \Rightarrow h/R = 2\gamma_{//}/\gamma_{\text{side}}$, which coincides with the optimal aspect ratio obtained in the above.

APPENDIX B

INTERDIFFUSION COEFFICIENT $\tilde{D} = C_B M G''$

The marker velocity can be given as,

$$v = (D_B - D_A)\frac{\partial X_B}{\partial x} = D_B\frac{\partial X_B}{\partial x} + D_A\frac{\partial X_A}{\partial x}$$

By substituting v into the equation of J_B, we have

$$J_B = j_B + C_B\left(D_B\frac{\partial X_B}{\partial x} + D_A\frac{\partial X_A}{\partial x}\right) = j_B - X_B(j_B + j_A)$$

In the analysis above, we have presented flux in terms of concentration gradient and we obtain a pair of equations of marker velocity v and interdiffusion coefficient, so that we can calculate the intrinsic diffusion coefficient of D_A and D_B. All the diffusion coefficients indicate that atomic diffusion are going with concentration gradient, that is, from high concentration to low concentration, However, in spinodal decomposition, it is against concentration gradient; the diffusion coefficient becomes negative in the field of concentration. To verify this, we recall that diffusion should be driven by chemical potential gradient. Below we present the atomic flux in terms of chemical potential gradient. We have

$$j = C < v >= CMF = CM\left(-\frac{\partial \mu}{\partial x}\right)$$

where μ is the chemical potential in the alloy and M is the mobility. We use chemical potential gradient instead of concentration gradient as the driving force of interdiffusion. We take

$$j_B = -C_B M_B\frac{\partial \mu_B}{\partial x} = -CX_B M_B\frac{\partial \mu_B}{\partial x}$$

$$j_A = -C_A M_A\frac{\partial \mu_A}{\partial x} = -CX_A M_A\frac{\partial \mu_A}{\partial x}$$

By substituting j_B and j_A into the equation of J_B, we have

$$J_B = -CX_B M_B \frac{\partial \mu_B}{\partial x} + CX_B \left[X_B M_B \frac{\partial \mu_B}{\partial x} + (1 - X_B) M_A \frac{\partial \mu_A}{\partial x} \right]$$

$$= -C \left[X_B M_B \frac{\partial \mu_B}{\partial x} - X_B^2 M_B \frac{\partial \mu_B}{\partial x} - (1 - X_B) M_A \frac{\partial \mu_A}{\partial x} \right]$$

$$= -C \left\{ X_B (1 - X_B) \left[M_B \frac{\partial \mu_B}{\partial x} - M_A \frac{\partial \mu_A}{\partial x} \right] \right\}$$

From Gibbs–Duhem equation, we have

$$X_A d\mu_A + X_B d\mu_B = 0$$

$$(1 - X_B) d\mu_A + X_B d\mu_B = 0$$

Thus we have

$$(1 - X_B) M_A \frac{d\mu_A}{dx} + X_B M_A \frac{d\mu_B}{dx} = 0 \quad \text{(B.1)}$$

$$(1 - X_B) M_B \frac{d\mu_A}{dx} + X_B M_B \frac{d\mu_B}{dx} = 0 \quad \text{(B.2)}$$

Now if we add Eq. (B.1) and subtract Eq. (B.2) in the bracket of J_B, in other words, if we just add a zero and subtract a zero from the bracket, we have

$$M_B \frac{\partial \mu_B}{\partial x} + (1 - X_B) M_A \frac{\partial \mu_A}{\partial x} + X_B M_A \frac{\partial \mu_B}{\partial x}$$

$$- M_A \frac{\partial \mu_A}{\partial x} - (1 - X_B) M_B \frac{\partial \mu_B}{\partial x} - X_B M_B \frac{\partial \mu_B}{\partial x}$$

$$= M_B \frac{\partial \mu_B}{\partial x} + M_A \frac{\partial \mu_A}{\partial x} - X_B M_A \frac{\partial \mu_A}{\partial x} + X_B M_A \frac{\partial \mu_B}{\partial x}$$

$$- M_A \frac{\partial \mu_A}{\partial x} - M_B \frac{\partial \mu_A}{\partial x} + X_B M_B \frac{\partial \mu_A}{\partial x} - X_B M_B \frac{\partial \mu_B}{\partial x}$$

$$= M_B \left(\frac{\partial \mu_B}{\partial x} - \frac{\partial \mu_A}{\partial x} \right) + X_B (M_A - M_B) \left(\frac{\partial \mu_B}{\partial x} - \frac{\partial \mu_A}{\partial x} \right)$$

$$= [(1 - X_B) M_B + X_B M_A] \left(\frac{\partial \mu_B}{\partial x} - \frac{\partial \mu_A}{\partial x} \right)$$

Thus we have

$$J_B = -C[X_B(1 - X_B)][(1 - X_B) M_B + X_B M_A] \left(\frac{\partial \mu_B}{\partial x} - \frac{\partial \mu_A}{\partial x} \right)$$

$$= -CX_B M \frac{\partial}{\partial x} (\mu_B - \mu_A)$$

where $M = (1 - X_B)[(1 - X_B) M_B + X_B M_A] = X_A [X_A M_B + X_B M_A]$.

By virtue of the definition of chemical potential,

$$dG = \mu_A dC_A + \mu_B dC_B = (\mu_B - \mu_A) dC_B$$

Or we have

$$\frac{dG}{dC_B} = \mu_B - \mu_A$$

Thus $J_B = -CX_B M(\partial/\partial y)(\partial G/\partial C_B) = -C_B M((\partial^2 G/\partial C_B^2)(\partial C_B/\partial y))$
$= -C_B MG''(\partial C_B/\partial x)$

Or we have $-J_B/(\partial C_B/\partial x) = C_B MG'' = \tilde{D}$, which is the interdiffusion coefficient.

We recall that we have already obtained $\tilde{D} = X_A D_B + X_B D_A$ as the interdiffusion coefficient. Also, as shown above, $\tilde{D} = C_B MG''$, so the interdiffusion coefficient takes the sign of G''. We note that G'' is negative within the spinodal region, so $\tilde{D}$ is negative, indicating that the diffusion is against concentration gradient, that is, an uphill diffusion. Outside the spinodal region, G'' is positive and the interdiffusion coefficient is positive as in the Darken's analysis.

APPENDIX C

NONEQUILIBRIUM VACANCIES AND CROSS-EFFECTS ON INTERDIFFUSION IN A PSEUDO-TERNARY ALLOY

One of the key assumptions in Darken's analysis of interdiffusion is the equilibrium vacancy concentration everywhere and forever. Physically, such assumption can be realized under a large density and intensity of vacancy sinks and sources (such as dislocation kinks, grain boundaries, and external surfaces). For example, it is reported recently that on a Cu having a high density of nanotwins, no Kirkendall void formation was observed in Cu–Sn reactions. Yet, as a quantitative characteristic for both density and effectiveness of vacancy sinks/sources, we may consider the "free length" of vacancies, L_V, and the corresponding relaxation time for vacancies, $\tau_V = L_V^2/D_V$, where D_V is the self-diffusion coefficient of vacancies. With respect to the "free length" of vacancies, Darken's assumption is realized, obviously, only at the coarse spatial scale of the diffusion processes, which means a physically large volume having a size much larger than L_V. In this case sinks/sources can be considered distributed uniformly all over the volume and quite effective. But if we investigate diffusion in a nanodiffusion zone of thickness 100 nm and "free length" $L_V \approx 1000$ nm, it is clear that we cannot apply Darken's assumption, and we should treat vacancies as an additional independent component.

The distribution of nonequilibrium vacancies must be taken into account in most diffusion-controlled processes with characteristic sizes (diffusion zone width, phase layer thickness, distance between precipitates, interlamellar distance, etc.) commensurate with vacancies' free length, and (or) with characteristic times commensurate with the relation of vacancies in relaxation time to vacancy concentration.

For example, in nanoshells, vacancies are in equilibrium only at the external and internal surfaces. Since these equilibria correspond to different vacancy concentrations, there should be a vacancy gradient inside the shell, in between these surfaces. Moreover, sinks and sources of vacancies are at the mentioned surfaces, and there are practically no effective vacancy sinks/sources within the nanoshell. Thus, within the nanoshell the vacancies are not at equilibrium and we should treat them as one more

Kinetics in Nanoscale Materials, First Edition. King-Ning Tu and Andriy M. Gusak.

component, V, besides the main components of A and B. So, with the nonequilibrium vacancies, the binary alloy AB should be treated as a pseudo-ternary alloy system of A–B–V. In the ternary system, two out of the three concentrations are independent and two out of the three fluxes are also independent:

$$X_A = 1 - X_B - X_V \ \ (\text{or}\ \ C_A = 1/\Omega - C_B - C_V)$$
$$j_A = -j_B - j_V$$

where X_B is the atomic fraction of B, from 0 to 1, C_B is the number of B atoms per unit volume, and Ω is the atomic volume. The same goes for A atoms as well.

The major difference in interdiffusion between binary and multicomponent (in particular, ternary) systems is the existence of cross-effects, as in irreversible processes, between fluxes and concentration gradients of different species. The flux of each species is determined not only by the concentration gradient of this very species, but also by the concentration gradient of another independent component. For example, in the pseudo-ternary system, we have

$$\begin{aligned} j_B &= -D_B \frac{\partial C_B}{\partial x} - D_{BV} \frac{\partial C_V}{\partial x} \\ j_V &= -D_V \frac{\partial C_V}{\partial x} - D_{VB} \frac{\partial C_B}{\partial x} \end{aligned} \tag{C.1}$$

In order to understand the physical meaning of direct terms and cross-terms and to analyze these processes mathematically, we need formulae for the components fluxes, taking into account nonequilibrium vacancies distribution. We illustrate below how these formulae are obtained for the simplest example of a cubic lattice.

Let concentration gradients of main components and vacancies of $(\partial C_B/\partial x), (\partial C_V/\partial x), (\partial C_A/\partial x) = -(\partial C_B/\partial x) - (\partial C_V/\partial x)$ be created in a simple cubic lattice along the <100> (x-axis) direction. Since vacancy concentration is much less than that of the main components, we have $\partial C_A/\partial x = -\partial C_B/\partial x$. Then, we can find, for instance, the flux of B atoms jumping across the saddle plane that is located at the position of x in the x-axis, as shown in Figure 3.12; the saddle plane is located between the two parallel atomic planes $x - d/2,\ \ x + d/2$, where d stands for interplanar spacing, which is equal to lattice parameter a in this very case.

To consider diffusion across the saddle plane, we suppose ν_{BV} is the frequency (probability per **unit time**) of a B atom jumping into a neighboring vacant **site** on the condition that it is vacant indeed. This frequency depends on the main components concentration in the place from where the atom jumps: $\nu_{BV} = \nu_{BV}(C_B(x \pm d/2))$. For the atom to jump, there must be a neighboring vacancy. A priori probability of this is that it is equal to vacant sites fraction that depends on the coordinate as well: $X_V = \Omega C_V = \Omega C_V(x \mp d/2)$. The number of B atoms in one plane (per unit of area) being able to jump through the **saddle** plane at x is equal to $1\,\text{cm}^2 \cdot d \cdot C_B(x \pm d/2)$, where $1\,\text{cm}^2$ times d is equal to volume, and C is the number of atoms per unit volume. Hence, the resulting atomic flux of B, which is defined as the difference between jumps quantity (per **unit time**) from the left plane to the right and from the right plane to the left, equals

$$j_{\mathrm{B}}(x) = d \cdot C_{\mathrm{B}}\left(x - \frac{d}{2}\right) \cdot \Omega C_{\mathrm{V}}\left(x + \frac{d}{2}\right) \nu_{\mathrm{BV}}\left(C_{\mathrm{B}}\left(x - \frac{d}{2}\right)\right) - d \cdot C_{\mathrm{B}}\left(x + \frac{d}{2}\right) \cdot \Omega C_{\mathrm{V}}\left(x - \frac{d}{2}\right) \nu_{\mathrm{BV}}\left(C_{\mathrm{B}}\left(x + \frac{d}{2}\right)\right) \quad \text{(C.2)}$$

In the case when all gradients being considered are not too large, we may neglect higher order terms of interplanar spacing at Taylor expansion:

$$\nu_{\mathrm{BV}}\left(x \pm \frac{d}{2}\right) C_{\mathrm{B}}\left(x \pm \frac{d}{2}\right) = \nu_{\mathrm{BV}}(x) C_{\mathrm{B}}(x) \pm \frac{d}{2} \frac{\partial(\nu_{\mathrm{BV}} C_{\mathrm{B}})}{\partial x}$$

$$C_{\mathrm{V}}\left(x \pm \frac{d}{2}\right) = C_{\mathrm{V}}(x) \pm \frac{d}{2} \frac{\partial C_{\mathrm{V}}}{\partial x}$$

Otherwise, we will come to nonlinear theory (see Chapter 2). Then by substituting them into Eq. (C.2), we have the following expression for the flux:

$$j_{\mathrm{B}}(x) = -d^2 X_{\mathrm{V}} \cdot \frac{\partial(\nu_{\mathrm{BV}} C_{\mathrm{B}})}{\partial x} + d^2 \nu_{\mathrm{BV}} \Omega C_{\mathrm{B}} \frac{\partial C_{\mathrm{V}}}{\partial x}$$

If we assume that the exchange frequency depends on the coordinate **implicitly**, that is, through concentration, we obtain

$$\frac{\partial(\nu_{\mathrm{BV}} C_{\mathrm{B}})}{\partial x} = \nu_{\mathrm{BV}} \frac{\partial C_{\mathrm{B}}}{\partial x} + C_{\mathrm{B}} \frac{\partial(\nu_{\mathrm{BV}})}{\partial x} = \nu_{\mathrm{BV}} \frac{\partial C_{\mathrm{B}}}{\partial x} + C_{\mathrm{B}} \frac{\partial(\nu_{\mathrm{BV}})}{\partial C_{\mathrm{B}}} \frac{\partial C_{\mathrm{B}}}{\partial x}$$

$$= \left(\nu_{\mathrm{BV}} + \frac{\partial(\nu_{\mathrm{BV}})}{\partial \ln C_{\mathrm{B}}}\right) \frac{\partial C_{\mathrm{B}}}{\partial x} = \nu_{\mathrm{BV}} \left(1 + \frac{\partial(\ln \nu_{\mathrm{BV}})}{\partial \ln C_{\mathrm{B}}}\right) \frac{\partial C_{\mathrm{B}}}{\partial x}$$

Hence,

$$j_{\mathrm{B}}(x) = -d^2 X_{\mathrm{V}} \cdot \nu_{\mathrm{BV}} \left(1 + \frac{\partial(\ln \nu_{\mathrm{BV}})}{\partial \ln C_{\mathrm{B}}}\right) \frac{\partial C_{\mathrm{B}}}{\partial x} + d^2 \nu_{\mathrm{BV}} \Omega C_{\mathrm{B}} \frac{\partial C_{\mathrm{V}}}{\partial x} \quad \text{(C.3)}$$

Comparing Eq. (C.3) with Eq. (C.1), we conclude that

$$D_{\mathrm{B}} = d^2 X_{\mathrm{V}} \cdot \nu_{\mathrm{BV}} \left(1 + \frac{\partial(\ln \nu_{\mathrm{BV}})}{\partial \ln C_{\mathrm{B}}}\right), \quad D_{\mathrm{BV}} = -d^2 \nu_{\mathrm{BV}} \Omega C_{\mathrm{B}} \quad \text{(C.4)}$$

The $d^2 X_{\mathrm{V}} \cdot \nu_{\mathrm{BV}}$ is the value that represents diffusion coefficient of tracer atoms B in alloy: $D_{\mathrm{B}}^* = d^2 X_{\mathrm{V}} \cdot \nu_{\mathrm{BV}}$. Thus, we can write the flux equation in the following form:

$$j_{\mathrm{B}}(x) = -D_{\mathrm{B}}^* \left(1 + \frac{\partial(\ln \nu_{\mathrm{BV}})}{\partial \ln C_{\mathrm{B}}}\right) \frac{\partial C_{\mathrm{B}}}{\partial x} + \frac{D_{\mathrm{B}}^* X_{\mathrm{B}}}{X_{\mathrm{V}}} \frac{\partial C_{\mathrm{V}}}{\partial x}$$

or

$$\Omega j_{\mathrm{B}}(x) = -D_{\mathrm{B}}^{*}\left(1 + \frac{\partial\left(\ln \nu_{\mathrm{BV}}\right)}{\partial \ln X_{\mathrm{B}}}\right)\frac{\partial X_{\mathrm{B}}}{\partial x} + \frac{D_{\mathrm{B}}^{*} X_{\mathrm{B}}}{X_{\mathrm{V}}}\frac{\partial X_{\mathrm{V}}}{\partial x} \tag{C.5}$$

The factor $(1 + (\partial(\ln \nu_{\mathrm{BV}})/\partial \ln X_{\mathrm{B}}))$ is usually interpreted as a thermodynamic factor in Darken's theory, which is equal to $\phi = (X_{\mathrm{B}}/kT)(\partial \mu_{\mathrm{B}}/\partial X_{\mathrm{B}}) = (X_{\mathrm{A}}/kT)(\partial \mu_{\mathrm{A}}/\partial X_{\mathrm{A}}) = (X_{\mathrm{A}} X_{\mathrm{B}}/kT)(\partial^2 g/\partial X_{\mathrm{B}}^2)$.

Thus, the cross-effects in binary systems with nonzero vacancy concentrations are described by the following equations for the fluxes of the main components and vacancies:

$$\Omega j_{\mathrm{A}}(x) = -D_{\mathrm{A}}^{*}\phi\frac{\partial X_{\mathrm{A}}}{\partial x} + \frac{D_{\mathrm{A}}^{*} X_{\mathrm{A}}}{X_{\mathrm{V}}}\frac{\partial X_{\mathrm{V}}}{\partial x} \tag{C.6A}$$

$$\Omega j_{\mathrm{B}}(x) = -D_{\mathrm{B}}^{*}\phi\frac{\partial X_{\mathrm{B}}}{\partial x} + \frac{D_{\mathrm{B}}^{*} X_{\mathrm{B}}}{X_{\mathrm{V}}}\frac{\partial X_{\mathrm{V}}}{\partial x} \tag{C.6B}$$

$$\Omega j_{\mathrm{V}}(x) = -\left(\Omega j_{\mathrm{A}}(x) + \Omega j_{\mathrm{B}}(x)\right) = \left(D_{\mathrm{B}}^{*} - D_{\mathrm{A}}^{*}\right)\phi\frac{\partial X_{\mathrm{B}}}{\partial x} - \frac{D_{\mathrm{A}}^{*} X_{\mathrm{A}} + D_{\mathrm{B}}^{*} X_{\mathrm{B}}}{X_{\mathrm{V}}}\frac{\partial X_{\mathrm{V}}}{\partial x} \tag{C.6V}$$

These flux equations are written in the lattice reference frame that, under the absence of working vacancy sinks/sources, coincide with the laboratory (Boltzmann–Matano) frame. The combination of $(D_{\mathrm{A}}^{*} X_{\mathrm{A}} + D_{\mathrm{B}}^{*} X_{\mathrm{B}})/X_{\mathrm{V}}$ before the vacancy gradient plays the role of vacancy self-diffusion coefficient D_{V} ("Frenkel relationship").

Thus,

$$\Omega j_{\mathrm{V}}(x) = (D_{\mathrm{B}}^{*} - D_{\mathrm{A}}^{*})\phi\frac{\partial X_{\mathrm{B}}}{\partial x} - D_{\mathrm{V}}\frac{\partial X_{\mathrm{V}}}{\partial x} \tag{C.7}$$

APPENDIX D

INTERACTION BETWEEN KIRKENDALL EFFECT AND GIBBS–THOMSON EFFECT IN THE FORMATION OF A SPHERICAL COMPOUND NANOSHELL

In the reaction of a concentric (spherical) or coaxial (cylindrical) nanoshell of a bilayer of A and B in forming an intermetallic compound of ApBq, we need to consider the interaction between Kirkendall effect and Gibbs–Thomson effect. Experimentally, the formation of nanoshells of CoO and Co_3S_4 by annealing nanospheres of Co in oxygen and sulfur ambient, respectively, has been reported. Kinetic analysis of the formation is presented here.

We analyze the interdiffusion in a pair of concentric spherical nanoshells of A and B to form an intermetallic compound layer between them, where A is the inside layer and B is the outside layer, as shown in Figure 3.10. We assume spherical symmetry and steady state process in analyzing the formation. Experimentally, we take the example of Co_3S_4 compound formation during the annealing of nanospheres of Co in sulfur ambient, in which A is the metal Co and B is the sulfur. To simplify the analysis, we assume that the total fluxes of A (diffuses out) and of B (diffuses in) through the compound layer do not depend on the distance from the center:

$$\Omega J_{\mathrm{B}}^{\mathrm{tot}} = 4\pi R^2 \Omega j_{\mathrm{B}}(R) = \text{const over } R$$
$$\Omega J_{\mathrm{A}}^{\mathrm{tot}} = 4\pi R^2 \Omega j_{\mathrm{A}}(R) = \text{const over } R \tag{D.1}$$

Here each flux density of $j_{\mathrm{B}}, j_{\mathrm{A}}$ can be expressed in terms of concentration times velocity ($j = Cv$), and, in turn, the velocity ($v = MF$) can be expressed in terms of mobility times chemical driving force (which is equal to the minus gradient of the corresponding chemical potential).

Kinetics in Nanoscale Materials, First Edition. King-Ning Tu and Andriy M. Gusak.

$$\Omega j_{\mathrm{B}}(R) = +X_{\mathrm{B}} v_{\mathrm{B}} = -X_{\mathrm{B}} \frac{D_{\mathrm{B}}^{*}}{kT} \frac{\partial \mu_{\mathrm{B}}}{\partial R} \tag{D.2a}$$

$$\Omega j_{\mathrm{A}}(R) = +X_{\mathrm{A}} v_{\mathrm{A}} = -X_{\mathrm{A}} \frac{D_{\mathrm{A}}^{*}}{kT} \frac{\partial \mu_{\mathrm{A}}}{\partial R} \tag{D.2b}$$

For almost all stoichiometric compounds, we can further take atomic fractions $X_{\mathrm{A}}, X_{\mathrm{B}}$ of B and A practically to be constant through the layer.

Now we make a mathematical rearrangement by using the constancy of the total fluxes over radius; we can multiply and divide simultaneously by the same integral $\int_{r_{\mathrm{i}}}^{r_{\mathrm{e}}} dR/(R^2)$, and use the possibility of taking a constant factor from outside into the integrand in Eq. (3.42):

$$\Omega J_{\mathrm{B}}^{\mathrm{tot}} = \frac{\int_{r_{\mathrm{i}}}^{r_{\mathrm{e}}} (\Omega J_{\mathrm{B}}^{\mathrm{tot}} dR)/R^2}{\int_{r_{\mathrm{i}}}^{r_{\mathrm{e}}} dR/R^2} = \frac{-4\pi X_{\mathrm{B}} D_{\mathrm{B}}^{*}}{kT \int_{r_{\mathrm{i}}}^{r_{\mathrm{e}}} dR/R^2} \int_{r_{\mathrm{i}}}^{r_{\mathrm{e}}} \frac{\partial(\mu_{\mathrm{B}})}{\partial R} dR$$

$$= -4\pi X_{\mathrm{B}} D_{\mathrm{B}}^{*} \frac{\mu_{\mathrm{B}}(r_{\mathrm{e}}) - \mu_{B}(r_{\mathrm{i}})}{kT((1/r_{\mathrm{i}}) - (1/r_{\mathrm{e}}))} \tag{D.3}$$

Thus, the total B-flux through the growing layer of compound (or through the outside interface of the compound layer) is

$$\Omega J_{\mathrm{B}}^{\mathrm{tot}} = -4\pi X_{\mathrm{B}} D_{\mathrm{B}}^{*} r_{\mathrm{i}} r_{\mathrm{e}} \frac{\mu_{\mathrm{B}}(r_{\mathrm{e}}) - \mu_{\mathrm{B}}(r_{\mathrm{i}})}{kT(r_{\mathrm{e}} - r_{\mathrm{i}})} \tag{D.4a}$$

Similarly, the total A-flux through the growing layer of compound (or through the inside interface of the compound layer) is

$$\Omega J_{\mathrm{A}}^{\mathrm{tot}} = -4\pi X_{\mathrm{A}} D_{\mathrm{A}}^{*} r_{\mathrm{i}} r_{\mathrm{e}} \frac{\mu_{\mathrm{A}}(r_{\mathrm{e}}) - \mu_{\mathrm{A}}(r_{\mathrm{i}})}{kT(r_{\mathrm{e}} - r_{\mathrm{i}})} \tag{D.4b}$$

Now we should write the condition of the flux balance for B or the metal (Co) at the outer boundary of the compound layer (since metal does not go outside of the nanoshell) and the flux balance for A or sulfur at the inner boundary of the compound layer (since sulfur does not go into the void).

$$4\pi r_{\mathrm{e}}^2 \frac{dr_{\mathrm{e}}}{dt} (X_{\mathrm{B}} - 0) = \Omega J_{\mathrm{B}}^{\mathrm{tot}} - 0 \tag{D.5}$$

$$4\pi r_{\mathrm{i}}^2 \frac{dr_{\mathrm{i}}}{dt} (X_{\mathrm{A}} - 0) = \Omega J_{\mathrm{A}}^{\mathrm{tot}} - 0 \tag{D.6}$$

Thus,

$$\frac{dr_{\mathrm{e}}}{dt} = D_{\mathrm{B}}^{*} \frac{r_{\mathrm{i}}}{r_{\mathrm{e}}} \frac{\mu_{\mathrm{B}}(r_{\mathrm{i}}) - \mu_{\mathrm{B}}(r_{\mathrm{e}})}{kT(r_{\mathrm{e}} - r_{\mathrm{i}})} \tag{D.7}$$

$$\frac{dr_i}{dt} = D_A^* \frac{r_e}{r_i} \frac{\mu_A(r_i) - \mu_A(r_e)}{kT(r_e - r_i)} \tag{D.8}$$

From standard thermodynamics, we have from Chapter 7, Section 7.2.4 that

$$\mu_B = g + X_A \frac{\partial g}{\partial X_B}, \quad \mu_A = g - X_B \frac{\partial g}{\partial X_B} \tag{D.9}$$

So,

$$\begin{aligned} \mu_B(r_i) - \mu_B(r_e) &= g(r_i) + X_A \left.\frac{\partial g}{\partial X_B}\right|_{ri} - g(r_e) - X_A \left.\frac{\partial g}{\partial X_B}\right|_{re} \\ &\cong X_A \left(\left.\frac{\partial g}{\partial X_B}\right|_{ri} - \left.\frac{\partial g}{\partial X_B}\right|_{re} \right) + (g(r_i) - g(r_e)) \end{aligned} \tag{D.10}$$

In Eqs (D.7) and (D.8), we should use Gibbs–Thomson potential for calculating chemical potentials at the curved boundaries. Gibbs energy per atom at the **internal** boundary is **lower** by Gibbs–Thomson term, and at **external** boundary is **higher** by Gibbs–Thomson term:

$$g(r_i) = g^{\text{compound}} - \frac{2\gamma\Omega}{r_i}, \quad g(r_e) = g^{\text{compound}} + \frac{2\gamma\Omega}{r_e}, \quad \text{so that}$$

$$g(r_i) - g(r_e) = -2\gamma\Omega \left(\frac{1}{r_i} + \frac{1}{r_e} \right) \tag{D.11}$$

Here we take into account that for an almost stoichiometric compound its composition is practically the same at both sides ($X_A(r_i) \cong X_A(r_e) = X_A^{\text{compound}}$), but the first derivatives of Gibbs energy are very different. They are determined by the common tangent rule (at inner boundary due to equilibrium between compound and remaining metallic core B; at external boundary due to equilibrium between compound and surrounding medium (in our case, sulfur A):

$$\begin{aligned} \left.\frac{\partial g}{\partial X_B}\right|_{ri} - \left.\frac{\partial g}{\partial X_B}\right|_{re} &= \frac{g_B - \left(g^{\text{compound}} - (2\gamma\Omega/r_i)\right)}{1 - X_B^{\text{compound}}} - \frac{\left(g^{\text{compound}} + (2\gamma\Omega/r_e)\right) - g_A}{X_B^{\text{compound}} - 0} \\ &= \frac{\Delta g}{X_B X_A} + 2\gamma\Omega \left(\frac{1}{X_A r_i} - \frac{1}{X_B r_e} \right) \end{aligned} \tag{D.12}$$

Here Δg is a Gibbs free energy (per atom) of compound formation from the mixture of pure A and B in necessary proportion (see Fig. 3.10). In other words, it is the thermodynamic driving force of solid state reaction to form the compound phase. Combining Eqs (D.11) and (D.12), we obtain for B component,

$$\mu_B(r_i) - \mu_B(r_e) = \frac{1}{X_B} \left(\Delta g - \frac{2\gamma\Omega}{r_e} \right) \tag{D.13a}$$

Similarly, we obtain for A component,

$$\mu_A(r_i) - \mu_A(r_e) = \frac{1}{X_A}\left(-\Delta g - \frac{2\gamma\Omega}{r_i}\right) \tag{D.13b}$$

Physically, we can understand Eq. (D.13) by referring to interdiffusion in a bulk sample of A/B to form an intermetallic compound between them. The chemical potential difference across the compound layer is Δg. In the case of nanospherical shells, the chemical potential is modified by $2\gamma\Omega/r$ at the inner as well as the outer surface of the shell. Now we can go back to the growth rate Eqs (D.7) and (D.8) and substitute the calculated driving forces in Eq. (D.13),

$$\frac{dr_e}{dt} = D_B^* \frac{r_i}{r_e} \frac{\Delta g - (2\gamma\Omega/r_e)}{X_B kT(r_e - r_i)} \tag{D.14}$$

$$\frac{dr_i}{dt} = -D_A^* \frac{r_e}{r_i} \frac{\Delta g + (2\gamma\Omega/r_i)}{X_A kT(r_e - r_i)} \tag{D.15}$$

Dividing Eq. (D.14) by Eq. (D.15), we have the equation for relative change of sizes

$$\frac{dr_e}{dr_i} = -\frac{D_B^* X_A r_i^2}{D_A^* X_B r_e^2} \frac{\Delta g^{\text{compound}} - (2\gamma\Omega/r_e)}{\Delta g^{\text{compound}} + (2\gamma\Omega/r_i)} \tag{D.16}$$

With further simplification of the boundary conditions and using the following nondimensional parameters,

$$x = \frac{r_i}{r_o}, \quad y = \frac{r_e}{r_o}, \quad G = \frac{2\gamma\Omega}{\Delta g r_o}$$

where r_o is the initial radius of nanoparticle, and Δg is the driving force of reaction and is equal to the formation energy of the sulfide. We have the main solution as

$$\frac{dy}{dx} = -\frac{x^3}{y^3} \frac{D_m^*}{D_o^*} \frac{1 - c_m}{c_m} \frac{y - G}{x + G} \tag{D.17}$$

This equation differs from a similar equation by Alivisatos et al. [8], in Chapter 3 with the last term containing $G \neq 0$ (responsible for Laplace pressure and Gibbs–Thomson effect both for vacancies and for main components). When both radii of r_e and r_i are large ($r_e, r_i \gg 2\gamma\Omega/\Delta g$, $G \gg 1$), the equations become similar.

It is evident from Eq. (D.17) that in the case where $G > 1$ the nanoshell formation is impossible (taking into account the fact that x and y start from almost 1). It means that in very small particles, $r_o < 2\gamma\Omega/\Delta g$, the reaction with void formation is impossible. It is interesting that this critical condition coincides with the critical radius for the nucleation of a new phase.

INDEX

Kinetics in Nanoscale Materials, First Edition. King-Ning Tu and Andriy M. Gusak.